T0251791

CHEMISTRY AND INDUSTRIAL TECHNIQUES FOR CHEMICAL ENGINEERS

Innovations in Physical Chemistry: Monograph Series

CHEMISTRY AND INDUSTRIAL TECHNIQUES FOR CHEMICAL ENGINEERS

Edited by
Lionello Pogliani, PhD
Suresh C. Ameta, PhD
A. K. Haghi, PhD

Apple Academic Press Inc.
4164 Lakeshore Road
Burlington ON L7L 1A4
Canada

Apple Academic Press, Inc.
1265 Goldenrod Circle NE
Palm Bay, Florida 32905
USA

© 2021 by Apple Academic Press, Inc.

First issued in paperback 2021

Exclusive worldwide distribution by CRC Press, a member of Taylor & Francis Group

No claim to original U.S. Government works

ISBN 13: 978-1-77463-513-1 (pbk)
ISBN 13: 978-1-77188-823-3 (hbk)

All rights reserved. No part of this work may be reprinted or reproduced or utilized in any form or by any electric, mechanical or other means, now known or hereafter invented, including photocopying and recording, or in any information storage or retrieval system, without permission in writing from the publisher or its distributor, except in the case of brief excerpts or quotations for use in reviews or critical articles.

This book contains information obtained from authentic and highly regarded sources. Reprinted material is quoted with permission and sources are indicated. Copyright for individual articles remains with the authors as indicated. A wide variety of references are listed. Reasonable efforts have been made to publish reliable data and information, but the authors, editors, and the publisher cannot assume responsibility for the validity of all materials or the consequences of their use. The authors, editors, and the publisher have attempted to trace the copyright holders of all material reproduced in this publication and apologize to copyright holders if permission to publish in this form has not been obtained. If any copyright material has not been acknowledged, please write and let us know so we may rectify in any future reprint.

Trademark Notice: Registered trademark of products or corporate names are used only for explanation and identification without intent to infringe.

Library and Archives Canada Cataloguing in Publication

Title: Chemistry and industrial techniques for chemical engineers / edited by Lionello Pogliani, PhD, Suresh C. Ameta, PhD, A.K. Haghi, PhD.
Names: Pogliani, Lionello, editor. | Ameta, Suresh C., editor. | Haghi, A. K., editor.
Series: Innovations in physical chemistry.
Description: Series statement: Innovations in physical chemistry: monographic series | Includes bibliographical references and index.
Identifiers: Canadiana (print) 20200177036 | Canadiana (ebook) 20200177060 | ISBN 9781771888233 (hardcover) | ISBN 9780429286674 (ebook)
Subjects: LCSH: Chemical engineering. | LCSH: Chemical engineering—Methodology. | LCSH: Chemical engineering—Vocational guidance.
Classification: LCC TP155 .C54 2020 | DDC 660—dc23

CIP data on file with US Library of Congress

Apple Academic Press also publishes its books in a variety of electronic formats. Some content that appears in print may not be available in electronic format. For information about Apple Academic Press products, visit our website at **www.appleacademicpress.com** and the CRC Press website at **www.crcpress.com**

ABOUT THE EDITORS

Lionello Pogliani, PhD
Professor of Physical Chemistry (retired),
University of Valencia-Burjassot, Spain

Lionello Pogliani, PhD, is a retired Professor of Physical Chemistry. He has contributed more than 200 papers in the experimental, theoretical, and didactical fields of physical chemistry, including chapters in specialized books and a book on numbers 0, 1, 2, and 3. A work of his has been awarded with the GM Neural Trauma Research Award. He is a member of the International Academy of Mathematical Chemistry, and he is on the editorial boards of many international journals. He is presently a part-time teammate at the Physical Chemistry Department of the University of Valencia, Spain. He received his postdoctoral training at the Department of Molecular Biology of the C.E.A. (Centre d'Etudes Atomiques) of Saclay, France, at the Physical Chemistry Institute of the Technical and Free University of Berlin, and at the Pharmaceutical Department of the University of California, San Francisco, USA. He spent his sabbatical years at the Technical University of Lisbon, Portugal, and at the University of Valencia, Spain.

Suresh C. Ameta, PhD
Dean, Faculty of Science, PAHER University, Udaipur, India

Suresh C. Ameta, PhD, is currently Dean, Faculty of Science at PAHER University, Udaipur, India. He has served as Professor and Head of the Department of Chemistry at North Gujarat University Patan and at M. L. Sukhadia University, Udaipur, and as Head of the Department of Polymer Science. He also served as Dean of Postgraduate Studies. Prof. Ameta has held the position of President of the Indian Chemical Society, Kotkata, and is now a life-long Vice President. He was awarded a number of prestigious awards during his career, such as national prizes twice for writing chemistry books in Hindi. He also received the Prof. M. N. Desai Award (2004), the Prof. W. U. Malik Award (2008), the National Teacher Award (2011), the Prof. G. V. Bakore Award (2007), a Life-Time Achievement Award by the Indian Chemical Society (2011) as well as the Indian Council of Chemist

(2015), etc. He has successfully guided 81 PhD students. Having more than 350 research publications to his credit in journals of national and international repute, he is also the author of many undergraduate- and postgraduate-level books. He has published three books with Apple Academic Press: *Chemical Applications of Symmetry and Group Theory; Microwave-Assisted Organic Synthesis*; and Green Chemistry: Fundamentals and Applications; and two with Taylor and Francis: Solar Energy Conversion and Storage and Photo-catalysis. He has also written chapters in books published by several other international publishers. Prof. Ameta has delivered lectures and chaired sessions at national conferences and is a reviewer of number of international journals. In addition, he has completed five major research projects from different funding agencies, such as DST, UGC, CSIR, and Ministry of Energy, Govt. of India.

A. K. Haghi, PhD
Professor Emeritus of Engineering Sciences, Former Editor-in-Chief, International Journal of Chemoinformatics and Chemical Engineering & Polymers Research Journal; Member, Canadian Research and Development Center of Sciences and Cultures

A. K. Haghi, PhD, is the author and editor of 165 books, as well as 1000 published papers in various journals and conference proceedings. Dr. Haghi has received several grants, consulted for a number of major corporations, and is a frequent speaker to national and international audiences. Since 1983, he served as professor at several universities. He is the former Editor-in-Chief of the *International Journal of Chemoinformatics and Chemical Engineering* and *Polymers Research Journal* and is on the editorial boards of many international journals. He is also a member of the Canadian Research and Development Center of Sciences and Cultures (CRDCSC), Montreal, Quebec, Canada.

INNOVATIONS IN PHYSICAL CHEMISTRY: MONOGRAPH SERIES

This book series offers a comprehensive collection of books on physical principles and mathematical techniques for majors, non-majors, and chemical engineers. Because there are many exciting new areas of research involving computational chemistry, nanomaterials, smart materials, high-performance materials, and applications of the recently discovered graphene, there can be no doubt that physical chemistry is a vitally important field. Physical chemistry is considered a daunting branch of chemistry—it is grounded in physics and mathematics and draws on quantum mechanics, thermodynamics, and statistical thermodynamics.

Editors-in-Chief

A. K. Haghi, PhD
Former Editor-in-Chief, *International Journal of Chemoinformatics* and *Chemical Engineering and Polymers Research Journal*; Member, Canadian Research and Development Center of Sciences and Cultures (CRDCSC), Montreal, Quebec, Canada
E-mail: AKHaghi@Yahoo.com

Lionello Pogliani, PhD
University of Valencia-Burjassot, Spain
E-mail: lionello.pogliani@uv.es

Ana Cristina Faria Ribeiro, PhD
Researcher, Department of Chemistry, University of Coimbra, Portugal
E-mail: anacfrib@ci.uc.pt

BOOKS IN THE SERIES

- **Applied Physical Chemistry with Multidisciplinary Approaches**
  Editors: A. K. Haghi, PhD, Devrim Balköse, PhD, and
  Sabu Thomas, PhD

- **Biochemistry, Biophysics, and Molecular Chemistry: Applied Research and Interactions**
  Editors: Francisco Torrens, PhD, Debarshi Kar Mahapatra, PhD, and
  A. K. Haghi, PhD

- **Chemistry and Industrial Techniques for Chemical Engineers**
  Editors: Lionello Pogliani, PhD, Suresh C. Ameta, PhD, and
  A. K. Haghi, PhD

- **Chemistry and Chemical Engineering for Sustainable Development: Best Practices and Research Directions**
  Editors: Miguel A. Esteso, PhD, Ana Cristina Faria Ribeiro, and
  A. K. Haghi, PhD

- **Chemical Technology and Informatics in Chemistry with Applications**
  Editors: Alexander V. Vakhrushev, DSc, Omari V. Mukbaniani, DSc,
  and Heru Susanto, PhD

- **Engineering Technologies for Renewable and Recyclable Materials: Physical-Chemical Properties and Functional Aspects**
  Editors: Jithin Joy, Maciej Jaroszewski, PhD, Praveen K. M.,
  and Sabu Thomas, PhD, and Reza Haghi, PhD

- **Engineering Technology and Industrial Chemistry with Applications**
  Editors: Reza Haghi, PhD, and Francisco Torrens, PhD

- **High-Performance Materials and Engineered Chemistry**
  Editors: Francisco Torrens, PhD, Devrim Balköse, PhD,
  and Sabu Thomas, PhD

- **Methodologies and Applications for Analytical and Physical Chemistry**
  Editors: A. K. Haghi, PhD, Sabu Thomas, PhD, Sukanchan Palit,
  and Priyanka Main

- **Modern Physical Chemistry: Engineering Models, Materials, and Methods with Applications**
  Editors: Reza Haghi, PhD, Emili Besalú, PhD,
  Maciej Jaroszewski, PhD, Sabu Thomas, PhD, and Praveen K. M.

- **Molecular Chemistry and Biomolecular Engineering: Integrating Theory and Research with Practice**

- Editors: Lionello Pogliani, PhD, Francisco Torrens, PhD, and
  A. K. Haghi, PhD

- **Modern Green Chemistry and Heterocylic Compounds: Molecular Design, Synthesis, and Biological Evaluation**
  Editors: Ravindra S. Shinde, and A. K. Haghi, PhD

- **Physical Chemistry for Chemists and Chemical Engineers: Multidisciplinary Research Perspectives**
  Editors: Alexander V. Vakhrushev, DSc, Reza Haghi, PhD, and J. V. de Julián-Ortiz, PhD

- **Physical Chemistry for Engineering and Applied Sciences: Theoretical and Methodological Implication**
  Editors: A. K. Haghi, PhD, Cristóbal Noé Aguilar, PhD, Sabu Thomas, PhD, and Praveen K. M.

- **Practical Applications of Physical Chemistry in Food Science and Technology**
  Editors: Cristóbal Noé Aguilar, PhD, Jose Sandoval Cortes, PhD, Juan Alberto Ascacio Valdes, PhD, and A. K. Haghi, PhD

- **Research Methodologies and Practical Applications of Chemistry**
  Editors: Lionello Pogliani, PhD, A. K. Haghi, PhD, and Nazmul Islam, PhD

- **Theoretical Models and Experimental Approaches in Physical Chemistry: Research Methodology and Practical Methods**
  Editors: A. K. Haghi, PhD, Sabu Thomas, PhD, Praveen K. M., and Avinash R. Pai

- **Theoretical and Empirical Analysis in Physical Chemistry: A Framework for Research**
  Editors: Miguel A. Esteso, PhD, Ana Cristina Faria Ribeiro, PhD, and A. K. Haghi, PhD

CONTENTS

CONTRIBUTORS

Tarek M. Aboul-Fotouh
Department of Mining and Petroleum Engineering, Al-Azhar University, Cairo, Egypt

Deewan Akram
Department of Chemistry, JRS College (Jamalpur), Munger, Bihar, India

Manawwer Alam
Research Centre-College of Science, King Saud University, P.O. Box 2455, Riyadh 11451, Saudi Arabia

Rakshit Ameta
Department of Chemistry, J. R. N. Rajasthan Vidyapeeth (Deemed to be University), Udaipur 313002, Rajasthan, India

Suresh C. Ameta
Department of Chemistry, PAHER University, Udaipur 313003, Rajasthan, India

Fatma H. Ashour
Department of Chemical Engineering, Faculty of Engineering, Cairo University, Giza 12613, Egypt

Sevdiye Atakul
Department of Chemical Engineering, Izmir Institute of Technology, Gulbahce Urla, Izmir, Turkey

Devrim Balkose
Department of Chemical Engineering, Izmir Institute of Technology, Gulbahce Urla, Izmir, Turkey

Jayesh Bhatt
Department of Chemistry, PAHER University, Udaipur 313003, Rajasthan, India

Gloria Castellano
Departamento de Ciencias Experimentales y Matemáticas, Facultad de Veterinaria y Ciencias Experimentales, Universidad Católica de Valencia San Vicente Mártir, Guillem de Castro-94, E-46001 València, Spain

Chin Kang Chen
The Indonesian Institute of Sciences, Indonesia Tunghai University, Taichung, Taiwan University of Brunei, Bandar Seri Begawan, Brunei
Computational Science, The Indonesian Institute of Sciences, Indonesia
Computer Science Department, Tunghai University, Taiwan
School of Business and Economics, Universiti Brunei Darussalam, Brunei

Hany A. Elazab
Department of Chemical Engineering, The British University in Egypt, El-Shorouk City, Cairo, Egypt

Mamdouh Gadalla
Department of Chemical Engineering, Faculty of Engineering, Port Said University, Egypt
Department of Chemical Engineering, The British University in Egypt, El-Shorouk City, Cairo, Egypt

Kanchan Kumari Jat
Department of Chemistry, PAHER University, Udaipur 313003, Rajasthan, India

T. K. Jumadilov
JSC Institute of Chemical Sciences After A.B. Bekturov, Almaty, The Republic of Kazakhstan

Neha Kapoor
Department of Chemistry, PAHER University, Udaipur 313003, Rajasthan, India

R. G. Kondaurov
JSC Institute of Chemical Sciences After A.B. Bekturov, Almaty, The Republic of Kazakhstan

Amjad Mumtaz Khan
Department of Chemistry, Faculty of Science, Aligarh Muslim University, Aligarh, India

Poonam Khullar
Department of Chemistry, B.B.K.D.A.V College for Women, Amritsar 143001, Punjab, India

Yahiya Kadaf Manea
Department of Chemistry, Faculty of Science, Aligarh Muslim University, Aligarh, India

Omar Mazen
Department of Chemical Engineering, The British University in Egypt, El-Shorouk City, Cairo, Egypt

Sukanchan Palit
43, Judges Bagan, Post-Office - Haridevpur, Kolkata-700082, India

Lionello Pogliani
Facultad de Farmacia, Dept. de Química Física, Universitat de Valencia, Av. V.A. Estellés s/n, 46100 Burjassot (València), Spain

Avinash Kumar Rai
Department of Chemistry, PAHER University, Udaipur 313003, Rajasthan, India

Eram Sharmin
Department of Pharmaceutical Chemistry, College of Pharmacy, Umm Al-Qura University, P.O. Box 715, 21955, Makkah Al-Mukarramah, Saudi Arabia
Materials Research Laboratory, Department of Chemistry, Jamia Millia Islamia, New Delhi-110025, India

Ravindra S. Shinde
Department of Chemistry, Dayanand Science College, Latur 413 512, Maharashtra, India

Heru Susanto
The Indonesian Institute of Sciences, Indonesia
Tunghai University, Taichung, Taiwan
University of Brunei, Bandar Seri Begawan, Brunei
Computational Science, The Indonesian Institute of Sciences, Indonesia
Computer Science Department, Tunghai University, Taiwan
School of Business and Economics, Universiti Brunei Darussalam, Brunei

Lavanya Tandon
Department of Chemistry, B.B.K.D.A.V College for Women, Amritsar 143001, Punjab, India

Francisco Torrens
Institut Universitari de Ciència Molecular, Universitat de València, Edifici d'Instituts de Paterna, PO Box 22085, E-46071 València, Spain

Sevgi Ulutan
Department of Chemical Engineering, Ege University, Bornova, 35100, Izmir, Turkey

Fahmina Zafar
Materials Research Laboratory, Department of Chemistry, Jamia Millia Islamia, New Delhi-110025, India

ABBREVIATIONS

AIDS	acquired immunodeficiency syndrome
ALP	alkaline phosphatase
ATP	adenosine triphosphate
BSA	bovine serum albumin
BT	bending ability
BWR	boiling water reactor
CC	climate change
CDs	cyclodextrins
CdTe	cadmium telluride
CNN	counterfeit neural network
CNQDs/PBA	biocompatible phenylboronic acid-functionalized graphitic carbon nitride quantum dots
CNT-FET	carbon nanotube-field effect transistor
CNTs	carbon nanotubes
CPFX	ciprofloxacin
CR	chain reaction
CTAB	cetyltrimethylammonium bromide
CTSs	centralized temporal stores
CVD	chemical vapor deposition
DGS	deep geological store
DOX	doxorubicin
DPV	differential pulse voltammetry
EDC	ethylcarbodiimide hydrochloride
FFs	fossil fuels
FRET	fluorescence resonance energy transfer
GEC	global ecological crisis
GHE	greenhouse effect
GTP	guanosine triphosphate
HLW	high-level waste
HV	hydroxyl value
HWRs	heavy water D_2O reactors
ICT	information communication technology
IRs	ionizing radiations
IRt	impact resistance

IS	information system
LCD	liquid crystal display
LILW	low/intermediate-level waste
MAA	mercaptoacetic acid
MF	microfiltration
MIBK	methyl isobutyl ketone
M-PU	metal-containing biobased polyester urethane
MRI	magnetic resonance imaging
MS	mesoporous silica
MW	molecular weight
MWCNT-PAH/SPE	multi-walled carbon nanotube-polyallylamine modified screen printed electrode
NB-CQDs	nitrogen and boron co-doped carbon quantum dots
NCDs	nitrogen-doped fluorescent carbon dots
NE	nuclear energy
NF	nuclear fuel
NGOs	nongovernmental organizations
NHS	N-hydroxysuccinimide
NIR	near infrared
NM	nuclear medicine
NP	nanoparticle
NPPs	nuclear power plants
NR	nuclear reactor
NS	nuclear safety
OAI-ORE	open archives initiative object reuse, and exchange
OPH	organophosphorus hydrolase
PAHs	polyaromatic hydrocarbons
PAN	phthalic anhydride
PDA	polydopamine
PDT	photodynamic therapy
PEG	polyethylene glycol
PET	photoinduced electron transfer
PET	positron emission tomography
PL	photoluminescence
PLS	partial least-square
PTE	periodic table of the elements
PTT	photothermal therapy
PTT-TPPDT	photothermal and two-photon photodynamic therapy
PWR	pressurized water reactor
QD	quantum dot
RITs	radioisotopes

RO	reverse osmosis
RP	radiological protection
RTP	room-temperature phosphorescence
RW	radioactive waste
SAR	structure activity relationship
SD	sustainable development
SH	scratch hardness
SKOS	simple knowledge organization
SWCNTs	single-walled carbon nanotubes
TA	tannic acid
TDI	toluylene-2,4-diisocyanate
TDS	total dissolved solids
TGA	thioglycolic acid
TIPS	thermally induced phase separation
TPPDT	two-photon photodynamic therapy
UF	ultrafiltration
VLLW	very low-level waste

PREFACE

This volume brings together innovative research, new concepts, and novel developments in the application of new tools for chemical and materials engineers. It is an immensely research-oriented, comprehensive, and practical work. Postgraduate chemistry students would benefit from reading this book as it provides a valuable insight into chemical technology and innovations. It should appeal most to the chemists and engineers in the chemical industry and research, who should benefit from the technological, scientific, and economic interrelationships and their potential developments. It contains significant research, reporting new methodologies and important applications in the fields of chemical engineering as well as the latest coverage of chemical databases and the development of new methods and efficient approaches for chemists.

This volume should also be useful to every chemist or chemical engineer involved directly or indirectly with industrial chemistry. With clear explanations, real-world examples, this volume emphasizes the concepts essential to the practice of chemical science, engineering, and technology while introducing the newest innovations in the field.

The book also serves a spectrum of individuals, from those who are directly involved in the chemical industry to others in related industries and activities. It provides not only the underlying science and technology for important industry sectors, but also broad coverage of critical supporting topics. Industrial processes and products can be much enhanced through observing the tenets and applying the methodologies covered in individual chapters.

This authoritative reference source provides the latest scholarly research on the use of applied concepts to enhance the current trends and productivity in chemical engineering. Highlighting theoretical foundations, real-world cases, and future directions, this book is ideally designed for researchers, practitioners, professionals, and students of materials chemistry and chemical engineering. The volume explains and discusses new theories and presents case studies concerning material and chemical engineering.

This book is an ideal reference source for academicians, researchers, advanced-level students, and technology developers seeking innovative research in chemistry and chemical engineering.

KEY FEATURES

- Presents key factors in industrial application of nanotechnology
- Comments on future research direction and ideas for membrane development
- Covers basic issues relative to chemistry and nanoscience
- Explores the key areas of chemistry needed for chemical engineers
- Provides an understanding of current pathways for professional development in chemical information technology
- Describes frequently used methods in mobile chemistry apps on teaching learning process in higher education
- Provides real-world examples of key issues

PART 1
New Techniques and Updates

CHAPTER 1

CONSTRUCTION AND USEFULNESS OF THE POURBAIX E-pH DIAGRAMS

LIONELLO POGLIANI

Facultad de Farmacia, Dept. de Química Física, Universitat de Valencia, Av. V.A. Estellés s/n, 46100 Burjassot (València), Spain

E-mail: liopo@uv.es

ABSTRACT

E-pH diagrams, first presented by Pourbaix in his 1963 "Atlas of Electrochemical Equilibria in Aqueous Solution," have become quite popular in many fields, mainly related to aqueous chemistry. They are widely used in geochemistry, environmental chemistry, metallurgy, in studies that concern immunity, corrosion, passivation, precipitation, adsorption for water treatment, leaching, and metal recovery for hydrometallurgy. Here we discuss the mathematics behind them, their general characteristics, a method to obtain the bidimensional iron–water and aluminum–water diagrams, their usefulness for the cited studies, but also their limits. Some new advances on the topic are discussed.

1.1 INTRODUCTION

Potential-pH (or E-pH) Pourbaix diagrams have a wide audience among geochemists, chemical engineers, electrochemists, metallurgists, and environmental chemists.[1–10] Everything started with Pourbaix,[11–14] whose fundamental work on the subject is honored by calling his diagrams Pourbaix diagrams.[15–17] It is nevertheless surprising to notice that even if these diagrams are based on thermodynamics and redox reactions they are practically overlooked not only among physical chemists but also among chemists.[15–17] Educational chemistry, instead, started quite early to emphasize their importance first with a paper co-signed by Pourbaix.[14] Soon, other educational papers followed[15-24] that extended the usefulness of the E-pH diagrams explaining some new applications. From the cited literature we notice that the diagrams are crucial in material and, especially, in corrosion studies. They allow reading on a potential-pH bidimensional plane the stability regions of diverse species of a given element either alone or in a multielement system. Recently, tridimensional, E-pH-pX, diagrams have been adopted in material and corrosion science, where they are extensively used for determining when metals are thermodynamically stable and when they are in a passive state. They have also been used to treat the portioning of ionic species at the interface between two immiscible electrolyte solutions.

1.2 STRUCTURE OF THE DIAGRAM

The normal Pourbaix either potential-pH or E-pH diagram is, a two-dimensional representation of aqueous phase electrochemical equilibria, and it shows water-stable phases as a function of potential, and pH, where, potential is defined with respect to the standard hydrogen electrode. In fact, some authors still use the symbol E_h. It was also Pourbaix, who proposed to use the regions of the diagram to indicate what aqueous species is predominant in terms of E and pH. Figure 1.1 shows the broad structure of such a diagram with its set of horizontal, vertical, and diagonal lines (boundaries) that define areas where different species are stable and where it is possible to read if the metal is undergoing either corrosion or passivity or immunity (consider the entire polygons inclusive the shaded area). The shaded area denotes the space where water is stable, that is, it encloses the region of interest in aqueous systems, which is the region of stability of the aqueous solvent to oxidation or reduction.

Normally, boundaries define transition from one stable phase (species) to another and where two species exist in equilibrium. The vertical lines separate species that are in a potential-independent acid–base equilibrium, that is, that undergo no redox reactions with exchange of electron. They concern pure acid–base pH-dependent reactions that undergo only exchange of protons and are completely reversible. The horizontal lines separate species that are involved in a pH-independent redox equilibrium, that is, they involve only redox reactions with exchange of electrons. At these lines, oxidation or reduction could occur and are completely reversible. Diagonal $E = f(pH)$ lines separate species whose redox equilibrium is pH-dependent, that is, they concern reactions that undergo exchange of both electrons and protons. The diagonal lines have normally negative slope because a positive slope would imply negative pH. The mathematics behind these considerations is explained in the next section.

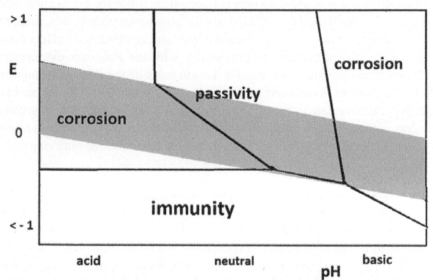

FIGURE 1.1 General structure of a Pourbaix potential(E)-pH diagram.

The shaded region, usually marked with dashed lines, as already told, encloses the region of stability of water and within this region, water does not undergo either oxidation or reduction. This region is of paramount importance to understand what happens to the different species in aqueous solutions. Outside this region, water breaks down, not the metal, as at high E values (strong oxidizing species, e.g., Cl_2) it is oxidized to O_2, whereas at sufficiently low E values (strong reducing species, e.g., Na) it is reduced to H_2 (more later on). In any point, P(E, pH), of the diagram, lines excluded, the most electrochemically (i.e., thermodynamically) stable form of the metal can be found, which means the most abundant form. The strong oxidizing agents or forms of the metal occur at the top of the diagram, while the strong reducing agents or forms of the metal occur at the bottom. If a species ranges from the top to the bottom of the diagram, at a given pH it will have neither oxidizing nor reducing properties (at that pH). If a species at a given potential ranges from highly acidic to highly basic values of the diagram, it will have neither acidic nor basic properties (at that E value). The aim to superimpose the aqueous E-pH diagram on a metal one allow showing (a) under what conditions a metal will corrode, (b) under what condition it is possible to stop the corrosion, that is, to passivate the metal, and (c) under what conditions it is stable, that is, it is immune to corrosion.

Many active metals, in presence of water, can corrode yet choosing the right conditions it is possible to prevent corrosion. Since the bronze and iron ages, we all depend on the use of metals for nearly everything; it is, thus, of paramount importance to understand how changing conditions affect their behavior. Today we can do so quite easily with the Pourbaix diagram as corrosion could be rapid (see Section 1.6) in parts of the Pourbaix diagram where the element is oxidized to a soluble, ionic product. However, solids, especially, oxides can form a protective coating on the metal that greatly prevents corrosion, a phenomenon called passivation. E-pH diagrams allow us also to understand why most metals do not exist in nature as pure elements.

1.3   THE MATHEMATICS OF THE DIAGRAM

Central for the present subject is the Nernst electrochemical equation, that is related to thermodynamics through the well-known expression: $n \cdot F \cdot E = -\Delta G$ [25]. The spontaneity of a reaction can be judged by the corresponding potential, E, as, if $\Delta G < 0$ then $E > 0$, and the reaction is spontaneous. The general redox reaction of a species in its oxidized, Ox, and reduced, Rd, forms, at some pH value (eq 1.1) and at T = 298.15 K, and 1 atm, obeys the Nernst eq (1.2), where a is the activity (E in V [Volt], even if it not explicitly written, $2.303 \cdot RT/F = 0.059$ V, and a is dimensionless),

$$qOx + mH^+ + ne^- = rRd + sH_2O \qquad (1.1)$$

$$E = E^0 + (0.059/n) \cdot \log\,[a^q(Ox) \cdot a^m(H^+)/a^r(Rd) \cdot a^s(H_2O)] \qquad (1.2)$$

This relation can be rearranged into the more useful eq (1.3), reminding that a for pure substances in condensed phases (solid or liquid) is taken as unit, $(a(H_2O) = 1)$, and $-\log a(H^+) = pH$,

$$E = E^0 + (0.059/n) \cdot \log\,[a^q(Ox)/a^r(Rd)] - (0.059m/n) \cdot pH \qquad (1.3)$$

For $a^q(Ox)/a^r(Rd) = 1$, this equation becomes

$$E = E^0 - (0.059m/n) \cdot pH \qquad (1.4)$$

Equations (1.3) and (1.4) determine the slope of the E = f(pH) line in a E versus pH diagram. If there is no redox reaction, that is, $n = 0$, the E = f(pH) line has infinite slope, that is, it is a vertical E-independent line at the pH value given by the equilibrium constant of a pure acid/base reaction. If, instead, $m = 0$, the E = f(pH) line has slope zero, that is, it is a horizontal pH-independent line at the E = E^0 value. If, instead, n and m differ from zero the E = f(pH) line is a diagonal with negative slope given by the coefficient 0.059m/n. For $a^q(Ox)/a^r(Rd) > 1$ the line displaces to the right of the line of eq (1.4) while for $a^q(Ox)/a^r(Rd) < 1$ it displaces to its left.

Thus, it is not at all surprising that the E-pH diagram is a set of crossing vertical, horizontal, and diagonal lines with different slopes. These lines define E-pH regions where chemical species are relatively stable. The diagram should also include the area for the stability of water that is essential to understand the behavior of the different species in water. This region is limited by two parallel lines encoded by eq (1.4) but with a different E^0 value. Usually, E^0 values (also known as *standard electrode potentials* or standard reduction potentials) are tabulated for a variety of redox couples at 298.15 K and P = 1.0 atm (values used in this paper). By convention, for comparison purposes, these potentials are for reactions written as a reduction.

These Tables do not cover any imaginable couple and for some of them E^0 may be unknown. In this case an apparent E^0_a should be derived by the aid of the line encoded by eq (1.4) (remind, $a^q(Ox)/a^r(Rd) = 1$) and the E-pH values of the point where it crosses another line. For instance, the couple $Fe(OH)_2/Fe^0$ obeys the following equation (see later on): E – E^0_a –0.059pH. The line encoded by this equation crosses the line for the couple Fe^{2+}/Fe^0 at P(E^0, pH) = P(–0.45, 5.95). Now, inserting these values into the previous equation we have $-0.45 = E^0_a - 0.059 \cdot 5.95 = E^0_a - 0.35$, and $E^0_a = -0.10$ V. Remind that the couple $Fe(OH)_2/Fe^{2+}$ is encoded by a vertical line (as $n = 0$) at pH = 5.95 that crosses the horizontal pH-independent ($m = 0$) line for the couple Fe^{2+}/Fe^0 at E = E^0 = –0.45 V (more in the paragraph for Fe).

Let us assume that S is a generic soluble species (for the nonsoluble species, $a = 1$) then, eq (1.3) can be written in the following way, where $-\log a(S) = pS$,

$$E = E^0 - (0.0591q/n) \cdot pS - (0.059m/n) \cdot pH \tag{1.5}$$

If both species are soluble, assuming $a^q(Ox)/a^r(Rd) = Y$, and $-\log Y = pY$, then,

$$E = E^0 - (0.0591/n) \cdot pY - (0.059m/n) \cdot pH. \tag{1.5'}$$

It is then possible to build a tridimensional diagram, $E = f(pOx \text{ or } pY, pH)$. In principle, it should also be possible, if both species are soluble, to compute a four-dimensional diagram that obeys the following relation;

$$E = E^0 - (0.0591q/n) \cdot pOx + (0.0591r/n) \cdot pRd - (0.059m/n) \cdot pH \quad (1.6)$$

Such 4D diagrams cannot be visualized; nevertheless, its set of tridimensional sections can be displayed and studied.

1.4 BUILDING THE BIDIMENSIONAL DIAGRAM

Throughout the following sections, we will discuss the equations that will allow constructing the bidimensional Pourbaix diagrams for Fe–water and Al–water systems. The following bidimensional diagrams can be considered sections of three-dimensional ones for either partial pressures or activities equal to one. The redox couples, their E^0 (from *standard electrode potentials*) or E^0_a calculated values as well as the concerned reactions and electrochemical equations for each couple for water, Fe, and Al will be collected throughout three different tables.

1.4.1 THE H₂O ITEM

Table 1.1 shows essential electrochemical information for water.

TABLE 1.1 Electrochemical Couples, E^0 (in V), Reactions, and Equations for H_2O at T = 298.15 K and P = 1 atm.

Couple	E^0	Chemical reactions	Equations*
$H^+/H_2(g)$	0.00	$2H^+ + 2e^- = H_2$	$E = -0.059 \text{ pH} - 0.0295 \log p(H_2)$ (1.7)
$O_2(g)/H_2O$	1.23	$O_2 + 4H^+ + 4e^- = 2H_2O$	$E = 1.23 - 0.059 \text{ pH} + 0.015 \log p(O_2)$ (1.8)

*The activity for pure water is equal to one.

Equations (1.7) and (1.8) encode two lines that enclose the area where water is stable. Outside this area, it undergoes either oxidation to $O_2(g)$ or reduction to $H_2(g)$. Notice: (a) With gases ($\equiv$ g) the partial pressure, $p(g)$, should be used, instead of the activity, a, and (b) this partial pressure should be read as a dimensionless quantity, that is, as $p(g)/P_0$, where $P_0 = 1$ atm.

Let us assume that for both gases, $p(g) = 1$ atm., then the simplified equations describe two $E = f(pH)$ parallel lines with slope -0.059 and intercept $E = 0$ V (the first one) and $E = 1.23$ V (the second one). Above the upper boundary, water is unstable with respect to oxygen, that is, for $E > 1.23 - 0.059 \cdot pH$ V, $O_2(g)$ is stable. Below the lower boundary, water is unstable relative to hydrogen gas, that is, for $E < -0.059 \cdot pH$ V, $H_2(g)$ is stable. In between liquid water is stable. Remind, the boundary at low E values is due to the $H^+/H_2(g)$ couple, whereas the one at high E values is due to the $O_2(g)/H_2O$ couple. Actually, reaction, $2H^+ + 2e- = H_2$, holds at low pH values, at higher pH values the reaction is $2H_2O + 2e- = H_2 + 2OH-$. In spite of this, the given equation still holds, as formally the electrochemical reaction has undergone no changes. Even the 2nd reaction at higher pH should be replaced by the reaction, $O_2 + 2H_2O + 4e- = 4OH^-$, but the corresponding equation still holds.

1.4.2 THE Fe ITEM

The data shown in Table 1.2 (at $T = 298.15$ K and $P = 1$ atm.) are taken from Ref. 14.

TABLE 1.2 Electrochemical Couples, E^0/E^0_a (in V), Reactions, and Equations for the Fe Species.

Couple	E^0/E^0_a	Chemical reactions	Equations*	
$Fe(OH)_3/Fe^{3+}$	–	$Fe(OH)_3 + 3H^+ = Fe^{3+} + 3H_2O$	$\log a(Fe^{3+}) = 4.62 - 3\ pH$	(1.9)
$Fe(OH)_2/Fe^{2+}$	–	$Fe(OH)_2 + 2H^+ = Fe^{2+} + 2H_2O$	$\log a(Fe^{2+}) = 13.2 - 2\ pH$	(1.10)
Fe^{3+}/Fe^{2+}	0.77	$Fe^{3+} + e- = Fe^{2+}$	$E = 0.77 + 0.059 \log [a(Fe^{3+})/ a(Fe^{2+})]$	(1.11)
Fe^{2+}/Fe^0	–0.44	$Fe^{2+} + 2e- = Fe^0$	$E = -0.44 + 0.0295 \log a(Fe^{2+})$	(1.12)
$Fe(OH)_3/Fe^{2+}$	1.04_a	$Fe(OH)_3 + 3H^+ + e- = Fe^{2+} + 3H_2O$	$E = 1.04 - 0.18\ pH - 0.059 \log a(Fe^{2+})$	1.13
$Fe(OH)_3/Fe(OH)_2$	0.26_a	$Fe(OH)_3 + H^+ + e- = Fe(OH)_2 + H_2O$	$E = 0.26 - 0.059\ pH$	(1.14)
$Fe(OH)_2/Fe^0$	-0.05_a	$Fe(OH)_2 + 2H^+ + 2e- = Fe^0 + 2H_2O$	$E = -0.05 - 0.059\ pH$	(1.15)

*The activities for water, and for the solid phases, Fe^0, $Fe(OH)_2$, and $Fe(OH)_3$ are equal to one.

With these data (every E^0_a has been independently confirmed) it is possible to build a simple diagram for Fe. In fact, to simplify things let us assume that all the activities are equal to one. Then, all equations simplify into the following set of relations (always check the corresponding reaction):

(eq 1.9) $4.62 - 3 \cdot pH = 0 \rightarrow pH = 1.54$: At this acidic pH value there is an E-independent vertical line that divides the (highly acidic) region where Fe^{3+} is stable from the less acidic region where $Fe(OH)_3$ is stable.

(eq 1.10) $13.2 - 2 \cdot pH = 0 \rightarrow pH = 6.6$: At this pH value there is an E-independent vertical line that divides the region where Fe^{2+} (acidic region) is stable from the region where $Fe(OH)_2$ is stable.

(eq 1.11) $E = 0.77$ V: At this E value there is a pH-independent horizontal line that divides the region where Fe^{3+} (higher E values) is stable from the region where Fe^{2+} is stable.

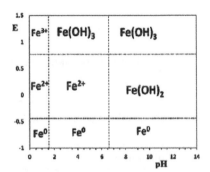 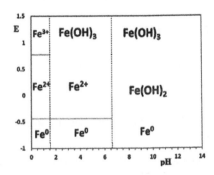

FIGURE 1.2   Left: Pourbaix diagram for Fe with horizontal and vertical lines only; right: Corrected Pourbaix diagram with meaningful horizontal and vertical lines only.

(eq 1.12) $E = -0.44$ V: At this E value there is a pH-independent horizontal line that parts the region where Fe^{2+} (higher E values) is stable from the region where Fe^0 is stable.

The diagram of Figure 1.2-left has been built with the newly defined vertical and horizontal lines. Actually, it is impossible to guess the boundaries that define the regions for the three couples: (a) $Fe(OH)_3/Fe^{2+}$, (b) $Fe(OH)_3/Fe(OH)_2$, and (c) $Fe(OH)_2/Fe^0$, as these boundaries concern pH-dependent redox reactions. This is why a more consistent figure should be like Figure 1.2-right. The voids in this figure will be transformed into different regions with the aid of the $E = f(pH)$ eqs (1.13)–(1.15) that define the respective boundaries. The three equations encode a diagonal line with slope -0.18 (eq 1.13), and two parallel lines with slope -0.059. The three lines have different intercepts but let us discuss them in detail.

(eq 1.13) $E = 1.04 - 0.18 \cdot pH$: This pH-dependent line concerns the equilibrium for the couple $Fe(OH)_3/Fe^{2+}$, where Fe^{2+} is stable in acidic media but decreases in stability with growing pH. Clearly, it is useless to calculate the E values for all pH values from zero to 14, as the meaningful pH values are only the ones for the region of this couple.

(eq 1.14) $E = 0.26 - 0.059 \cdot pH$: This pH-dependent line concerns the equilibrium for the couple $Fe(OH)_3/Fe(OH)_2$, where $Fe(OH)_3$ increases its stability with growing pH, that is, in less acidic media. The meaningful pH values are only the ones belonging to the region of this couple.

(eq 1.15) $E = -0.05 - 0.059 \cdot pH$: This pH-dependent line concerns the equilibrium for the couple $Fe(OH)_2/Fe^0$, where $Fe(OH)_2$ increases its stability with growing pH, that is, in less acidic media. The pH values concerned are, even here, only the ones belonging to the region of the present couple.

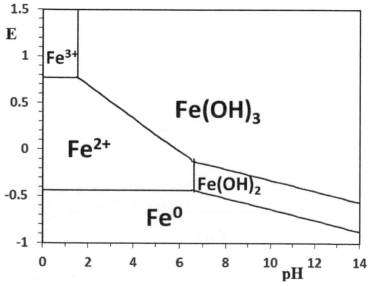

FIGURE 1.3   The final Pourbaix diagram for the different iron species.

Inserting these three lines in the three voids of Figure 1.2-right (respectively), we obtain the new regions for the final iron Pourbaix diagram shown in Figure 1.3.

The last step in the construction of a generalized iron Pourbaix diagram, shown in Figure 1.4, is to superimpose to the diagram of Figure 1.3 the region of the stability of water defined by the two lines discussed in the previous section (Table 1.1). The H_2O region (shaded area) allows understanding the real behavior of iron in presence of water, which is the main condition found on Earth. The diagram of Figure 1.4 explains why the inner core of the Earth is made of pure iron, Fe^0 (the outer core of pure Ni^0 also), and why pure Fe^0 cannot be found in nature. For an active metal such as iron, the region where the pure element is stable is below the $H^+/H_2(g)$ line for water. This means that iron metal is unstable in water and it undergoes a set of oxidation-pH-dependent reactions. In solution and over the $O_2(g)/H_2O$ line Fe can only exist either as $Fe(OH)_3$ or as Fe^{3+} at quite low pH values. At slightly acidic pH, in poorly oxygenated water, iron as such is unstable, it loses immunity and corrodes into Fe^{2+}. In well-oxygenated water and at not too low pH values, the system gets closer to the highest H_2O/O_2 line, where O_2 is reduced to H_2O and where the stable species is $Fe(OH)_3$ that could protect the surface but due to porosity and poor adherence passivation fails.

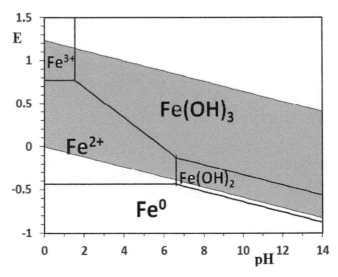

FIGURE 1.4 The generalized Pourbaix diagram for the iron–water system at 298.15 K, P = 1 atm.

Actually, iron can be better protected by connecting it with a more active metal, such as Mg ($E^0 = -2.70$ V) or Zn ($E^0 = -0.76$ V), that should periodically be replaced as both undergo corrosion. It becomes now clear why most active metals are not found pure in nature but mainly as oxides.

Let us now try, with the acquired expertise, to do the same for aluminum, Al.

1.4.3   THE Al ITEM

All the data needed (taken from Ref. 24) to build the Pourbaix diagram for Al are shown in Table 1.3 (at T = 298.15 K and P = 1 atm.). As already done with water and iron we assume that all activities are equal to unit. With this assumption eqs (1.16)–(1.20) simplify into the following set of relations (always check the corresponding reaction in Table 1.3):

(eq 1.16) $10.2 - 3 \cdot pH = 0 \rightarrow pH = 3.4$: At this acidic pH value there is an E-independent vertical line that divides the (acidic) region where Al^{3+} is stable from the less acidic region where $Al(OH)_3$ is stable.

(eq 1.17) $12.4 - pH = 0 \rightarrow pH = 12.4$: at this quite basic pH value there is an E-independent vertical line that divides the region where $Al(OH)_3$ is stable from the more basic region where $Al(OH)_4^-$ is stable.

(eq 1.18) $E = -1.68$ V: At this E value, there is a pH-independent horizontal line that parts the region where Al^{3+} (higher E values) is stable from the region where Al^0 is stable.

TABLE 1.3   Electrochemical Couples, E^0/E^0_a (in V), Reactions and Equations for the Al Species.

Couple	E^0/E^0_a	Chemical reactions	Equations*
$Al(OH)_3/Al^{3+}$	–	$Al(OH)_3 + 3H^+ = Al^{+++} + 3H2O$	$\log a(Al^{3+}) = 10.2 - 3\ pH$ (1.16)
$Al(OH)_4^-/$ $Al(OH)_3$	–	$Al(OH)_4^- + H^+ = Al(OH)_3 + H_2O$	$\log a[Al(OH)_4^-] = 12.4 - pH$ (1.17)
Al^{3+}/Al^0	–1.68	$Al^{3+} + 3e^- = Al^0$	$E = -1.68 + 0.020 \log a(Al^{3+})$ (1.18)
$Al(OH)_3/Al^0$	-1.47_a	$Al(OH)_3 + 3e^- + 3H^+ = Al^0 + 3H_2O$	$E = -1.47 - 0.059\ pH$ (1.19)
$Al(OH)_4^-/Al^0$	-1.23_a	$Al(OH)_4^- + 3e^- + 4H^+ = Al^0 + 4H_2O$	$E = -1.23 - 0.079\ pH + 0.02 \log a[Al(OH)_4^-]$ (1.20)

*The activities for water, and for the solid phases, Al^0 and $Al(OH)_3$ are equal to one.

There are then two horizontal lines and a vertical line in the first draft of the Pourbaix diagram for Al shown in Figure 1.5. Nevertheless, as already told for iron in Figure 1.3, the boundaries between the regions for the couples, $Al(OH)_3/Al^0$ and $Al(OH)_4^-/Al^0$ are ill-defined as they concern boundaries where $E = f(pH)$, as we face two pH-dependent redox reactions.

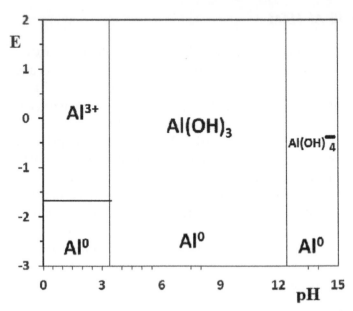

FIGURE 1.5 Corrected Pourbaix diagram for Al with meaningful horizontal and vertical lines only.

In the middle and right regions, the horizontal line at $E = -1.68$ has been omitted. These two undefined regions will be fixed by the boundaries encoded with the following equations:

(eq 1.19) $E = -1.47 - 0.059 \cdot pH$: This pH-dependent boundary concerns the equilibrium for the couple $Al(OH)_3/Al^0$, where $Al(OH)_3$ increases its stability with growing pH. The meaningful pH values here are only the ones for the boundary between the two regions of the present couple.

(eq 1.20) $E = -1.23 - 0.079 \cdot pH$: This pH-dependent boundary concerns the equilibrium for the couple $Al(OH)_4^-/Al^0$, where $Al(OH)_4^-$ increases its stability with growing pH, that is, in highly alkaline media. The pH values concerned are only the ones for the boundary between the two regions of the present couple.

The final diagram with the shaded region for water stability is shown in Figure 1.6.

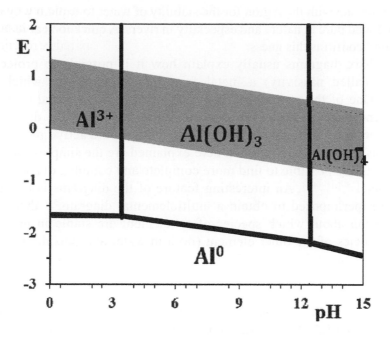

FIGURE 1.6   The generalized Pourbaix diagram for the Al–water system at 298.15 K, p = 1 atm.

This diagram makes it clear why Al^0 cannot exist immune in a nature rich in water. Pure elemental aluminum at low pH values is corroded and at higher pH values it undergoes passivity, thanks to a rather homogeneous formation of a thin protective layer of $Al(OH)_3$. The diagram tells also that in highly alkaline media passivity is threatened and corrosion starts again with formation of the soluble species $Al(OH)_4^-$. The diagram captures quite well the amphoteric character of Al with corrosion under highly acidic and highly alkaline conditions and protection with a hydroxide film throughout the middle pH region.

1.5   OTHER TYPES OF POURBAIX DIAGRAMS

From what has been told about the Pourbaix diagrams of Fe and Al it is not only possible to understand why some metals are never found pure (immune) in nature but also why some others are actually found that way. The reader can surely imagine that the region for Au^0 in an E-pH diagram should overlap with the region for the stability of water to explain why gold can be found pure in nature and especially in rivers. A quick look at its E-pH diagram[26] confirms this guess.

Pourbaix diagrams usually explain how it is possible to protect (a process called passivity) a metal from corrosion and in which pH region. One of the approaches is the deposition on the metal of a thin homogeneous protective layer whose function is to avoid that water (and especially aerated water) gets in touch with the active metal. The diagrams for iron and aluminum here explained are the simplest ones, in literature, it is possible to find more complete and complex diagrams for both metals[5–7,11–14,24]. An interesting feature of the diagrams is that they can be superimposed to obtain a multielemental diagram,[5–7] that gives information about which species of an element are stable in presence of the species of the other element (both in water and outside it). With the multielement diagram, it becomes possible, for example, to predict if Zn (or Mg) can really protect Fe from corrosion and in under what conditions.

Recently, tridimensional Pourbaix diagrams, E-pH-pS (S = soluble species, see eq 1.5), have started to be appreciated among practitioners[1–3,24]. With these 3D diagrams, it becomes possible, after inspection, to choose meaningful bidimensional sections for specific pS values.

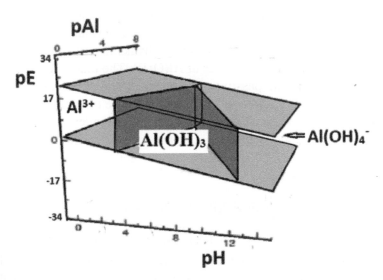

FIGURE 1.7   The water stability portion of the tridimensional Pourbaix diagram for Al.

Figure 1.7 (data are from Ref. 14) is an example of a 3D for Al, that is, for $S = Al^{3+} \equiv Al$ and pS = pAl. Contrasting Figures 1.6 and 1.7, we notice that, thanks to the third variable, pAl, the boundaries got transformed into planes and the regions into volumes, and that at high pAl values (low Al^{3+} concentration), and at mid-pH values, the amphoteric character of Al disappears as Al^{3+} goes directly into $Al(OH)_4^-$. Some authors, as in Figure 1.7, have started to use pE instead of the electrical potential E.[5] To understand what it is, divide eq (1.3) by 0.059 V to obtain,

$$E/0.059 = E^0/0.059 + (1/n) \cdot \log \left[a^q(Ox)/a^r(Rd) \right] - (m/n) \cdot pH \qquad (1.21)$$

Now, with pE = E/0.059 (or = 16.95·E) and $pE^0 = E^0/0.059$, we obtain, under standard conditions, the fully adimensional equation (pE, like pH, is adimensional)

$$pE = pE^0 + (1/n) \cdot \log \left[a^q(Ox)/a^r(Rd) \right] - (m/n) \cdot pH \qquad (1.22)$$

For $a^q(Ox)/a^r(Rd) = 1$, we obtain the simple relation: $pE = pE^0 - (m/n) \cdot pH$ and for the couple $H^+/H_2(g)$ we have pE = −pH (the line is a real diagonal with slope-1).

1.6 THE LIMITS OF THE DIAGRAM

The limits of these diagrams are mainly due to the fact that their predictions are based on thermodynamics that could be of marginal help with kinetic controlled reactions. For instance, reaction kinetics not only could affect the corrosion rate but could also yield different products, like oxides instead of hydroxides (Fe_2O_3 instead of $Fe(OH)_3$, and Al_2O_3 instead of $Al(OH)_3$). It should also be noticed that due to kinetic control multielement diagrams are not always good at predicting reactions between different species for two different elements. Concerning passivity, as already told for iron, not all insoluble products are protective, as porosity, thickness, and adherence of the protective layer to substrate could affect it. Concerning water, the potentials for its decomposition could be different from the experimental ones due to the different overpotentials required for the oxygen production on metals. Finally, the diagrams do not provide any insight into the electrochemical stability of metastable materials.

1.7   CONCLUSION

The advantages of the Pourbaix diagrams fairly compensate for all its limitations.

KEYWORDS

- **Pourbaix diagrams**
- **mathematics of the diagram**
- **Fe–water diagram**
- **Al–water diagram**
- **usefulness**
- **other types of diagrams**
- **limits**

REFERENCES

1. Minguzzi, A.; Fan, F.-R. F.; Vertova, A.; Rondinini, S.; Bard, A. J. Dynamic Potential–pH Diagrams Application to Electrocatalysts for Water Oxidation. *Chem. Sci.* **2012**, *3*, 217–229.
2. Ichige, Y.; Ouchi, M.; Mishima, K.; Haga, K.; Kondo, S. *Three Dimensional pH - Potential Diagram of Cobalt Slurry with Corrosion Rate.* 2015 International Conference on Planarization/CMP Technology (ICPT), IEEE Xplore Digital Library, Feb 25 2016, ISBN: 978-1-6195-6510-4.
3. Huang, H. -H. The Eh-pH Diagram and Its Advances. *Metals* **2016**, *6* (23), 1–30.
4. Wills, L. A.; Qu, X.; Chang, I-Y.; Mustard, T. J. L.;, Keszler, D. A.; Persson, K. A.;Cheong P. H. -Y., Group Additivity-Pourbaix Diagrams Advocate Thermodynamically Stable Nanoscale Clusters in Aqueous Environments. *Nature Commun.* **2017**, *8*, article n. 15852, 1–7, DOI: 10.1038.
5. Yokokawa, H.; Sakai, S.; Kawada, T.; Dokiya, M. Generalized Electrochemical Potential Diagrams for Complex Aqueous (M-X-H-O-e⁻) Systems. *J. Electrochem. Soc.* **1990**, *137*, 388–398.
6. Sai Jayaraman, S.; Singh, A. K. The Materials Project. https://materialsproject.org/janrain/loginpage/?next=/#apps/pourbaixdiagram. (accessed Oct 27, 2018).
7. Thompson, W. T.; Kaye, M. H.; Bale, C. W.; Pelton, A. D. Pourbaix Diagrams for Multielement Systems; In *Uhlig's Corrosion Handbook*, 3rd ed.; Revie R. W., Ed.; John Wiley & Sons, Inc.: Hoboken, NJ, 2011.
8. Ahmad, Z. *Principles of Corrosion Engineering and Corrosion Control*; Butterworth-Heinemann, 2006.
9. Ashby, M. F.; Jones, D. R. H. *Engineering Materials 1*; 4th ed; Butterworth-Heinemann, Elsevier: Oxford, 2012.

10. Azumi, K.; Elsentriecy H. H.; Tang, J. *Corrosion Prevention of Magnesium Alloys*; Woodhead Publishing, Elsevier: Oxford, 2013.

11. Pourbaix, M. J. N. *Thermodynamique des Solutions Aqueuses Diluees: Representation Graphique du Role du pH e du Potential*; Ph.D., Delft, Delft: The Netherlands, 1945; translation by Agar, J. N. published as *Thermodynamics of Dilute Aqueous Solutions*; Arnold: London, 1948.

12. Pourbaix, M. J. N. *Atlas d'equilibres electrochimiques*; Gauthier-Villars: Paris, 1963.

13. Pourbaix, M.; Franklin, J. A. *Atlas of Electrochemical Equilibria in Aqueous Solutions*; Pergamon Press: Oxford, England, 1966.

14. Delahay, P.; Pourbaix, M.; Van Rysselberghe, P. J. Potential-pH Diagrams. *J. Chem. Educ.* **1950**, *27*, 683–688.

15. Barnum, D. W. Potential-pH Diagrams. *J.Chem. Ed.* **1982**, *59*, 809–812.

16. Napoli, A.; Pogliani, L. Potential-pH Diagrams. *Educ. in Chem.* **1997**, *34*, 51–52.

17. Napoli, A.; Pogliani, L. Diagrammi Potenziale-pH un Percorso Didattico. *La Chimica nella Scuola* **1996**, Anno XVIII, *5*, 152–215.

18. Campbell, J. A.; Whiteker, R. A. A Periodic Table Based on Potential-pH Diagrams. *J. Chem. Educ.* **1969**, *46*, 90–92.

19. Williams, B. C.; Patrick Jr. W. H. A Computer Method for the Construction of Eh-pH Diagrams. *J. Chem. Educ.* **1977**, *54*, 107.

20. Atkinson, G. F. Precipitation Titrations in Terms of the Pourbaix Diagrams. *J. Chem. Educ.* **1977**, *54*, 109.

21. Powell, D.; Cortez, J.; Mellon, E. K. A Laboratory Exercise Introducing Students to the Pourbaix Diagram for Cobalt. *J. Chem. Educ.* **1987**, *64*, 165–167.

22. Alonso, C.; Ocon, P. Passivation of Copper in Acid Medium. *J. Chem. Educ.* **1987**, *64*, 459–460.

23. Solorza, O.; Olivares, L. Experimental Demonstration of Corrosion Phenomena. *J. Chem. Educ.* **1991**, *68*, 175–177.

24. Pesterfield, L. L.; Maddox, J. B.; Crocker, M. S.; Schweitzer, G. K. Pourbaix (E–pH-M) Diagrams in Three Dimensions. *J. Chem. Educ.* **2012**, *89*, 891–899.

25. Castellan, G. W. *Physical Chemistry*, 3rd ed; Addison-Wesley: London, 1983.

26. Van Muylder, J.; Pourbaix, M. *Atlas of Electrochemical Equilibria in Aqueous Solutions*; Pourbaix, M., Eds.; 2nd ed; National Association of Corrosion Engineers: New York, 1974.

CHAPTER 2

BIOMEDICAL APPLICATIONS OF CARBON NANOTUBES

AVINASH KUMAR RAI[1], NEHA KAPOOR[1], JAYESH BHATT[1], RAKSHIT AMETA[2*], and SURESH C. AMETA[1]

[1]Department of Chemistry, PAHER University, Udaipur 313003, Rajasthan, India

[2]Department of Chemistry, J. R. N. Rajasthan Vidyapeeth (Deemed to be University), Udaipur 313002, Rajasthan, India

*Corresponding author. E-mail: ameta_sc@yahoo.com

ABSTRACT

Carbon nanotubes (CNTs) are cylinders of graphene sheets. These are associated with wide range of electronic, thermal, and structural properties, depending on diameter, length, chirality, and twist. CNTs are large macromolecules having unique size and shape with some remarkable physical properties such as mechanical strength, aspect ratio, electrical, and thermal conductivity. These have a wide range of industrial applications such as air filtration, catalytic application, water purification, electronics, composite materials, energy storage, super capacitors, biosensors, and chemical sensors. Nanotechnology has recently been applied in biomedical fields for detection, diagnosis, imaging, therapies, etc. CNTs have received great attention in biomedical fields because of their unique structures and properties, including high aspect ratios, large surface areas, rich surface chemical functionalities, and size stability on the nanoscale and found varied applications such as cancer diagnosis and therapy, targeted drug delivery, fertility, contrast agents in magnetic resonance imaging (MRI), sensors (bio and chemical), photothermal therapy, and photodynamic therapy. The field of CNTs in biomedical field has been reviewed here.

2.1 INTRODUCTION

The carbon nanotube (CNT) represents one of the most unique inventions in the field of nanotechnology. Since their discovery in 1991,[1] they became one of the major component of the nanotechnology era.[2,3] CNTs have been studied closely over the last two decades by many researchers around the world due to their great potential in different fields. CNTs are allotropes of carbon with a nanostructure that can have a length-to-diameter ratio greater than 1,000,000. CNTs are long and thin cylinders of carbon. These have a higher tensile strength more than even steel. It is all due to sp^2 bonds between the individual carbon atoms. These are large macromolecules having unique size and shape with some remarkable physical properties such as mechanical strength, aspect ratio, electrical, thermal conductivity, etc. They can be considered as a sheet of graphite rolled into a cylinder. Nanotubes are associated with wide range of electronic, thermal, and structural properties, depending on its diameter, length, chirality, and twist. Single-walled carbon nanotubes (SWCNTs) have single cylinder of graphene sheet, while nanotubes having multiple walls are called multi-walled carbon nanotubes (MWCNTs) having cylinders inside other cylinder. A CNT is a tube-shaped material of carbon, which has a diameter of few nanometers. A nanometer is 1-billionth part of a meter, or it is about 10,000 times smaller than a human hair. CNTs are unique in feature as the bonding between the carbon atoms is very strong and the tubes can have extreme aspect (length/diameter) ratios.

The production of CNTs consists in the transformation of a carbon source into nanotubes, usually at high temperature and low pressure; however, the conditions of synthesis influence the characteristics of the final product. The synthesis conditions can be modified to obtain either MWCNTs or SWCNTs to fulfill different needs. Generally, three techniques are used for producing CNTs and these are:

- Carbon arc-discharge technique,[4-9]
- Laser-ablation technique, [10-13] and
- Chemical vapor deposition (CVD) technique.[14-22]

There are various applications of CNTs, where advantage of unique properties of CNTs may be utilized. Combining the electrical, thermal, and mechanical properties of CNTs with polymers could find a vast range of potential applications. A number of studies have carried out to use these properties for industrial applications such as air filtration,[23,24] graphene-based noble metals,[25] catalytic application, and water purification.[26-30] CNTs find several other applications in field of engineering also such as, electronics, composite materials, energy storage, super capacitors,[31-34] and chemical sensors.[35-38]

Nanotechnology is the manipulation of matter, including the synthesis, assembly, control, and measurement, on the atom and molecular level. Since the concept of nanomanipulation was established, this technology has been used in a number of applications in various fields such as electronics, mechanics, chemistry, and biology. Furthermore, nanotechnology has been applied in biomedical fields for detection, diagnosis, imaging, and therapy. CNTs have received great attention in biomedical fields because of their unique structures and properties, including high aspect ratios, large surface areas, rich surface chemical functionalities, and size stability on the nanoscale.

2.2 CANCER THERAPY

Cancer is defined as the uncontrolled growth of cells that destroys normal tissues and organs. Various mutations and uncontrolled growth and division of cancerous cells allow cancer cells to acquire properties such as self-sufficiency in growth signals, unlimited proliferation potential, and resistance to signals that stop proliferation or induce apoptosis by normal cells. Tumors have evolved to utilize additional supports via interactions with surrounding stromal cells, promotion of angiogenesis, evasion of immune detecting systems, and metastasis to other organs.

CNTs have high aspect ratios and are very small, and thus they have high specific surface areas, related to their needle-like shapes, enabling them to adsorb onto or conjugate with various therapeutic molecules. The needle-like shape of CNTs also enables their internalization into target cells. Therefore, CNTs are considered as promising nanocarriers for the delivery of drugs, genes, and proteins. However, because vesicle-based carriers such as liposomes have decreased other diseases other than cancer, CNT-based nanocarriers have been widely studied for the delivery of anticancer agents.

Particularly, CNTs are attractive transporters for drug delivery and mediators of noninvasive therapy. Appropriate functionalization enables the use of CNTs as nanocarriers to transport anticancer drugs such as DOX, CPT, CP, CDDP, PTX, and DTX. CNTs have been used as carriers for genes such as pDNA, siRNA, ODNs, and RNA/DNA aptamers. CTNs can also deliver proteins and immunotherapy components. A combination of light energy such as near infrared (NIR) and CNTs enables their use in noninvasive therapeutic techniques. CNTs have been used as mediators for photothermal therapy (PTT) and photodynamic therapy (PDT) to directly destroy cancer cells without severely damaging normal tissue.

Although many studies have shown promising results for CNT-based therapies in vitro and in vivo, there are several limitations to the clinical applications of these methods. First, safety issues in the human body have not been adequately addressed. Although many in vitro tests showed the safety of f-CNTs, most in vivo toxicity tests were conducted over a relatively short time. Long-term safety has gained increasing attention, and in vivo studies related to long-term toxicity and external excretion of CNTs have shown some progress. In addition, efforts have been made to minimize toxicity via the purification and surface functionalization of CNTs. Second, the size uniformity of synthesized CNTs and uniformity of the loading amount at drug–CNT complexes must be improved. Many strategies including control of the synthetic catalyst, growth temperature, environmental gas pressure, flux, and composition of the feedstock gas have been proposed to increase the uniformity of CNTs and studied over the past several decades. Various functionalization methods using covalent or non-covalent surface binding of molecules have been engrafted to CNT-based drug loading systems to increase loading uniformity. The accuracy of targeting cancer cells and controllability of loaded drugs should be also improved. Different CNT functionalization strategies using various molecules and materials have been reported to enhance the activity and stability of drug–CNTs. Additional studies are required to improve these methods for clinical application, and then only CNTs will become one of the strongest tools in various other biomedical fields as well as cancer therapy.

Liu et al.[39] fabricated polyethylene glycol (PEG) modified mesoporous silica (MS) coated SWCNTs. They used them as a multifunctional platform for imaging guided combination therapy of cancer. A model chemotherapy drug, doxorubicin (DOX), could be loaded into the mesoporous structure of the obtain SWCNT@MS-PEG nano-carriers and shows high efficiency. Photothermally triggered drug release from DOX loaded SWCNT@MS-PEG was observed upon stimulation under NIR light inside cells, which resulted in a synergistic cancer cell killing effect. A combination therapy (in vivo) using this agent was demonstrated in a mouse tumor model, where a remarkable synergistic anti-tumor effect was observed better than that obtained by mono-therapy.

SWCNTs were coated with a shell of polydopamine (PDA) by Zhao et al.,[40] which is further modified by PEG. The PDA shell in SWCNT@

PDA-PEG could from complex with Mn^{2+}, which along with metallic nanoparticulate impurities anchored on SWCNTs offer enhanced both T1 and T2 contrasts under magnetic resonance imaging (MRI). The radionuclide ^{131}I could be easily labeled onto SWCNT@PDA-PEG, utilizing the PDA shell; thus, enabling nuclear imaging and radioisotope cancer therapy. Efficient tumor accumulation of SWCNT@PDA-^{131}I-PEG was observed after systemic administration into mice evident from MR and gamma imaging. NIR-triggered PTT in combination with ^{131}I-based radioisotope therapy showed a remarkable synergistic antitumor therapeutic effect as compared to monotherapies.

The biodistribution of radio-labeled SWCNTs in mice has been investigated by Lin et al.[41] in vivo positron emission tomography (PET), ex vivo biodistribution and Raman spectroscopy. It is found that SWCNTs functionalized with phospholipids bearing PEG were quite stable in vivo. The effect of PEG chain length was also studied on the biodistribution and circulation of the SWCNTs. It was revealed that PEGylated SWCNTs exhibited relatively long blood circulation times and also low uptake by the reticuloendothelial system (RES). They achieved efficient targeting of integrin positive tumour in mice using SWCNTs coated with PEG chains, which were linked to an arginine–glycine–aspartic acid (RGD) peptide.

Liu et al.[42] reported that chemically functionalized SWCNT have a great potential in tumor-targeted accumulation in mice. These exhibited biocompatibility, excretion, and little toxicity. Paclitaxel (PTX), a widely used cancer chemotherapy drug, was conjugate to branched PEG chains on SWCNTs via a cleavable ester bond; just to have a water-soluble SWCNT-PTX conjugate. It afforded higher efficacy in suppressing tumor growth than clinical taxol in a murine 4T1 breast cancer model, because of prolonged blood circulation and 10-fold higher tumor PTX uptake by SWCNT delivery. It may be due to enhanced permeability and retention. It was reported that drug molecules carried into the RES are released from SWCNTs and these are excreted via biliary pathway without causing any toxic effects to normal organs.

A method has been presented by Heister et al.,[43] for the triple functionalization of oxidised SWCNTs with an anti-cancer drug DOX, a monoclonal antibody, and fluorescent marker at non-competing binding sites. This proposed methodology allowed for the targeted delivery of the anti-cancer drug to site and the visualization of the cellular uptake of SWCNTs by confocal microscopy. It was revealed that complex is efficiently taken up by cancer cells with subsequent intracellular release of DOX, which then translocated to the nucleus while the nanotubes remained in the cytoplasm.

Hampel et al.[44] presented CNTs as carriers for carboplatin, a therapeutic agent for cancer treatment. Carboplatin was introduced into CNTs to prove that these are suited as nanocontainers and nanocarriers; thus can release this drug to initialize its medical virtue. The effect on cell proliferation and cytotoxicity of the carboplatin-filled CNT was also investigated. It was revealed that the structure of carboplatin incorporated into the CNTs was retained. It was also found that carboplatin-filled CNTs inhibited growth of bladder cancer cells, but unfilled, opened CNTs hardly affected cancer cell growth.

2.3 FERTILITY

The fast pace of human life has presented various challenges in the area of healthcare. Diabetes, hypertension, depression, cancers, several other infectious diseases, etc. are just some examples of common outcomes resulting due to speed stress-filled lifestyle of human beings. Early diagnosis has been the goal for prompt arrest and management of these health conditions. Use of microchips, biosensors, nanorobots, nano-identification of single-celled structures, and micro electromechanical systems are current techniques being developed for use in nanodiagnostics. The intrinsic properties of CNTs have huge potential to provide accurate medical imaging and medical therapy (magnetic hyperthermia, targeted drug delivery, chemotherapy, etc.) even at the cellular level.

Jha et al.[45] demonstrated the loading of silver nanoparticle (AgNP) inside the MWCNT and their targetability to the intracellular part of the sperm cell for its further application in biosensing-based infertility diagnosis. They synthesized photosynthesized AgNP using *Ocimum tenuiflorum* (tulsi extract), and observed that spherical shaped (5–40 nm) AgNP were obtained, which show surface plasmonic resonance at 430 nm. The loading was ascertained with different methods. A shifting at 3450 (-OH stretching) and 1615 cm^{-1} (CNT back bone) supported the binding of AgNP with MWCNT. It was revealed that heat flow Ag loaded MWCNT had greater stability than AgNP. It was observed that the loading of AgNP inside MWCNT increases surface height of MWCNT from 22 to 32 nm, which established the encapsulation of AgNP (10 nm) inside the nanotubes as evident from AFM study. DNA fragmentation and morphological examination was also carried out, which confirmed the binding and targetability of AgNP to the sperm nucleus. This AgNP- MWCNT composite can be considered suitable in fertility diagnosis as it has improved targeting efficiency and biosenssing ability.

2.4 MAGNETIC RESONANCE IMAGING (MRI)

It is a noninvasive medical diagnostic technique. Although certain endogenous contrast can be achieved in the process of nuclei excitation and relaxation of magnetic spins, specific exogenous contrast agents (CAs) are often required to give an acceptable MRI image with high spatiotemporal resolution, sensitivity, specificity, and volumetric coverage. Paramagnetic complexes (Ga^{3+} or Mn^{2+}-based chelates), paramagnetic ion nanoparticles (Gd_2O_3 and MnO), and superparamagnetic iron oxide (SPIO) nanoparticles (Fe_3O_4, FeCO, and $MnFe_2O_4$) are commonly used CAs for MRI. The major advantage of MRI is its high spatial resolution (25–100 µm level) and the excellent tissue contrast. MRI overruns other imaging approaches available till date, and it is available for both; morphological as well as functional assessments, which requires only a certain quantity of CAs for the relatively time-consuming imaging process.

A nanoscale loading and confinement of aquated Gd^{3+} n-ion clusters within ultra-short SWCNTs (US-tubes) was reported by Sitharaman et al.[46.] These Gd^{3+} n@US-tube species were linear superparamagnetic molecular magnets with MRI, which has efficacies 40–90 times larger than any Gd^{3+} -based CA in present clinical use.

Superparamagnetic Gd^{3+}-ion clusters (1×5 nm) confined within ultra-short (20–80 nm) SWCNT capsules have been reported by Hartmen et al.[47] These gadonanotubes are high-performance T_1-weighted CAs for MRI. The r_1 relaxivity (ca. 180 mM^{-1} s^{-1} per Gd^{3+} ion) of gadonanotubes at 1.5 T, 37°C, and pH 6.5 was found to be 40 times greater than Gd^{3+} ion-based clinical agent used at present. Gadonanotubes are also ultrasensitive pH-smart probes with their r_1/pH response from pH 7.0–7.4, which is an order of magnitude greater than for any other MR CA. Gadonanotubes are excellent candidates for the early detection of cancer, where the extracellular pH of tumors can drop to pH $\leq$ 7 or below.

Wu et al.[48] loaded Fe_3O_4 nanoparticles in situ on the surface of MWCNTs by a solvothermal method. They used diethylene glycol and diethanolamine as solvents and complexing agents, respectively. As-prepared MWCNT/ Fe_3O_4 hybrids were having excellent hydrophilicity, superparamagnetic property at room temperature, and a high T_2 relaxivity of 175.5 mM^{-1} s^{-1} in aqueous solutions. It was revealed that MWCNT/Fe_3O_4 exhibited excellent MRI enhancement effect on cancer cells. They also displayed low cytotoxicity and neglectable hemolytic activity. After intravenous administration, the T_2-weighted MRI signal decreased significantly in the liver and spleen of mice, which suggests potential application of these hybrids as MRI CAs.

Zavaleta et al.[49] used an optimized noninvasive Raman microscope to observe tumor targeting and localization of SWCNTs in mice. Raman images were obtained in two groups of tumor-bearing mice. It was reported that control group received plain-SWCNTs, while the experimental group received tumor targeting RGD-SWCNTs intravenously. Raman imaging commenced over the next 72 h, revealed an increased accumulation of RGD-SWCNTs in tumor ($p < 0.05$) as compared to plain-SWCNTs. This may pave a path for the development of a new preclinical Raman imager.

Non-ionizing whole-body high field MRI was used by Marangon et al.[50] to follow the distribution of water-dispersible non-toxic functionalized CNTs. Oxidized CNTs have positive MRI contrast properties, if there are covalently functionalization with the chelating ligand diethylenetriaminepentaacetic-dianhydride (DTPA), followed by chelation to Gd^{3+}. They also evaluated structural and magnetic properties, MR relaxivities, cellular uptake, and application for MRI cell imaging of Gd-CNTs as compared to the oxidized CNTs. It was observed that anchoring of paramagnetic gadolinium onto the nanotube sidewall allowed efficient T_1 contrast and MR signal enhancement, which was preserved after CNT internalization by cells in spite of the intrinsic T_2 contrast of oxidized CNTs internalized in macrophages. Thus, Gd-CNTs have the potential to produce positive contrast in vivo following injection into the blood stream due to their high dispersibility. The uptake of Gd-CNTs in the liver as well as spleen and rapid renal clearance of extracellular Gd-CNTs was also assessed using MRI.

SPIO nanoparticles within the inner cavity of MWCNTs were fabricated by lin et al.,[51] which have high mechanical stability. A simple, effective, and self-assembled coating with RAFT diblock copolymers was done to have a high dispersion stability under physiological conditions in SPIO-MWCNTs. In vivo acute tolerance testing was also carried out with a high tolerance dose up to 100 mg kg^{-1} in mice. It was reported that a 55% increase in tumor to liver contrast ratio was observed after administration of the material with in vivo MRI measurements as compared to the preinjection image with better detection of the tumor.

2.5 SENSORS

Biosensors are analytical devices that combine biological molecules, or biorecognition systems like enzymes, binding proteins, nucleic acid, bacteria, cells, or even whole tissues of higher organisms and the physical transducers or detectors, which convert a biological response or biorecognition into digital electronic signals. Biosensors can be classified based on biorecognition events or signal transduction. In the case of transducers, biosensors can be classified as piezoelectric (such as quartz crystal microbalance or QCM), optical (such as surface plasmon resonance or SPR), or electrochemical sensors.

Choi et al.[52] developed a SWCNT-based biosensor to detect *Staphylococcus aureus*. The specificity of 11 bacteria and polyclonal anti-*Staphylococcus aureus* antibodies (pAbs) was determined and indirect ELISA was used for this purpose. Hybridization of 1-pyrenebutanoic acid succinimidyl ester (PBASE) was followed by immobilization of pAbs onto sensor platform. The difference in resistance (ΔR) was calculated using a potentiostat. The optimum concentration of SWCNTs on the platform was found to be 0.1 mg mL^{-1}. It was observed that binding of pAbs with *S. aureus* resulted in a significant increase in resistance value of the biosensor ($p < 0.05$). Specific binding of *S. aureus* on the biosensor was confirmed by SEM images. It was reported that SWCNT-based biosensor could detect *S. aureus* with a limit of detection (LOD) of 4 log CFU mL^{-1}.

SWCNTs and immobilized antibodies were integrated by Kara et al.[53] into a disposable bio-nano combinatorial junction sensor which can be used for detection of *Escherichia coli* K-12. They aligned gold tungsten wires (50 μm diameter) coated with polyethylenimine (PEI) and SWCNTs forming a crossbar junction. It was functionalized with streptavidin and biotinylated antibodies to allow for higher specificity toward targeted microbes. They monitered changes in electrical current (ΔI) after bioaffinity reactions between bacterial cells (*E. coli* K-12) and antibodies on the SWCNT surface, just to evaluate the performance of sensor. The averaged ΔI was found to be increased from 33.13 to 290.9 nA in the presence of SWCNTs in a 108 CFU mL^{-1} concentration of *E. coli*. It was reported that current decreased as cell concentrations increases, may be due to increased bacterial resistance on the bio-nano modified surface. The detection limit of this sensor was 102 CFU mL^{-1} with a detection time (< 5 min) with nanotubes.

Zhu et al.[54] designed a novel brush-like electrode based on CNT nano-yarn fiber. It was used for electrochemical biosensor applications and its efficacy as an enzymatic glucose biosensor was observed. CNT nano-yarn fiber was spun directly from a CVD gas flow reaction using a mixture of ethanol and acetone as the carbon source and an iron nano-catalyst. The fiber (28 μm diameter) was made of bundles of double-walled CNTs (DWNTs), which were concentrically compacted into multiple layers forming a nano-porous network structure. It was revealed that a superior electrocatalytic activity for

CNT fiber was there as compared to the commonly used Pt–Ir coil electrode. The electrode end tip of the CNT fiber was freeze-fractured, which gave a unique brush-like nano-structure resembling a scale-down electrical "flex." Here, glucose oxidase (GOx) enzyme can be immobilized using glutaraldehyde crosslinking in the presence of bovine serum albumin (BSA). An outer epoxy-polyurethane (EPU) layer was used as semi-permeable membrane. It was reported that sensitivities, linear detection range, and linearity for glucose detection for this CNT fiber electrode was found to be better than that reported for a Pt–Ir coil electrode. It was revealed that thermal annealing of the CNT fiber at 250°C for 30 min before fabrication of the sensor showed a 7.5-fold increase in glucose sensitivity. As-spun CNT fiber-based glucose biosensor was stable for 70 days. Gold coating of the electrode connecting end of the CNT fiber has enhanced the glucose detection limit to 25 μM.

A novel vertically aligned CNT-based electrical cell impedance sensing biosensor (CNT-ECIS) was reported by Abdolahad et al.[55] as a more rapid, sensitive, and specific device for detecting cancer cells. This biosensor is based on the fast entrapment of cancer cells on vertically aligned CNT arrays, which leads to mechanical and electrical interactions between CNT tips and entrapped cell membranes; thus, impedance of the biosensor is changed. This biosensor was fabricated through a photolithography process on $Ni/SiO_2/Si$ layers. CNT arrays were grown on 9 nm thick patterned Ni microelectrodes. SW48 colon cancer cells were passed over the surface of CNT covered electrodes. CNT arrays act as both; adhesive as well as conductive agents. The impedance changes occurred as fast as 30 s (for whole entrapment and signaling processes). The present biosensor was able to detect the cancer cells with the concentration as low as 4000 cells cm^{-2} on its surface with a sensitivity of 1.7×10^{-3} Ω cm^2.

Thayyath and Alexander[56] developed a novel potentiometric sensor with high selectivity and sensitivity for the determination of lindane. A MWCNT was grafted using glycidyl methacrylate (GMA). The reaction of MWCNT with GMA produces MWCNT-g-GMA. Epoxide ring of GMA reacts with allylamine to produce the vinylated MWCNT (MWCNT-CH=CH$_2$). MWCNT-based imprinted polymer (MWCNT-MIP) was synthesized using methacrylic acid (MAA) as the monomer, ethylene glycol dimethacrylate (EGDMA) as the cross linker, and α,α′-azobisisobutyronitrile (AIBN) as the initiator. Lindane, an organochlorine pesticide molecule, was used as the template. The optimizations of operational parameters were also done. The sensor responds to lindane in the range $1.0 \times 10^{-10} - 1.0 \times 10^{-3}$ M and the detection limit was 1.0×10^{-10} M.

A "green" synthesis method was used by Hu et al.[57] to prepare Pt@BSA nanocomposite. It was reported that electrocatalytic activity toward oxygen reduction was enhanced due to the excellent bioactivity of anchored GOD and superior catalytic performance of platinum nanoparticles, which was gradually restrained with the addition of glucose. They developed a sensitive glucose biosensor upon the restrained oxygen reduction peak current. Differential pulse voltammetry (DPV) was used to determine the performance of the enzyme biosensor, which gave a linear response range from 0.05 to 12.05 mM with an optimal detection limit of 0.015 mM. As-proposed sensing technique revealed high selectivity, satisfactory storage stability, durability, and favorable fabrication reproducibility with the RSD of 3.8%. This glucose biosensor had a good detection accuracy of analytical recoveries within 97.5 to 104.0% in human blood serum samples.

Oh et al.[58] fabricated CNT-based biosensor with a field-effect transistor structure, which can detect hepatitis B. For real-time measurement, they mounted microfluidic channel on the device and the presence of hepatitis B antigen was detected by measurement of electrical conductance as a function of time. It was reported that when hepatitis B antigen was exposed to the CNT biosensor with hepatitis B antibody immobilized, the conductance was found to increase, but it almost became constant within 10 min. As the antigen concentration increases, the conductance was also increased further. Thus, it is possible to detect hepatitis B in real-time using the CNT biosensor.

A highly sensitive flow injection amperometric biosensor was designed by Lui and Lin[59] to detect organophosphate pesticides and nerve agents. It was based on self-assembled acetylcholinesterase (AChE) on a CNT-modified glassy carbon (GC) electrode. AChE was immobilized on the negatively charged CNT surface by assembling alternatively a layer of cationic poly(diallyldimethylammonium chloride) (PDDA) and AChE. Formation of layer-by-layer nanostructures on carboxyl-functionalized CNTs was evident from TEM. The immobilization of AChE on the CNT/PDDA surface was also confirmed. The electrocatalytic activity of CNT leads to a significantly improved electrochemical detection of the thiocholine product generated enzymatically, which includes a low oxidation overvoltage, higher sensitivity, and stability. As-developed PDDA/AChE/PDDA/CNT/GC biosensor integrated into a flow injection system was successfully used to monitor organophosphate pesticides and nerve agents, like paraoxon. This biosensor was used to measure even very low concentration of paraoxon (0.4 pM) with only 6 min inhibition time. It has excellent operational lifetime stability with no decrease in the activity of enzymes for more than 20 repeated measurements.

Wang et al.[60] developed deoxyribonucleic acid (DNA) biosensor, which was based on a CNT-modified electrode and used it as an anticancer drug screening device for rapid electrochemical detection of cyclophosphamide. The interaction of cyclophosphamide with double stranded (ds) calf thymus DNA was detected with this biosensor and it was compared with the carbon paste-based biosensor. It was indicated that the developed CNT-based DNA biosensor had a faster response and a higher detection reproducibility as compared to the carbon paste-based biosensor.

A simple and sensitive method for the real-time detection of a prostate cancer marker (PSA-ACT complex) has been reported by Kim et al.[61] using label-free protein biosensors based on a carbon nanotube field effect transistor (CNT-FET). They functionalized CNT-FET with a solution containing various linker-to-spacer ratios. The binding event of the target PSA-ACT complex onto the receptor was detected by monitoring the gating effect caused by charges in the target PSA-ACT complex. As a buffer solution is used, it was difficult to control the distance between the receptors through introduction of linkers and spacers. As a result, the charged target PSA-ACT complex could easily approach the CNT surface within the Debye length and a large gating effect is observed. CNT-FET biosensors modified with only linkers was unable to detect target proteins, until a very high concentration of the PSA-ACT complex solution (~500 ng mL^{-1}) was injected. On the other hand, biosensor modified with a 1:3 ratio of linker-to-spacer could detect 1.0 ng mL^{-1} without any pretreatment. This linker and spacer-modified CNT-FET could successfully block non-target proteins and detect the target protein in human serum quite selectively.

A fast, sensitive, and label-free biosensor for the selective determination of *Salmonella Infantis* has been reported by Villamizar et al.[62] It is based on a field effect transistor (FET), here a network of single-walled carbon nantotubes acts as the conductor channel. Anti-*Salmonella* antibodies were adsorbed onto the SWCNTs and then the SWCNTs were protected with Tween 20 to prevent the non-specific binding of other bacteria or proteins. They exposed this FET device to increasing concentrations of *S. Infantis* and could detect at least 100 CFU mL^{-1} in 1 h. *Streptococcus pyogenes* and *Shigella sonnei* were also tested as potential competing bacteria for *Salmonella*. It was observed that *Streptococcus* and *Shigella* did not interfere with the detection of *Salmonella* at a concentration of 500 CFU mL^{-1}.

Bacterial food poisoning is a common threat, but it can be prevented with due care and proper handling of food products. Viswanathana et al.[63] developed a disposable electrochemical immunosensor for the simultaneous measurements of common food pathogenic bacteria namely *Escherichia coli* O157:H7 (E. coli), *campylobacter*, and *salmonella*. They fabricated this immunosensor by immobilizing the mixture of anti-*E. coli, anticampylobacter*, and *anti-salmonella* antibodies (1:1:1) on the surface of the multi-walled carbon nanotube-polyallylamine modified screen printed electrode (MWCNT-PAH/SPE). Bacteria suspension was attached to the immobilized antibodies, when the immunosensor was incubated in liquid samples. The calibration curves for three selected bacteria were found in the range of 1 × 103 × 105 cells mL^{-1} with the LOD 400 for *salmonella*, 400 for *campylobacter*, and 800 cells mL^{-1} for *E. coli*.

Novel TiO_2/CNT nanocomposites have been prepared by Shen et al.[64] doped on the carbon paper as the modified electrodes. Then, the redox behavior of the ferricyanide probe and the surface properties of the cancer cells coated on the modified electrodes were investigated using electro-chemical and contact angle measurements. Significantly enhanced electro-chemical signals on the modified electrodes covered with cancer cells have been observed as compared with electrochemical signals on bare carbon paper and nanocomposite modified substrates. It was revealed that different leukemia cells (i.e., K562/ ADM cells and K562/B.W. cells) could also be recognized because of their different electrochemical behavior as well as hydrophilic/hydrophobic features on the modified electrodes.

A single nanotube field effect transistor array, functionalized with IGF1R-specific and Her2-specific antibodies, has been reported by Shao et al.[65] It has highly sensitive and selective sensing of live, intact MCF7 and BT474 human breast cancer cells in blood. This biosensor showed 60% decrease in conductivity upon interaction with BT474 or MCF7 breast cancer cells in two µL drops of blood while non-specific antibodies or with MCF10A control breast cells produced only < 5% decrease in electrical conductivity. Free energy change upon cell–antibody binding, the stress exerted on the nanotube, and the change in conductivity are particularly specific to a specific antigen–antibody interaction and as such, these properties can be used as a fingerprint for the molecular sensing of circulating cancer cells. It was revealed that the binding of a single cell to a single nanotube field effect transistor changes electrical conductivity. A nanoscale oncometer with single cell sensitivity with a diameter 1000 times smaller than a cancer cell is possible that functions in a drop of fresh blood.

A prototype 4-unit electrochemical immunoarray based on SWCNT forests has been reported by Chikkaveeraiah et al.[66] for the simultaneous detection of multiple protein biomarkers for prostate cancer. They designed immunoarray procedures to measure simultaneously prostate specific antigen (PSA), prostate specific membrane antigen (PSMA), platelet factor-4 (PF-4), and interleukin-6 (IL-6) in a single serum sample. All of these proteins are elevated in serum of patients with prostate cancer, but these have quite different relative levels of serum concentration.

An electrochemical biosensor was developed by Bareket et al.[67] for the detection of formaldehyde in aqueous solution. It was based on the coupling of the enzyme formaldehyde dehydrogenase and a CNT-modified screen-printed electrode (SPE). The amperometric response to formaldehyde released from U251 human glioblastoma cells situated in the biosensor chamber was monitored in response to treatment with various anticancer prodrugs of formaldehyde and butyric acid. It was revealed that current response was higher for prodrugs that release two molecules of formalde-hyde (AN-193) than those prodrugs that release only single molecule of formaldehyde (AN-1, AN-7). It was reported that homologous prodrugs that release one (AN-88) or two (AN-191) molecules of acetaldehyde, showed no signal. This sensor was found to be fast, sensitive, selective, inexpensive, and disposable, and also easy to manufacture and operate.

2.6 DRUG DELIVERY

CNTs can also act as delivery vehicles for drugs; thus, these are suitable for therapeutics in regenerative medicine. One of the most desirable property of a drug delivery system is the ability to have a system, which responds to external stimuli and programmable triggers followed by drug release that can be adjusted according to therapeutic requirements. Programmable external triggering by ultrasound, electrical field, magnetic fields, light, enzymes, pH or temperatures can direct the optimized controllable drug release from a drug eluting system. SWCNTs are having a great potential as vehicles for cancer diagnostics and chemotherapies because of their properties, such as high drug-carrying capacities, remarkable cell membrane penetrability, pH-dependent therapeutic unloading, prolonged circulating times and intrinsic fluorescent, photoacoustic, photothermal, and Raman properties.

Sirivisoot and Pareta[68] reported that MWCNTs grown out of anodized nanotubular titanium (MWCNT-Ti) is useful as a sensing electrode for various biomedical applications. These sensors detected the redox reactions of proteins deposited by osteoblasts (bone-forming cells) in extracellular matrix bone formation. Penicillin or Streptomycin and drug dexamethasone were immobilized on and released from nanostructured polypyrrole-drug-coated Ti on demand using electrical stimuli. These coatings were helpful in treating bacterial infections, reduce inflammation, promote bone growth and also to reduce fibroblast functions, which will enhance the use of such materials as implant biosensing and therapeutic devices.

A new family of folate-decorated and CNT-mediated drug delivery system has been reported by Huang et al.,[69] which involves uniquely combining CNTs with anticancer drug (DOX) for controlled drug release. The synthesis of such nanocarrier involved attachment of DOX to CNT surface via π–π stacking interaction, which was followed by encapsulation of CNTs with folic acid-conjugated chitosan. The π–π stacking interaction allowed controlled release of drug. Apart from it, encapsulation of CNTs enhances the stability of the nanocarrier in aqueous medium, because of the hydrophilicity and cationic charge of chitosan. A unique integration of drug targeting and visualization showed great promises toward current challenges in cancer therapy.

Heister et al.[70] designed a drug delivery system based on CNTs for the anti-cancer drugs DOX and mitoxantrone. It was found stable under biological conditions, which allowed sustained release, and promotes selectivity through an active targeting scheme. A systematic approach to PEG conjugation has been selected in order to create a formulation of stable and therapeutically effective CNTs. This type of drug delivery system may have improved cancer treatment modalities by reducing drug-related side effects.

Bhirde et al.[71] reported in vivo killing of cancer cells using a drug-single wall CNT bioconjugate. They also demonstrated that these are superior as compared to nontargeted bioconjugates. First line anticancer agent cisplatin and epidermal growth factor (EGF) were attached to SWCNTs to specifically target squamous cancer. SWCNT-cisplatin without EGF was used for nontargeted control HNSCC for comparison. It was revealed that SWCNT-Qdot-EGF bioconjugates internalized rapidly into the cancer cells. While, limited uptake was there for control cells without EGF. The uptake was blocked by siRNA knockdown of EGFR in cancer cells, which revealed the importance of EGF-EGFR binding. Head and neck squamous carcinoma cells (HNSCC) treated with SWCNT-cisplatin-EGF were also killed selectively, but control systems did not influence cell proliferation as there was no EGF-EGFR binding. It was observed that regression of tumor growth was fast in mice treated with targeted SWCNT-cisplatin-EGF as compared to nontargeted SWNT-cisplatin.

A method for the triple functionalization of oxidised SWCNTs has been reported by Heister et al.[72] with the anti-cancer drug DOX, a monoclonal antibody, and a fluorescent marker at non-competing binding sites. It allowed for the targeted delivery of the anti-cancer drug to cancer cells. It was shown that the complex is efficiently taken up by cancer cells, which will release DOX, and translocate to the nucleus, while the nanotubes remained in the cytoplasm.

Zhang et al.[73] reported a targeted drug delivery system, which is triggered by changes in pH based on SWCNTs, derivatized with carboxylate groups and coated with a polysaccharide material. It can be loaded with the anticancer drug DOX. It was observed that drug binds at physiological pH (pH 7.4) and it is only released at a lower pH (lysosomal pH and the pH characteristic of certain tumor environments). Loading efficiency and release rate of the associated DOX can be controlled, if surface potentials of the modified nanotubes were modified by modification of the polysaccharide coating. It was revealed that folic acid (FA), which is a targeting agent for many tumors, can be additionally tethered to the SWCNTs to deliver DOX selectively into the lysosomes of HeLa cells with a significantly higher efficiency than free DOX. The DOX released from these modified nanotubes was found to damage nuclear DNA and inhibited the cell proliferation.

A targeted drug-delivery system was developed by Jeyamohan et al.[74] for selective killing of breast cancer cells with PEG biofunctionalized and DOX-loaded SWCNTs conjugated with folic acid. In vitro drug-release studies were carried out, which showed that the drug DOX binds at physiological pH (7.4) and it is released only at a lower pH, that is, lysosomal pH (4.0), a characteristic pH of the tumor environment. A sustained release of DOX from the SWCNTs was observed for a period of 3 days. Optical properties of SWCNTs provide an avenue for selective photothermal ablation in cancer therapy. In vitro studies showed that laser was effective in destroying the cancer cells, while it spares the normal cells. An enhanced killing of breast cancer cells was achieved on combining laser effect with DOX-conjugated SWCNTs. Thus, this nanodrug-delivery system (laser, drug, and SWCNTs) looks to be a promising candidate with high treatment efficacy and low side effects for cancer therapy.

Singh et al.[75] developed a novel multifunctional hybrid nanomaterial, magnetic CNTs, which was ensheathed with MS. They used it for the simultaneous applications of drug delivery and imaging. Magnetic nanoparticles

(MNPs) were decorated onto the multi-walled CNTs, which was then layered with mesoporous silica ($mSiO_2$) to facilitate the loading of bioactive molecules to a large quantity showing magnetic properties. This hybrid nanomaterial had a high mesoporosity due to the surface-layered $mSiO_2$, and excellent magnetic properties (MRI in vitro and in vivo). These nanocarriers showed high loading capacity for therapeutic molecules including drug gentamicin and protein cytochrome C. The genetic molecule siRNA was effectively loaded, which was released over a period of days to a week. Hybrid nanocarriers exhibited a high cell uptake rate through magnetism, while eliciting favourable biological efficacy within the cells.

As the majority of side effects are there in present chemotherapies like excessive dosing of anticancer drugs, it is essential to minimize the amount of drug but with maximize drug efficacy just to increase the life-quality of chemotherapy patients. lee et al.[76] demonstrated that the intracellular delivery of amide linked DOX on CNT can nullify the efflux of cancer cells through prolonged endolysosome delivery. It can induce burst release of DOX in an acidic hydrolase environment, resulting into reduction in the amount of anti-cancer drug by ten folds as compared to conventional effective drug dose. The clearance of accumulated CNTs in the liver was over after 4 weeks. The analysis of liver toxicity markers showed almost no changes in GOT and GPT levels and release of pro-inflammatory cytokines across both; short-term as well as long-term periods.

A new type of drug delivery system (DDS) was constructed by Ji et al.,[77] which involves chitosan (CHI) modified SWCNTs for controllable loading/release of anti-cancer drug DOX. First CHI was non-covalently wrapped around SWCNTs, which imparts water-solubility and biocompatibility to the nanotubes. Folic acid (FA) was also bound to the outer CHI layer so as to have selective killing of tumor cells. It was reported that targeting DDS could effectively kill the HCC SMMC-7721 cell lines and it depresses the growth of liver cancer in nude mice; thus, show better pharmaceutical efficiency to free DOX. It was revealed that the targeting DDS had negligible in vivo toxicity. Thus, it is promising for high treatment efficacy and low side effects for future cancer therapy.

Shao et al.[78] developed a lipid-drug approach for efficient drug loading onto CNT. A long chain lipid molecule was conjugated to the drug molecule so that this can be loaded directly onto CNT through binding of tail of lipid in the drug molecule to CNT surfaces through hydrophobic interactions. PTX and folic acid was conjugated with a non-toxic lipid molecule docosanol for functionalization with CNT for targeted drug delivery. A high level of drug loading onto SWCNT could be achieved by this approach. It was reported that conjugation of FA to SWCNT-lipid-PTX led to enhanced cell penetration capacity. Targeted SWCNT-lipid-PTX showed much improved drug efficacy in vitro as compared to free drug taxol while non-targeted SWCNT-lipid-PTX at 48 h (78.5% *vs.* 31.6% and 59.1% in cytotoxicity, respectively, $p < 0.01$). A human breast cancer xenograft mouse model also confirmed that drug efficacy has improved. This targeted SWCNT-lipid-PTX was found to be non-toxic .

Fahrenholtz et al.[79] developed a three-component drug-delivery system, which consists MWCNTs coated with a non-classical platinum chemotherapeutic agent ([PtCl(NH$_3$)$_2$(L)]Cl (P3A1 where L = N-(2-(acridin-9-ylamino) ethyl)-N-methylproprionimidamide) and 1,2-distearoyl-sn-glycero-3-phosphoethanolamine-N-[amino(PEG)-5000] (DSPE-mPEG). The optimized P3A1-MWCNTs were found to be colloidally stable in physiological solution and were able to deliver more P3A1 into breast cancer cells than free drug. These are cytotoxic to several cell models of breast cancer and induce S-phase cell cycle arrest and non-apoptotic cell death in breast cancer cells. It was suggested that delivery of P3A1 to cancer cells using MWCNTs (as a drug carrier) is better for combination of cancer chemotherapy and PTT.

A multifunctional anti-cancer prodrug system has been reported by Fan et al.[80] which is based on water-dispersible CNT. This type of prodrug system features properties like active targeting, pH-triggered drug release, and photodynamic therapeutic properties. They incorporated an anti-cancer drug, DOX onto CNT *via* a cleavable hydrazone bond. Folic acid (a targeting ligand) was also coupled onto CNT. It was reported that this prodrug preferably enters folate receptor (FR)-positive cancer cells undergoing intracellular release of the drug triggered by the lower pH. CNT carrier exhibited photodynamic therapeutic action; and as such, the cell viability of FR-positive cancer cells can be further reduced upon light irradiation. This dual effect of drug release triggered by pH and PDT increase the therapeutic efficacy of the DOX–CNT prodrug.

A novel SWCNT-based tumor-targeted drug delivery system (DDS) has been developed by Chen et al.[81] It consisted of a functionalized SWCNT linked to tumor-targeting modules as well as prodrug modules. There are three key features of this nanoscale DDS and these are:

- Use of functionalized SWCNTs as a biocompatible platform for the delivery of therapeutic drugs or diagnostics,

- Conjugation of prodrug modules of an anticancer agent (taxoid with a cleavable linker) that is activated to its cytotoxic form inside the tumor cells upon internalization and in situ drug release, and
- Attachment of tumor-recognition modules (biotin and a spacer) to the nanotube surface.

Three fluorescent and fluorogenic molecular probes were designed, synthesized, characterized, and subjected to the analysis of the receptor-mediated endocytosis and drug release inside the cancer cells (L1210FR leukemia cell line) by means of confocal fluorescence microscopy just to prove the efficacy of this DDS. The specificity and cytotoxicity of the conjugate have also been assessed and compared with L1210 and human noncancerous cell lines.

2.7 PHOTODYNAMIC AND PHOTOTHERMAL THERAPY

PTT is considered a promising technique for cancer treatment because it is selective and have localized therapeutic effect by laser irradiation. This therapy damages malignant cells by heat converted from light by an agent. PDT also uses photosensitizers, which become cytotoxic upon irradiation with laser light at a particular excitation wavelength. Although phototherapy only has certain limitations, it can be combined with PDT for better output.

Zhang et al.[82] developed Ru(II) complex-functionalized SWCNTs (Ru@ SWCNTs) as nanotemplates for bimodal photothermal and two-photon photodynamic therapy (PTT-TPPDT) as SWCNTs have the ability to load a large amount of Ru(II) complexes (Ru1 or Ru2) *via* noncovalent π–π interactions. Such loaded Ru(II) complexes are effectively released by the photothermal effect of irradiation (808 nm diode laser; 0.25 W cm^{-2}). This released Ru(II) complexes produce singlet oxygen species upon two-photon laser irradiation,which can be used as a two-photon photodynamic therapy (TPPDT) agent. A combination of PTT and TPPDT, will give Ru@SWCNTs greater anticancer efficacies than individual PDT using Ru(II) complexes or PTT using SWCNTs in two-dimensional (2D) cancer cell and three-dimensional (3D) multicellular tumor spheroid (MCTS) models. An in vivo tumor ablation was also achieved with excellent treatment efficacy under diode laser irradiation for 5 min.

Construction of a nanosystem has been described by Marangon et al.,[83] which is based on MWCNTs and a photosensitizer m-tetrahydroxyphenylchlorin (mTHPC) for cancer treatment. It is a combination of PDTs and PTTs. The study of photoactivity of complex mTHPC/MWCNT revealed quenching of the photosensitizer associated to CNT and an almost total release of the photosensitizer was achieved into the cell cytoplasm after uptake by SKOV3 ovarian cancer cells. Cytotoxicity had a correlation at the cell level with the uptake of mTHPC/MWCNT. PDT and PTT treatment induced different signaling pathways lead to cell apoptosis.

Zinc monocarboxyphenoxyphthalocyanine was modified chemically with ascorbic acid via an ester bond to give ZnMCPPc-AA. Complexes were coordinated to SWCNTs via $\pi-\pi$ interaction to give ZnMCPPc-AA-SWCNT and ZnMCPPc-SWCNT respectively. Later these complexes showed comparatively better photophysical properties: such as improved triplet lifetimes and quantum yields, and singlet oxygen quantum yields, when compared to first one. The PDT activities of all the four complexes were tested by Ogbodu and Nyokong[84] in vitro on MCF-7 breast cancer cells. Ascorbic acid was found to suppress the PDT effect of first due to its ability to reduce oxidative DNA damage because of potent reducing properties. They observed highest phototoxicity for fourth complex, which resulted in 77% decrease in cell viability, followed by third complex which showed 67% decrease in cell viability.

An improved PDT effect of zinc monocarboxyphenoxyphthalo-cyanine (ZnMCPPc) was also reported by Ogbodu et al.,[85] upon conjugation to spermine (via amide bond) as a targeting molecule on MCF-7 breast cancer cells. ZnMCPPc-spermine conjugate was adsorbed onto SWCNTs (ZnMCPPc-spermine-SWCNT). No change was observed in the fluorescence quantum yield of first complex following formation of second and third one, which showed improved photophysical properties; that is, over 50% increases in triplet and singlet oxygen quantum yields compared to first complex. All the three complexes were relatively non-toxic to MCF-7 cancer cells, when incubated with 5–40 μM of each complex for 24 h in the dark. It was revealed that first complex (40 μM) resulted in only 64% decrease in cell viability, while rest two improved the PDT effect of first complex to 97% and 95% decrease in cell viability at 40 μM, respectively.

Shi et al.[86] synthesized CNT derivative, hyaluronic acid-derivatized CNTs (HA-CNTs) having high aqueous solubility, neutral pH, and tumor-targeting activity. A new PDT agent, hematoporphyrinmonomethyl ether (HMME), was adsorbed onto the functionalized CNTs to afford HMME-HA-CNTs. Inhibition of tumor growth was investigated both; in vivo and in vitro by a combination of PTT and PDT using HMME-HA-CNTs. HMME-HA-CNT nanoparticles are able to combine local specific PDT with external NIR PTT;

as a result significant improvement in the therapeutic efficacy of cancer treatment was achieved. The combined treatment had a synergistic effect and as a consequence, higher therapeutic efficacy was observed without obvious toxic effects to normal organs as compared with PDT or PTT alone.

Hashida et al.[87] evaluated the photothermal activity of a novel SWCNTs composite with a designed peptide having a repeated structure of H-(-Lys-Phe-Lys-Ala-)$_7$-OH [(KFKA)$_7$] against tumor cells in vitro as well as in vivo. It was demonstrated that SWCNT-(KFKA)$_7$ composite had high aqueous dispersibility that enabled SWCNTs to be used in tumor ablation. It was observed that 3 min of NIR irradiation of a colon 26 or HepG2 cell culture incubated with SWCNT-(KFKA)$_7$ had remarkable cell damage, while it was only moderate by single treatment with SWCNT-(KFKA)$_7$ or NIR irradiation alone. The intra-tumoral injection of SWCNT-(KFKA)$_7$ solution followed by NIR irradiation resulted in a rapid rise of temperature to 43°C in the subcutaneously inoculated colon 26 tumor, where remarkable suppression of tumor growth was observed as compared with treatment with only SWCNT-(KFKA)$_7$ injection or NIR irradiation.

The dual application of intravenously injected SWCNTs as photoluminescent agents has been reported by Robinson et al.[88] for in vivo tumor imaging in the 1.0–1.4 µm emission region and as NIR absorbers and heaters at 808 nm for photothermal tumor elimination at the lowest injected dose (70 µg of SWCNT/mouse, equivalent to 3.6 mg kg^{-1}) and laser irradiation power (0.6 W cm^{-2}) It was reported that complete tumor elimination was achieved for large numbers of photothermally treated mice without any toxic side effects even after more than six months post-treatment. The performance of SWCNTs and gold nanorods (AuNRs) at an injected dose of 700 µg of AuNR/mouse (equivalent to 35 mg kg^{-1}) was observed in NIR photothermal ablation of tumors in vivo. It was revealed that there was effective tumor elimination with SWCNTs at ten times lower injected doses and lower irradiation powers than for AuNRs.

2.8   CONCLUSION

CNTs have numerous applications, but their applications are limited. CNTs have found use in detecting and target drug delivery for cancer, biosensing of certain pathogens, in CAs in MRI, combination therapy, etc. Time is not far off, when CNTs will find much interesting applications and will occupy a prominent application in different areas of biomedical field.

KEYWORDS

- carbon nanotubes
- cancer
- targeted drug delivery
- photothermal therapy
- photodynamic therapy
- biosensors
- chemical sensors

REFERENCES

1. Iijima, S. Helical Microtubules of Graphitic Carbon. *Nature* **1991**, *354*, 56–58.
2. Dresselhaus, M. S.; Avouris, P. *Carbon Nanotubes: Synthesis, Structure, Properties and Applications*, Berlin: Springer, 2001.
3. Baughman, R. H.; Zakhidov, A. A.; de Heer, W. A. Carbon Nanotubes—The Route Toward Applications. *Science* **2002**, *297*, 787–792.
4. Ijima, S.; Ajayan, P. M.; Ichihashi, T. Growth Model for Carbon Nanotubes. *Phys. Rev. Lett.* **1992**, *69* (21), 3100–3103.
5. Ebbesen, T. W.; Ajayan, P. M. Large Scale Synthesis of Carbon Nanotubes. *Nature* **1992**, *358*, 220–221.
6. Bethune, D. S.; Kiang, C. H.; de Vries, M. S.; Gorman, G.; Savoy, R.; Vazquez, J.; Beyers, R. Cobalt-Catalyzed Growth of Carbon Nanotubes with Single-Atomic-Layer Walls. *Nature* **1993**, *363* (6430), 605–607.
7. Liu, C.; Cong, H. T.; Li, F.; Tan, P. H.; Cheng, H. M.; Lu, K.; Zhou, B. L. Semi-Continuous Synthesis of Single-Walled Carbon Nanotubes by a Hydrogen Arc Discharge Method. *Carbon* **1999**, *37* (11), 1865–1868.
8. Antisari, M. V.; Marazzi, R.; Krsmanovic, R. Synthesis of Multiwall Carbon Nanotubes by Electric Arc Discharge in Liquid Environments. *Carbon* **2003**, *41* (12), 2393–2401.
9. Li, H.; Guan, L.; Shi Z.; Gu, Z. Direct Synthesis of High Purity Single-Walled Carbon Nanotube Fibers by Arc Discharge. *J. Phys. Chem. B* **2004**, *108* (15), 4573–4575.
10. Thess, A.; Lee, R.; Nikolaev, P.; Dai, H. J.; Petit, P.; Robert, J.; Xu, C. H.; Lee, Y. H.; Kim, S. G.; Rinzler, A. G.; Colbert, D. T.; Scuseria, G. E.; Tomanek, D.; Fischer, J. E.; Smalley, R. E. Crystalline Ropes of Metallic Carbon Nanotubes. *Science* **1996**, *273* (5274), 483–487.
11. Braidy, N.; El Khakani, M. A.; Botton, G. A. Carbon Nanotubular Structures Synthesis by Means of Ultraviolet Laser Ablation. *J. Mater. Res.* **2002**, *17* (9), 2189–2192.
12. Takahashi, S.; Ikuno, T.; Oyama, T.; Honda, S. I.; Katayama, M.; Hirao, T.; Oura, K. Synthesis and Characterization of Carbon Nanotubes Grown on Carbon Particles by Using High Vacuum Laser Ablation. *J. Vac. Soc. Jpn.* **2002**, *45* (7), 609–612.

13. Vanderwal, R. L.; Berger, G. M.; Ticich, T. M. Carbon Nanotube Synthesis in a Flame Using Laser Ablation for In Situ Catalyst Generation. *Appl. Phys. A* **2003**, *77* (7), 885–889.

14. Yacaman, M. J.; Yoshida, M. M.; Rendon, L.; Santiesteban, J. G. Catalytic Growth of Carbon Microtubules with Fullerene Structure. *Appl. Phys. Lett.* **1993**, *62*, 202–204.

15. Li, W. Z.; Xie, S. S.; Qian, L. X.; Chang, B. H.; Zou, B. S.; Zhou, W. Y.; Zhao, R. A.; Wang, G. Large-Scale Synthesis of Aligned Carbon Nanotubes. *Science* **1996**, *274* (5293), 1701–1703.

16. Qin, L. C. CVD Synthesis of Carbon Nanotubes. *J. Mater. Sci. Lett.* **1997**, *16* (6), 457–459.

17. Choi, Y. C.; Bae, D. J.; Lee, Y. H.; Lee, B. S.; Park, G. S.; Choi, W. B.; Lee, N. S.; Kim, J. M. Growth of Carbon Nanotubes by Microwave Plasma-Enhanced Chemical Vapor Deposition at Low Temperature. *J. Vac. Sc. Technol. A* **2000**, *18* (4), 1864–1868.

18. Varadan, V. K.; Xie, J. Large-Scale Synthesis of Multiwalled Carbon Nanotubes by Microwave CVD. *Smart Mater. Struct.* **2002**, *11* (4), 610–616.

19. Chatterjee, A. K.; Sharon, M.; Banerjee, R.; Neumann-Spallart, M. CVD Synthesis of Carbon Nanotubes Using a Finely Dispersed Cobalt Catalyst and Their Use in Double Layer Electrochemical Capacitors. *Electrochim. Acta* **2003**, *48* (23), 3439–3446.

20. Park, D.; Kim, Y. H.; Lee, J. K. Synthesis of Carbon Nanotubes on Metallic Substrates by a Sequential Combination of PECVD and Thermal CVD. *Carbon* **2003**, *41* (5), 1025–1029.

21. Chaisitsak, S.; Yamada, A.; Konagai, M. Hot Filament Enhanced CVD Synthesis of Carbon Nanotubes by Using a Carbon Filament. *Diamond Relat. Mater.* **2004**, *13* (3), 438–444.

22. Seidel, R.; Duesberg, G. S.; Unger, E.; Graham, A. P.; Liebau, M.; Kreupl, F. Chemical Vapor Deposition Growth of Single-Walled Carbon Nanotubes at 600°C and a Simple Growth Model. *J. Phys. Chem. B* **2004**, *108* (6), 1888–1893.

23. Jung, J. H.; Hwang, G. B.; Lee, J. E.; Bae, G. N. Preparation of Airborne Ag/CNT Hybrid Nanoparticles Using an Aerosol Process and Their Application to Antimicrobial Air Filtration. *Langmuir* **2011**, *27* (16), 10256–10264.

24. Yang, S.; Zhu, Z.; Wei, F.; Yang, X. Carbon Nanotubes/Activated Carbon Fiber Based Air Filter Media for Simultaneous Removal of Particulate Matter and Ozone. *Building Environ.* **2017**, *125*, 60–66.

25. Tan, C.; Huang, X.; Zhang, H. Synthesis and Applications of Graphene-Based Noble Metal Nanostructures. *Materials Today* **2013**, *16* (1–2), 29–36.

26. Prakash, J., Sun, S., Swart, H. C., Gupta, R. K. Noble Metals-TiO$_2$ Nanocomposites: From Fundamental Mechanisms to Photocatalysis, Surface Enhanced Raman Scattering and Antibacterial Applications. *Appl. Mater. Today* **2018**, *11*, 82–135.

27. Qiu, B.; Xing, M.; Zhang, J. Recent Advances in Three-Dimensional Graphene Based Materials for Catalysis Applications. *Chem. Soc. Rev.* **2018**, *47* (6), 2165–2216.

28. Liu, Y.; Liu, X.; Zhao, Y.; Dionysiou, D. D. Aligned α-FeOOH Nanorods Anchored on a Graphene Oxide-Carbon Nanotubes Aerogel Can Serve as an Effective Fenton-Like Oxidation Catalyst. *Appl. Catal. B: Environ.* **2017**, *213*, 74–86.

29. Piccinino, D.; Abdalghani, I.; Botta, G.; Crucianelli, M.; Passacantando, M.; Di Vacri, M. L.; Saladino, R. Preparation of Wrapped Carbon Nanotubes Poly(4-Vinylpyridine)/MTO Based Heterogeneous Catalysts for the Oxidative Desulfurization (ODS) of Model and Synthetic Diesel Fuel. *Appl. Catal. B: Environ.* **2017**, *200*, 392–401.

30. Xu, J.; Cao, Z.; Zhang, Y.; Yuan, Z.; Lou, Z.; Xu, X.; Wang, X. A Review of Functionalized Carbon Nanotubes and Graphene for Heavy Metal Adsorption from Water: Preparation, Application, and Mechanism. *Chemosphere* **2018**, *195*, 351–364.

31. Ma, S.-B.; Nam, K.-W.; Yoon, W.-S.; Yang, X.-Q.; Ahn, K.-Y.; Oh, K.-H.; Kim, K.-B. Electrochemical Properties of Manganese Oxide Coated onto Carbon Nanotubes for Energy-Storage Applications. *J. Power Sources* **2008**, *178* (1), 483–489.

32. Ma, P.-C.; Siddiqui, N. A.; Marom, G.; Kim, J.-K. Dispersion and Functionalization of Carbon Nanotubes for Polymer-Based Nanocomposites: A Review. *Composites Part A: Appl. Sci. Manufact.* **2010**, *41* (10), 1345–1367.

33. Koti, V.; George, R.; Koppad, P.; Murthy, K. V. S.; Shakiba, A. Friction and Wear Characteristics of Copper Nanocomposites Reinforced with Uncoated and Nickel Coated Carbon Nanotubes. *Mater. Res. Express* **2018**, *5* (9), doi:10.1088/2053-1591/aad8fa.

34. Zhu, T.; Xia, B.; Zhou, L.; Wen (David) Lou, X. Arrays of Ultrafine CuS Nanoneedles Supported on a CNT Backbone for Application in Supercapacitors. *J. Mater. Chem.* **2012**, *22* (16), 7851–7855.

35. Chappanda, K. N.; Shekhah, O.; Yassine, O.; Patole, S. P.; Eddaoudi, M.; Salama, K. N. The Quest for Highly Sensitive QCM Humidity Sensors: The Coating of CNT/MOF Composite Sensing Films as Case Study. *Sensors Actuators B: Chem.* **2018**, *257*, 609–619.

36. Chaitongrat, B.; Chaisitsak, S. Fast-LPG Sensors at Room Temperature by α-Fe2O3/CNT Nanocomposite Thin Films. *J. Nanomater.* **2018**, *2018*, 1–11.

37. Wu, L.; Zhang, X.; Wang, M.; He, L.; Zhang, Z. Preparation of Cu2O/CNTs Composite and Its Application as Sensing Platform for Detecting Nitrite in Water Environment. *Measurement* **2018**, *128*, 189–196.

38. Goldoni, A.; Alijani, V.; Sangaletti, L.; D'Arsiè, L. Advanced Promising Routes of Carbon/Metal Oxides Hybrids in Sensors: A Review. *Electrochim. Acta* **2018**, *266*, 139–150.

39. Liu, J.; Wang, C.; Wang, X.; Wang, X.; Cheng, L.; Li, Y.; Liu, Z. Mesoporous Silica Coated Single-Walled Carbon Nanotubes as a Multifunctional Light-Responsive Platform for Cancer Combination Therapy. *Adv. Funct. Mater.* **2015**, *25* (3), 384–392.

40. Zhao, H.; Chao, Y.; Liu, J.; Huang, J.; Pan, J.; Guo, W.; Wu, J.; Sheng, M.; Yang, K.; Wang, J.; et al. Polydopamine Coated Single-Walled Carbon Nanotubes as a Versatile Platform with Radionuclide Labelling for Multimodal Tumor Imaging and Therapy. *Theranostics* **2016**, *6* (11), 1833–1843.

41. Liu, Z.; Cai, W.; He, L.; Nakayama, N.; Chen, K.; Sun, X.; Chen, X.; Dai, H. In Vivo Biodistribution and Highly Efficient Tumor Targeting of Carbon Nanotubes in Mice. *Nat. Nanotechnol.* **2007**, *2*, 47–52.

42. Liu, Z.; Chen, K.; Davis, C.; Sherlock, S.; Cao, Q.; Chen, X.; Dai, H. Drug Delivery with Carbon Nanotubes for In Vivo Cancer Treatment. *Cancer Res.* **2008**, *68* (16), 6652–6660.

43. Heister, E.; Neves, V.; Tîlmaciu, C.; Lipert, K.; Beltrán, V. S.; Coley, H. M.; Silva S. R. P.; McFadden, J. Triple Functionalisation of Single-Walled Carbon Nanotubes with Doxorubicin, a Monoclonal Antibody, and a Fluorescent Marker for Targeted Cancer Therapy. *Carbon* **2009**, *47* (9), 2152–2160.

44. Hampel, S., Kunze, D., Haase, D., Krämer, K., Rauschenbach, M., Ritschel, M., Büchner, B., et al. Carbon Nanotubes Filled with a Chemotherapeutic Agent: A Nanocarrier Mediates Inhibition of Tumor Cell Growth. *Nanomedicine* **2008**, *3* (2), 175–182.

45. Jha, P. K.; Jha, R. K.; Rout, D.; Gnanasekar, S.; Rana, S. V. S.; Hossain, M. Potential Targetability of Multi-Walled Carbon Nanotube Loaded with Silver Nanoparticles Photosynthesized from *Ocimum tenuiflorum* (Tulsi Extract) in Fertility Diagnosis. *J. Drug Target.* **2017**, *25* (7), 616–625.

46. Sitharaman, B.; Kissell, K. R.; Hartman, K. B.; Tran, L. A.; Baikalov, A.; Rusakova, I.; Sun, Y.; Khant, H. A.; Ludtke, S. J.; Chiu, W.; Laus, S; To′th, E.; Helm, L.; Merbachd A. E.; Wilson, L. J. Superparamagnetic Gadonanotubes are High-Performance MRI Contrast Agents. *Chem. Commun.* **2005**, *31*, 3915–3917.

47. Hartman, K. B.; Laus, S.; Bolskar, R. D.; Muthupillai, R.; Helm, L.; Toth, E.; Merbach, A. E.; Wilson, L. J. Gadonanotubes as Ultrasensitive pH-Smart Probes for Magnetic Resonance Imaging. *Nano Lett.* **2008**, *8* (2), 415–419.

48. Wu, H.; Liu, G.; Zhuang, Y.; Wu, D.; Zhang, H.; Yang, H.; He H.; Yang, S. The Behavior after Intravenous Injection in Mice of Multiwalled Carbon Nanotube/Fe$_3$O$_4$ Hybrid MRI Contrast Agents. *Biomaterials* **2011**, *32* (21), 4867–4876.

49. Zavaleta, C.; de la Zerda, A.; Liu, Z.; Keren, S.; Cheng, Z.; Schipper, M., Chen, X.; Dai, H.; Gambhir, S. S. Noninvasive Raman Spectroscopy in Living Mice for Evaluation of Tumor Targeting with Carbon Nanotubes. *Nano Lett.* **2008**, *8* (9), 2800–2805.

50. Marangon, I.; Ménard-Moyon, C.; Kolosnjaj-Tabi, J.; Béoutis, M. L.; Lartigue, L.; Alloyeau, D.; Pach, E.; Ballesteros, B.; Autret, G.; Ninjbadgar, T.; et al. Covalent Functionalization of Multi-Walled Carbon Nanotubes with a Gadolinium Chelate for Efficient T1-Weighted Magnetic Resonance Imaging. *Adv. Funct. Mater.* **2014**, *24* (45), 7173–7186.

51. Liu, Y.; Muir, B. W.; Waddington, L. J.; Hinton, T. M.; Moffat, B. A.; Hao, X.; Qiu, J.; Hughes, T. C. Colloidallystabilized Magnetic Carbon Nanotubes Providing MRI Contrast in Mouse Liver Tumours. *Biomacromolecules* **2015**, *16* (3), 790–797.

52. Choi, H. K.; Lee J.; Park, M. K.; Oh, J. H. Development of Single-Walled Carbon Nanotube-Based Biosensor for the Detection of *Staphylococcus Aureus*. *J. Food Quality* **2017**, *2017*, 1–8, doi: 10.1155/2017/5239487.

53. Kara, Y.; Kim, C. T.; Kim, J. H.; Chung, J. H.; Lee, H. G.; Jun, S. Single-Walled Carbon Nanotube-Based Junction Biosensor for Detection of *Escherichia coli*. *J. Pone. Org.* **2014**, *9* (9), doi.:10.1371/journal.pone.0105767.

54. Zhu, Z.; Song, W. S.; Burugapalli, K.; Moussy, F.; Li, Y. L.; Zhong, X. H. Nano-Yarn Carbon Nanotube Fiber Based Enzymatic Glucose Biosensor. *Nanotechnol.* **2010**, *21* (16), doi:10.1088/0957-4484/21/16/165501.

55. Abdolahad, M.; Taghinejad, M.; Hossein Taghinejad, H.; Janmaleki M.; Shams Mohajerzadeh, S. A Vertically Aligned Carbon Nanotube-Based Impedance Sensing Biosensor for Rapid and High Sensitive Detection of Cancer Cells. *Lab Chip.* **2012**, *12* (6), 1183–1190.

56. Thayyath, S.; Alexander, A. S. Design and Fabrication of Molecularly Imprinted Polymer-Based Potentiometric Sensor from the Surface Modified Multiwalled Carbon Nanotube for the Determination of Lindane (γ-Hexachlorocyclohexane), an Organochlorine Pesticide. *Biosensor. Bioelectron.* **2015**, *64*, 586–593.

57. Hu, C.; Yang, D. P.; Zhu, F.; Jiang, F.; Shen, S. Enzyme-Labeled Pt@BSA Nanocomposite as a Facile Electrochemical Biosensing Interface for Sensitive Glucose Determination. *ACS Appl. Mater. Interfaces* **2014**, *6* (6), 4170–4178.
58. Oh J.; Yoo S.; Chang Y. W.; Lim K.; Yoo K.-H. Carbon Nanotube-Based Biosensor for Detection Hepatitis B. *Curr. Appl. Phys.* **2009**, *9* (4), e229–e231, doi: 10.1016/j. cap.2009.06.045.
59. Liu G.; Lin Y. Biosensor Based on Self-Assembling Acetylcholinesterase on Carbon Nanotubes for Flow Injection/Amperometric Detection of Organophosphate Pesticides and Nerve Agents. *Anal. Chem.* **2006**, *78* (3), 835–843.
60. Wang, S.; Wang, R.; Sellin, P. J.; Chang, S. X. Carbon Nanotube Based DNA Biosensor for Rapid Detection of Anti-Cancer Drug of Cyclophosphamide. *Curr. Nanosci.* **2009**, *5* (3), 312–317.
61. Kim, J. P.; Lee, B. Y.; Lee, J.; Hong, S.; Sima, S. J. Enhancement of Sensitivity and Specificity by Surface Modification of Carbon Nanotubes in Diagnosis of Prostate Cancer Based on Carbon Nanotube Field Effect Transistors. *Biosens. Bioelectron.* **2009**, *24*, 3372–3378.
62. Villamizar, R. A.; Marotoa, A.; Rius, F. X.; Inza, I.; Figueras, M. J. Fast Detection of *Salmonella Infantis* with Carbon Nanotube Field Effect Transistors. *Biosens. Bioelectron.* **2008**, *24*, 279–283.
63. Viswanathana, S.; Rani, C.; Ho, J. A. Electrochemical Immunosensor for Multiplexed Detection of Food-Borne Pathogens using Nanocrystal Bioconjugates and MWCNT Screen-Printed Electrode. *Talanta* **2012**, *94*, 315–319.
64. Shen, Q.; You, S.-K.; Park, S.-G.; Jiang, H.; Guo, D.; Chen, B.; Wang, X. Electrochemical Biosensing for Cancer Cells Based on TiO_2/CNT Nanocomposites Modified Electrodes. *Electroanalysis* **2008**, *20* (23), 2526–2530.
65. Shao, N.; Wickstrom, E.; Panchapakesan, B. Nanotube–Antibody Biosensor Arrays for the Detection of Circulating Breast Cancer Cells. *Nanotechnology* **2008**, *19* (46), doi: 10.1088/0957-4484/19/46/465101.
66. Chikkaveeraiah, B. V.; Bhirde, A.; Malhotra, R.; Patel, V.; Gutkind, J. S.; Rusling, J. F. Single-Wall Carbon Nanotube Forest Arrays for Immunoelectrochemical Measurement of Four Protein Biomarkers for Prostate Cancer. *Anal. Chem.* **2009**, *81* (21), 9129–9134.
67. Bareket, L.; Rephaeli, A.; Berkovitch, G.; Nudelman, A.; Rishpon, J. Carbon Nanotubes Based Electrochemical Biosensor for Detection of Formaldehyde Released from a Cancer Cell Line Treated with Formaldehyde-Releasing Anticancer Prodrugs. *Bioelectrochemistry* **2010**, *77* (2), 94–99.
68. Sirivisoot, S.; Pareta, R. A. Orthopedic Carbon Nanotube Biosensors for Controlled Drug Delivery. *Nanomedicine* **2012**, 149–179, doi:10.1533/9780857096449.2.149.
69. Huang, H.; Yuan, Q.; Shah, J. S.; Misra, R. D. K. A New Family of Folate-Decorated and Carbon Nanotube-Mediated Drug Delivery System: Synthesis and Drug Delivery Response. *Adv. Drug Delive. Rev.* **2011**, *63* (14–15), 1332–1339.
70. Heister, E.; Neves, V.; Lamprecht, C.; Silva, S. R. P.; Coley, H. M.; McFadden, J. Drug Loading, Dispersion Stability, and Therapeutic Efficacy in Targeted Drug Delivery with Carbon Nanotubes. *Carbon* **2012**, *50* (2), 622–632.
71. Bhirde, A. A.; Patel, V.; Gavard, J.; Zhang, G.; Sousa, A. A.; Masedunskas, A.; Leapman, R. D.; Weigert, R.; Gutkind, J. S.; Rusling, J. F. Targeted Killing of Cancer Cells In Vivo and In Vitro with EGF-Directed Carbon Nanotube-Based Drug Delivery. *ACS Nano* **2009**, *3* (2), 307–316.

72. Heister, E.; Neves, V.; Tîlmaciu, C.; Lipert, K.; Beltrán, V. S.; Coley, H. M.; Silva, S. R. P.; McFadden, J. Triple Functionalisation of Single-Walled Carbon Nanotubes with Doxorubicin, a Monoclonal Antibody, and a Fluorescent Marker for Targeted Cancer Therapy. *Carbon* **2009,** *47* (9), 2152–2160.

73. Zhang, X.; Meng, L.; Lu, Q.; Fei, Z.; Dyson, P. J. Targeted Delivery and Controlled Release of Doxorubicin to Cancer Cells Using Modified Single Wall Carbon Nanotubes. *Biomaterials* **2009,** *30* (30), 6041–6047.

74. Jeyamohan, P.; Hasumura, T.; Nagaoka, Y.; Yoshida, Y.; Maekawa, T.; Kumar, D. S. Accelerated Killing of Cancer Cells Using a Multifunctional Single-Walled Carbon Nanotube-Based System for Targeted Drug Delivery in Combination with Photothermal Therapy. *Int. J. Nanomed.* **2013,** *8,* 2653–2667.

75. Singh, R. K.; Patel, K. D.; Kim, J. J.; Kim, T. H.; Kim, J. H.; Shin, U. S.; Lee, E. J.; Knowles, J. C.; Kim, H. W. Multifunctional Hybrid Nanocarrier: Magnetic CNTs Ensheathed with Mesoporous Silica for Drug Delivery and Imaging System. *ACS Appl. Mater. Interfaces* **2014,** *6* (4), 2201–2208.

76. Lee, Y. K.; Choi, J.; Wang, W.; Lee, S.; Nam, T. H.; Choi, W. S.; Kim, C. J.; Lee, J. K.; Kim, S. H.; Kang, S. S.; et al. Nullifying Tumor Efflux by Prolonged Endolysosome Vesicles: Development of Low Dose Anticancer-Carbonnanotube Drug. *ACS Nano* **2013,** *7* (10), 8484–8497.

77. Ji, Z.; Lin, G.; Lu, Q.; Meng, L.; Shen, X.; Dong, L.; Fu, C.; Zhang, X. Targeted Therapy of SMMC-7721 Liver Cancer In Vitro and In Vivo with Carbon Nanotubes Based Drug Delivery System. *J. Colloid Interface Sci.* **2012,** *365* (1), 143–149.

78. Shao, W.; Paul, A.; Zhao, B.; Lee, C.; Rodes, L.; Prakash, S. Carbon Nanotube Lipid Drug Approach for Targeted Delivery of a Chemotherapy Drug in a Human Breast Cancer Xenograft Animal Model. *Biomaterials* **2013,** *34* (38), 10109–10119.

79. Fahrenholtz, C. D.; Ding, S.; Bernish, B. W.; Wright, M. L.; Zheng, Y.; Yang, M.; Yao, X.; Donati, G. L.; Gross, M. D.; Bierbach, U.; et al. Design and Cellular Studies of a Carbon Nanotube-Based Delivery System for a Hybrid Platinum-Acridine Anticancer Agent. *J. Inorg. Biochem.* **2016,** *165,* 170–180.

80. Fan, J.; Zeng, F.; Xu, J.; Wu, S. Targeted Anti-Cancer Prodrug Based on Carbon Nanotube with Photodynamic Therapeutic Effect and pH-Triggered Drug Release. *J. Nanopart. Res.* **2013,** *15* (1911), 1–15.

81. Chen, J.; Chen, S.; Zhao, X.; Kuznetsova, L. V.; Wong, S. S.; Ojima, I. Functionalized Single-Walled Carbon Nanotubes as Rationally Designed Vehicles for Tumor-Targeted Drug Delivery. *J. Amer. Chem. Soc.* **2008,** *130* (49), 16778–16785.

82. Zhang, P.; Huang, H.; Huang, J.; Chen, H.; Wang, J.; Qiu, K.; Zhao, D.; Ji, L.; Chao, H. Noncovalent Ruthenium (II) Complexes–Single-Walled Carbon Nanotube Composites for Bimodal Photothermal and Photodynamic Therapy with Near-Infrared Irradiation. *ACS Appl. Mater. Interfaces* **2015,** *7* (41), 23278–23290.

83. Marangon, I.; Ménard-Moyon, C.; Silva, A. K.; Bianco, A.; Luciani, N.; Gazeau, F. Synergic Mechanisms of Photothermal and Photodynamic Therapies Mediated by Photosensitizer/Carbon Nanotube Complexes. *Carbon* **2016,** *97,* 110–123.

84. Ogbodu, R. O.; Nyokong, T. The Effect of Ascorbic Acid on the Photophysical Properties and Photodynamic Therapy Activities of Zinc Phthalocyanine-Single Walled Carbon Nanotube Conjugate on MCF-7 Cancer Cells. *Spectrochim. Acta Part A: Mol. Biomol. Spectrosc.* **2015,** *151,* 174–183.

85. Ogbodu, R. O.; Limson, J. L.; Prinsloo, E.; Nyokong, T. Photophysical Properties and Photodynamic Therapy Effect of Zinc Phthalocyanine-Spermine-Single Walled Carbon Nanotube Conjugate on MCF-7 Breast Cancercell Line. *Synth. Metals* **2015**, *204*, 122–132.

86. Shi, J.; Ma, R.; Wang, L.; Zhang, J.; Liu, R.; Li, L.; Liu, Y.; Hou, L.; Yu, X.; Gao, J.; et al. The Application of Hyaluronic Acid-Derivatized Carbon Nanotubes in Hematoporphyrin Monomethyl Ether-Based Photodynamic Therapy for In Vivo and In Vitro Cancer Treatment. *Int. J. Nanomed.* **2013**, *8*, 2361.

87. Hashida, Y.; Tanaka, H.; Zhou, S.; Kawakami, S.; Yamashita, F.; Murakami, T.; Umeyama, T.; Imahori, H.; Hashida, M. Photothermal Ablation of Tumor Cells Using a Single-Walled Carbon Nanotube–Peptide Composite. *J. Controll. Release* **2014**, *173*, 59–66.

88. Robinson, J. T.; Welsher, K.; Tabakman, S. M.; Sherlock, S. P., Wang, H., Luong, R.; Dai, H. High Performance In Vivo Near-IR (>1 μm) Imaging and Photothermal Cancer Therapy with Carbon Nanotubes. *Nano Res.* **2010**, *3* (11), 779–793.

CHAPTER 3

QUANTUM DOTS AND THEIR SENSING APPLICATIONS

JAYESH BHATT[1], KANCHAN KUMARI JAT[2], AVINASH KUMAR RAI[1], RAKSHIT AMETA[3], and SURESH C. AMETA[1*]

[1]*Department of Chemistry, PAHER University, Udaipur-313003, Rajasthan, India*

[2]*Department of Chemistry, M. L. Sukahdia University, Udaipur-313001 Rajasthan, India*

[3]*Department of Chemistry, J. R. N. Rajasthan Vidyapeeth (Deemed to be University), Udaipur-313001 Rajasthan, India*

*Corresponding author. E-mail: ameta_sc@yahoo.com

ABSTRACT

Quantum dots are nanoparticles in their lower range of size < 10 nm. They are called dots because they are considered zero-dimensional and termed as quantum, because they follow quantum mechanical rules (quantum confinement). These dots have found wide range of applications, which include photodetector, photovoltaic device, photocatalysis, biosensor, chemical sensor, display, photodynamic therapy, diode laser, etc. With their superior transport and optical properties, it is anticipated that these will find many more new applications in coming decades. Hence, the sensing applications of different quantum dots are presented.

3.1 INTRODUCTION

A quantum dot (QD) is a nanocrystal made of semiconductor material that is small enough to exhibit quantum mechanical properties. Specifically, its excitons are confined in all three spatial dimensions. The electronic properties of these materials are intermediate between those of bulk semi-conductors and discrete molecules. Quantum dots were first discovered by Alexey Ekimov in 1981[1] in a glass matrix and then in colloidal solutions by Louis Brus in 1985.[2] The term "quantum dot" was coined by Reed and others[3] in 1988. Quantum dot is a conducting island of a size comparable to the Fermi wavelength in all spatial directions. A semiconductor quantum dot, however, is made out of roughly a million atoms with an equivalent number of electrons. Virtually, all electrons are tightly bound to the nuclei of the material; however, the number of free electrons in the dot can be very small; between one and a few hundred. The de Broglie wavelength of these electrons is comparable to the size of the dot, and the electrons occupy discrete quantum levels (similar to atomic orbitals in atoms) and have a discrete excitation spectrum. A quantum dot has another character-istic, usually called the charging energy, which is analogous to the ionization energy of an atom. This is the energy required to add or remove a single electron from the dot. QDs are often called the artificial atoms.

3.2 APPLICATIONS

Being zero-dimensional, quantum dots have a sharper density of states than higher-dimensional structures. As a result, they have superior transport and optical properties. They have potential uses in diode lasers, photodetectors, photovoltaic devices, chemical sensors, displays, photodynamic therapy, photocatalysis, biosensors, amplifiers, chemical sensors, etc.

3.2.1 METALS

Metal ions are major sources of water contamination as some of them are very toxic. This has encouraged researchers to develop novel low cost metal ion sensors to detect their presence even at trace levels. Although some of the heavy metals such as copper (Cu), iron (Fe) and zinc (Zn) are biologi-cally essential for living organisms, but they can be hazardous at higher concentrations and can cause serious concerns to human health. Mercury is highly toxic and lead to even death. These metal ion pollutants can be detected using some sophisticated and expensive instruments, but it takes considerable time for measurement while quantum dots like CdSe, CdS, CdTe, ZnS, etc. can be used as low cost heavy metal ion sensors, which are less time consuming.

Luminescent CdSe-ZnS quantum dots were modified by Xie et al.[4] using bovine serum albumin (BSA). They used it as selective copper ion probe. The fluorescence of the water-soluble QDs can be quenched only by the presence of Cu^{2+} and Fe^{3+} in physiological buffer solution. It was revealed that as Zn^{2+}, Na^+, and K^+ have no effect on the fluorescence. Addition of F^- to form the colorless complex FeF_6^{3-} was found to eliminate the interference of Fe^{3+}. The detection limit of Cu^{2+} ions was determined as 10 nM. The results were explained in terms of strong binding of copper ions onto the surface of CdSe which resulted in a chemical displacement of cadmium ions and the formation of CuSe on the surface of the QDs.

A new fluorescent nanosensor family for Zn^{2+} determination has been reported by Ruedas-Rama et al.,[5] which is based on azamacrocycle derivatization of CdSe/ZnS core/shell quantum dot nanoparticles. It was claimed that these are the first zinc ion sensors using QD nanoparticles in a host-guest and receptor-fluorophore system. Three azamacrocycles were used as receptors:

- TACN (1,4,7-triazacyclononane),
- Cyclen (1,4,7,10-tetraazacyclododecane), and
- Cyclam (1,4,8,11-tetraazacyclotetradecane).

Azamacrocycles conjugated to QDs via an amide link interact directly with one of the photoinduced QD charge carriers. They may transfer a hole in the QDs to the azamacrocycle, and as a result, the radiative recombination process is disrupted. As zinc ion enters the aza-crown, the lone pair electrons present on nitrogen atom become involved in the coordination and hence, proper energy level is no longer available for the hole-transfer mechanism, switching on the QD emission. Thus, an increase of the fluorescence intensity allowed the detection of even low concentrations of zinc ions. Three zinc ion sensors based on CdSe-ZnS core-shell QD nanoparticles showed a very good linearity in the range 5–500 µM. The detection limits were lower than 2.4 µM. In a study of interferences, the zinc-sensitive QDs had good selectivity as compared to other physiologically important cations and transition metals. A very good applicability in the determination of zinc ion in physiological samples was also demonstrated.

Luminescent and stable CdSe/ZnS core/shell quantum dots capped with sulfur calixarene were prepared by Li et al.[6] They used these QDs for the selective determination of mercury ions in acetonitrile with high sensitivity.

Shang et al.[7] prepared water-soluble luminescent CdSe quantum dots surface-modified with triethanolamine (TEA-CdSe-QDs) with high stability. It was revealed that fluorescence of the TEA-CdSe-QDs was greatly quenched only, when Hg^{2+} and I^- coexisted in the solution, while there was no noticeable effect on the fluorescence emission on addition of Hg^{2+} or I^-alone. It was reported that such a unique quenching effect could be used for recognition of mercury (II) ions and/or iodide anions in aqueous solution with high selectivity and sensitivity. The detection limits of Hg^{2+} or I^- ion were 1.9×10^{-7} mol L^{-1} and 2.8×10^{-7} mol L^{-1}, respectively. They showed that I^- could bridge between TEA-CdSe-QDs and Hg^{2+} to form a stable complex (QDs–I^-–Hg^{2+}) and then an effective electron transfer from the QDs to the Hg^{2+} may be there. As a result, fluorescence quenching of QDs was observed.

Water-soluble rhodamine B assisted graphene quantum dots (RhB-GQDs) were synthesized by Yang et al.[8] via one-pot hydrothermal method, using rhodamine B. Ethylenediamine was used as nitrogen source. As-synthesized RhB-GQDs have few layers configuration (average size of 4.5 nm) with good distribution. They applied fluorescence quenching approach for the sensing of mercury ions in water. Rhodamine B was absorbed on the surface of GQDs, and it can prevent RhB-GQDs to contact interfering cations. It was revealed that presence of N atoms reduces the ability of interfering cations to form stable complex with carboxyl groups; thus, giving RhB-GQDs good selectivity for mercury ions. It was reported that fluorescence quenching efficiency of as-synthesized RhB-GQDs had a good linear relationship with concentration even at low concentration. The fluorescence of RhB-GQDs was found to be quenched by nonradiative electron transfer between Hg^{2+} and the excited state of RhB-GQDs because of strong affinity of Hg^{2+} to amino and carboxyl on the surface of RhB-GQDs. It was revealed that as-synthesized RhB-GQDs can serve as fluorescent platform for the sensitive and selective detection of Hg^{2+} with a detection limit of 0.16 nM.

3.2.2 ANIONS

Anions like F^-ion are responsible for fluorosis while the presence of dichromate in polluted water is also harmful due to its oxidizing properties. This is known that cynide ion is deadly toxic. Detection of harmful ions is therefore important even in traces. Quantum dots are successful in detecting the presence of such anions.

Water-soluble luminescent CdSe quantum dots were synthesized by Jin et al.,[9] which were surface-modified with 2-mercaptoethane sulfonate. They used this probe for the selective determination of free cyanide in aqueous solution with high sensitivity (detection limit of 1.1×10^{-6} M), via analyte-induced changes in their photoluminescence after photoactivation.

Water-soluble carbon dots (CDs) have been developed by Qiao et al.[10] via precursors of 1,3-phenylenediamine and citric acid. The quantum dots (3–4 nm) were prepared based on a simple one-pot ultrasonic irradiation method, which shows bright blue emission at 440 nm. It was reported that CDs can be remarkably quenched by the presence of dichromate due to inner filter effect. Carbon dots possess highly selective and sensitive responses to dichromate ion through color changes. This type of sensing modes (fluorescence as well as colorimetric detection) can be used for the detection of dichromate ion and the limits of detection were found to be as low as 140 and 410 nM, respectively. The concentration of dichromate ion has been determined in real water and river samples also just to establish this method of detection.

Lu et al.[11] developed a novel dual-ligand strategy to synthesize fluoride-responded $CH_3NH_3PbBr_3$ PQDs emploing n-octylamine (OA) and 6-amino-1-hexanol (AH) as two capping ligands. It was reported that usual ligand, OA, stabilizes the PQDs emitting intense fluorescence; while the new ligand, AH, could interact with fluoride through hydrogen bonding between hydroxyl group and fluoride. It leads to the growth and fluorescence quenching of the PQDs. This fluorescent nanosensor was applied successfully to the detection of fluoride with high sensitivity and excellent selectivity. The limit of detection was found to be 3.2 μM (0.061 mg L$^-$) which is quite lower than the WHO guidelines. Spot plate test was also carried out, which indicated that the nanosensor could be used for visual detection of fluoride.

3.2.3 INSECTICIDES AND HERBICIDES

Quantum dots are basically semiconductor fluorescent nanoparticles and these can be used for environmental monitoring with high sensitivity. Levels of pesticides and herbicides used in agricultural practices have reached an alarming situation. Pesticides, herbicides and insecticides are toxic and harmful to living beings, even if present at very low concentrations. Therefore, there is an urgent need for their rapid, sensitive and specific detection of these in different samples like food and environment.

A new biosensor has been reported by Ji et al.[12] for the detection of para-oxon using (CdSe)ZnS core–shell quantum dots (QDs) and an organophosphorus hydrolase (OPH) bioconjugate. First, OPH was coupled to (CdSe)ZnS core–shell QDs through an electrostatic interaction between negatively charged QDs surfaces and the positively charged protein side chain and ending amino groups. The activity of OPH was preserved after the bioconjugation as evident from circular dichroism (CD) spectroscopy. Detectable secondary structure changes were observed by CD spectroscopy when the OPH/QDs bioconjugate was exposed to organophosphorus compounds like paraoxon. It was also reported that photoluminescence intensity of the OPH/QDs bioconjugate was also quenched in the presence of paraoxon. It was indicated that the quenching of PL intensity is due to conformational changes in the enzyme. The limit of detection for paraoxon concentration using this bioconjugate was about 10^{-8} M. An increase in OPH molar ratio in the bioconjugates was found to increase slightly sensitivity of this biosensor, but no further increase of sensitivity was achieved, when the molar ratio of OPH to QDs was >20 as at this level, the surface of QDs was saturated by OPH.

Vinayaka et al.[13] developed a reliable and rapid method for analysis and detection of 2,4-diphenoxy acetic acid (2,4-D, a herbicide). They used cadmium telluride quantum dot nanoparticles (CdTe QDs) for this purpose. Fluorescent property of quantum dots was used utilizing fluoroimmunoassay along with immunoassay to detect the presence of 2,4-D. CdTe capped with mercaptopropionic acid was conjugated using N-(3-dimethylaminopropyl)-N-ethylcarbodiimide hydrochloride (EDC) and a coupling reagent like N-hydroxysuccinimide (NHS) to alkaline phosphatase (ALP), which was then conjugated to 2,4-D molecule. It was possible to detect very low concentration of 2,4-D, upto 250 pg mL^{-1}.

The development of advanced tools for sensing specific materials remains an ongoing challenge. A new quantum dot based sensor was developed by Kashpova et al.[14] via supramolecular interactions, which demonstrates a novel simplicity of design a sensitive QDs and avoiding their covalent cross-linking. A simple label-free and turn-off method for the detection of paraoxon and its degradation products in aqueous media was proposed by them using the fluorescent QD/surfactant/cyclodextrin supramolecular system. This nanocomposite was prepared from 3-mercaptopropionic acid-capped CdTe QDs coated with cetyltrimethylammonium bromide (CTAB) through electrostatic self-assembly. Nanocomposite modification was made by β-cyclodextrin (β-CD) using hydrophobic interaction between cetyl tails of surfactants and inner cavity of macrocycle, which contributed to higher emission intensity and stability in aqueous solution. It was reported that strong fluorescence of CdTe/CTAB/β-CD nanocomposite can be quenched effectively by the addition of paraoxon. It is all due to the host–guest complexation between β-CD cavity and paraoxon degradation product.

They also tested functionality of the paraoxon sensor with blood samples of paraoxon-poisoned rats.

3.2.4 NUCLEOTIDES AND NUCLEIC ACIDS

Semiconductor quantum dots have drawn considerable attention and find their applicability in numerous fields within the life sciences. Nucleotides and nucleic acids can also be detected using quantum dots as sensors.

Two new classes of quantum dot (QD)–mediated biosensing methods have been reported by Yeh et al.[15] to detect specific DNA sequences in a separation-free format. Both these methods use two target-specific oligo-nucleotide probes for recognition of a specific target. It is based on cross-linking of 2 QDs with distinct emission wavelengths caused by probe-target hybridization. Here, QDs are used as both; fluorescent tags and nanoscaffolds that capture multiple fluorescently labeled hybridization products, resulting in amplified target signals.

The presence of targets was determined according to spatiotemporal coincidence of two different wavelength fluorescent signals, which were emitted from the QD/DNA/probe complexes. The fluorescent signals can be measured with single-dot/molecule sensitivity, when a single wavelength-excitation, dual wavelength-emission confocal spectroscopic system is used. The present methods have advantages in simplicity, testing speed, and multi-plexed applications as compared to other nanoparticle-based, separation-free assays.

Zhao et al.[16] presented a DNA fluorescence probe system, which is based on fluorescence resonance energy transfer (FRET) from CdTe quantum dot (QD) donors to Au nanoparticle (AuNP) acceptors. CdTe QDs, (2.5 nm diameter), were prepared in water as energy donors. Au nanoparticles (16 nm diameter) were prepared as energy acceptors from gold chloride by reduction. CdTe QDs were linked to 5'-NH_2-DNA through 1-ethyl-3-(dimethylaminopropyl)carbodiimide hydrochloride (EDC) as a linker 3'-SH-DNA was self-assembled onto the surface of AuNPs. The FRET distance of CdTe QDs and Au nanoparticles was determined by hybridiza-tion of complementary double stranded DNA (dsDNA) bound to the QDs and AuNPs (CdTe-dsDNA-Au). The fluorescence of CdTe-DNA-Au conju-gates was found to decrease extremely as compared to the fluorescence of CdTe-DNA, indicating that the FRET has occurred between CdTe QDs and Au nanoparticles. The fluorescence change of this conjugate depends on the ratio of Au-DNA to CdTe-DNA. When this ratio was kept 10:1, the FRET efficiency reached to its maximum. This probe system had a certain degree of fluorescence recovery, when a complementary single stranded DNA was introduced into this system, which showed that the distance between CdTe QDs and Au nanoparticles was increased.

Commercially available CdSe-ZnS quantum dots have been modified by Callen et al.[17] They exchange the hydrophobic surface ligands with (2-mercaptoethyl)-trimethylammonium chloride. Resulting water soluble conjugate was titrated with solutions of different phosphates (adenosine triphosphate (ATP), adenosine diphosphate, adenosine monophosphate, guanosine triphosphate (GTP), guanosine diphosphate and guanosine monophosphate) in 0.01 M 4-(2-hydroxyethyl)-1-piperazineethanesulfonic acid buffer (pH 7.4). A strong fluorescence quenching of about 80% and 25% was observed for ATP and GTP respectively. Virtually no effect was indicated for other phosphates. The quenching effect of ATP and GTP may be due to the high negative charge density associated with these substrates resulting in a strong attraction to the QD surface, which enabled them to engage in electron transfer with the excited QD. It was also reported that lack of fluorescence quenching associated with the other nucleotides may be attributed to their reduced charge density resulting in a lower affinity for the QD surface.

3.2.5   AMINO ACIDS

Amino acids are building blocks of proteins. There has been a continued interest in the development of reliable methods for the determination of amino acids for diagnostic and research purposes.

Huang et al.[18] developed a simple and sensitive method for l-cysteine detection, which is based on the increase in fluorescence intensity of mercaptoacetic acid-capped CdSe/ZnS quantum dots (QDs) in aqueous solution. It was reported that under the optimized conditions, a linear range of QDs fluorescence intensity versus the concentration of l-cysteine was 10–800 nmol L^{-1}, and a limit of detection of 3.8 nmol L^{-1}. It was also observed that there was no interference of coexisting foreign substances including some common ions, carbohydrates, nucleotide acids and other amino acids. The present method had advantages of simplicity, rapidity and sensitivity. The results were found satisfying, when synthetic amino acid samples, medicine sample together with human urine samples were analyzed by this method.

Water-soluble and stable semiconductor CdSe/ZnS quantum dots were prepared by Han and Li.[19] They used β- and α-cyclodextrins (CDs) as surface-coating agents in a simple sonochemical method. It allowed highly enantioselective fluorescent recognition of amino acids. are treated with L-amino acids. It was revealed that a much greater fluorescence enhancement was observed with CD-QDs as compared with D-amino acids.

3.2.6 PROTEINS

The quantitative and selective detection of proteins is significant because some diseases can be diagnosed based on their detection by the presence of blood-borne antibodies against the infection. Quantum dots help us in detecting the protein and related substance also.

It was reported by Choi et al.[20] that aptamer-capped near-infrared PbS quantum dots can detect a target protein based on selective charge transfer. These water-soluble QDs were synthesized with thrombin-binding aptamer, which retained the secondary quadruplex structure necessary for binding to thrombin. These QDs were 3-6 nm in diameter and show fluorescence around 1050 nm. When the aptamer-functionalized QD binds to its target, then a fluorescence quenching was observed, which may be due to charge transfer from amine groups on the protein to the QD. It was possible to detect thrombin within a min and that too with a detection limit of approximately 1 nM. It was also reported that this selective detection is observed even in the presence of high background concentrations of interfering proteins (either negatively or positively charged, which suggests that aptamer-capped QDs could proved to be useful for label-free protein assays.

3.2.7 GLUCOSE

Glucose sensing in diabetes diagnosis is of significant importance because of ever increasing prevalence of diabetes all over the world. This sensing is also critical in the food and drug industries. Quantum dots integrated with fluorescent technique has allowed for the development of novel glucose sensors, which has superior sensitivity and convenience.

The ratiometric fluorescence analysis of enzyme activities and their substrates is possible after coupling of oxidases with fluorophore-labeled CdSe/ZnS quantum dots by the interaction between the biocatalytically generated H_2O_2 and the quantum dots. Gill et al.[21] applied this method to the analysis of glucose and the inhibition of acetylcholine esterase.

Duan et al.[22] reported a new fluorescence method for detection of glucose using tungsten disulfide quantum dots (WS_2 QDs). This method is based on the transformation of Fe^{2+}/Fe^{3+} and enzymatic reaction. It was observed that the fluorescence for WS_2 QDs can be quenched via photoinduced electron transfer (PET), in the presence of Fe^{3+} ions but Fe^{2+} does not influence the fluorescence of WS_2 QDs. Glucose is known to produce hydrogen peroxide in the presence of glucose oxidase (GOx), and Fe^{2+} can be transferred to Fe^{3+} via oxidation by H_2O_2. Thus, a sensitive and simple strategy for glucose detection can be developed based on different influences of Fe^{2+} and Fe^{3+} on the fluorescence intensity of WS_2 QDs. It was found that there was a linear response for glucose in the range of 1–60 $\mu mol\ L^{-1}$ with a detection limit 0.30 $\mu mol\ L^{-1}$ under the optimized conditions. They also applied present method for glucose detection in serum sample, which gave satisfactory results.

Highly biocompatible phenylboronic acid-functionalized graphitic carbon nitride quantum dots (g-CNQDs/PBA) were prepared by Ngo et al.[23] via a simple hydrothermal process. They used these as glucose sensing fluorescence material. The quantum yield of g-CNQDs/PBA was found to be as high as 67%, when measured using quinine sulfate as a standard. It is reported to be the highest QY value for fluorescent g-CNQDs. The g-CNQDs/PBA exhibited two wide linear regions in the range of 25 nM^{-1} μM and 1 μM^{-1} mM and had a detection limit as low as 16 nM. It also showed excellent selectivity in the presence of different interfering reagents and very low cellular toxicity with excellent bio-imaging properties. This fabricated g-CNQDs/PBA can be a promising material for a range of clinical diagnostics and biomedical applications.

3.2.8 DRUGS

Quantum dots have proved themselves as powerful fluorescent probes as integrated targeting, imaging and therapeutic functionalities make them excellent material to study drug delivery and detection of some drugs even at lower concentrations.

Wang et al.[24] reported the use of pH-sensitive cadmium telluride (CdTe) quantum dots as proton probes for determination of tiopronin. They proposed a simple, rapid and specific quantitative method based on the fluorescence quenching of CdTe QDs. A calibration plot of $\ln(F_0/F)$ with concentration of tiopronin was found to be linear in the range of 0.15–20 $\mu g\ mL^{-1}$ (0.92–122.5 $\mu mol\ L^{-1}$) under the optimal conditions, with a good correlation coefficient.

The limit of detection (LOD) ($3\sigma/k$) was reported to be 0.15 μg mL^{-1}(0.92 μmol mL^{-1}). The content of tiopronin was determined in pharmaceutical tablet by this method and the results were in good agreement with those obtained from other methods like oxidation–reduction titration method.

Liang et al.[25] proposed quenching of the fluorescence of CdSe quantum dots by spironolactone. They used a simple, rapid and specific method for spironolactone determination. Spironolactone concentration versus quantum dot fluorescence gave a linear response with an excellent correlation coefficient (0.997), between 2.5 and 700 mg mL^{-1} (6.0–1680 μmol L^{-1}) and the limit of detection (S/N = 3) was 0.2 μg mL^{-1} (0.48 μmol L^{-1}) under the optimum conditions. The contents of spironolactone in some tablets were determined by this method and the results agreed well with the claimed values.

A little attention has been paid to the phosphorescence properties of QDs and their possible potential for phosphorescence detection as compared to quantum dots based fluorescence sensors. He et al.[26] explored the phosphorescence property of Mn-doped ZnS QDs to develop a novel room-temperature phosphorescence (RTP) method for the facile, rapid, cost-effective, sensitive, and selective detection of enoxacin in biological fluids. It was reported that Mn-doped ZnS QDs-based RTP method does not require the use of deoxidants and other inducers. It allowed the detection of enoxacin in biological fluids without any interference from autofluorescence and the scattering light of the matrix. It was revealed that Mn-doped ZnS QDs offers excellent selectivity for detection of enoxacin even in the presence of the some metal ions in biological fluids, biomolecules, and other kinds of antibiotics. Quenching of the phosphorescence emission due to the addition of enoxacin at 1.0 μM was found to be unaffected by 5000-fold excesses of Na$^+$ and 10,000-fold excesses of K$^+$, Mg^{2+}, and Ca^{2+} ions. Amino acids like tryptophan, histidine, and l-cysteine at 1000-fold concentration of enoxacin do not affect this detection. Glucose was also found not to affect the detection even at 10000-fold concentration of enoxacin. Typical coadministers (ceftezole, cefoperazone and oxacillin) are permitted at 50-, 10-, and 100-, fold excesses, respectively, without any significant interference. The detection limit for enoxacin was found to be 58.6 nM. The recovery of spiked enoxacin in human urine and serum samples was in the range of 94 –104%. This developed method was also employed to monitor the time-dependent concentration of enoxacin in urine from a healthy volunteer after the oral medication of enoxacin.

Carbon quantum dots (CQDs) were prepared from wood soot by Zhong et al.[27] It was observed that the CQDs could catalyze the oxidation of 3,

3', 5, 5'-tetramethylbenzidine (TMB) in the presence of H_2O_2 to form oxidized TMB (ox-TMB), which has an absorption peak at 652 nm. It was reported that the introduction of glutathione (GSH) could reduce ox-TMB. As a result, a decrease of the absorbance was observed at 652 nm. Thus, a highly sensitive colorimetric sensor for detection of GSH was fabricated. Effects of various operational parameters such as pH, temperature, and the concentrations of H_2O_2, CQDs and TMB were investigated. The response was found to be linearly proportional to the concentration of GSH within the range of 0.05 to 20 µM with a low detection limit of 0.016 µM at pH 3.5, temperature 35°C, H_2O_2 1.0 mM, CQDs 2.5 µg mL^{-1} and TMB 0.5 mM. The present method was successfully applied by them for determination of GSH in human serum samples with the recovery of 95.7–103.6%.

The graphene quantum dots (GQD) are quite unique for applications as sensor as they provide a platform for large surface area, where sensing material can be attached. Ahamed et al.[28] reported a new analytical method for sensing ciprofloxacin (CPFX), an antibiotic, using GQD electrode in differential pulse voltammetry (DPV). It is based on interaction of ferric ion with CPFX. It was observed that ferric ion undergoes a well defined one electron reduction at GQD electrode at Ep = 0.310 V vs saturated calomel electrode with a peak width of 0.100 V. When very low concentration (nM-mM) of CPFX is present in the electrolytic bath, the ferric ion reduction peak decreases with the appearance of three new peaks at 0.200, 0.050, and −0.085 V. These three peaks were attributed to the three stages of binding of CPFX with three positive charges of ferric ion. It was reported that decrease of the ferric ion peak at 0.31 V was found to be proportional to the concentration of CPFX. CPFX bound ferric ion showed enhanced currents as compared to glassy carbon electrode due to large surface area of GQD.

3.2.9 MISCELLANEOUS

Apart from these applications, quantum dots also find useful applications in detection of other materials and as biomarkers such as cancer, tuberculosis, urea, PAHs, tannic acid etc.

Tuberculosis (TB) is a major pulmonary disease, and it can be diagnosed using breath analysis techniques. A number of sensing methods have been developed to detect TB-VOCs. A low cost and less time consuming method is very much needed. With this view in mind, Bhattacharyya et al.[29] synthesized stable colloidal suspensions of CdSe QDs as well as carbon dots and used as a viable photoluminescent platform for detection of TB biomarkers. CdSe/

carbon dots based sensing solution acted as a fluorescent probe for detection of TB biomarkers due to tunable excitation and emission properties.

Liu et al.[30] suggested that as photoluminescence of mercaptoacetic acid (MAA)-capped CdSe/ZnSe/ZnS semiconductor nanocrystal quantum dots in SKOV-3 human ovarian cancer cells is pH-dependent, it can find applications, where QDs served as intracellular pH sensors. It was observed that in both fixed and living cells, the fluorescence intensity of intracellular MAA-capped QDs (MAA QDs) was found to increase monotonically with increasing pH. It was also revealed that electrophoretic mobility of MAA QDs also increases with pH, which indicates an association between surface charging and fluorescence emission. MAA dissociates from the ZnS outer shell at lower pH, resulting in aggregation and loss in solubility, MAA QD fluorescence changes were observed in the intracellular environment.

An original and novel assay system has been developed by Huang et al.[31] using urease as a catalyst and CdSe/ZnS quantum dots (QDs) as an indicator for quantitative analysis of urea. It was reported that determination of urea can be performed in a quantitative manner by mixing urease and QDs. It is based on the enhancement of QD photoluminescence (PL) intensity, which depends to the enzymatic degradation of urea. By selecting a particular buffer concentration and pH, PL enhancement due to the degradation of urea was found to be linear in the concentration of urea ranging from 0.01 to 100 mM. Urease/QDs system can be considered a promising urea-biosensing system. This system is a superior design and has many advantages, like simple preparation, low cost, no requirement of enzyme immobilization, high flexibility, and good sensitivity.

The presence of cyclic polyaromatic hydrocarbons (PAHs) is harmful in the environment. These PAHs are known to cause cancer and may affect the eyes, kidneys, and liver. Qu et al.[32] developed a simple fluorescence (FL) analysis method to detect the presence of polycyclic aromatic hydrocarbons using supramolecular nano-sensitizers, a combination of CdTe quantum dots and cyclodextrins (CDs) as an additive. This new fluorescence analysis pathway utilizes cyclodextrins and thioglycolic acid (TGA)-modified CdTe QDs, which are easy to prepare. It was revealed that the fluorescent response of CdTe QDs was there towards different PAHs in the presence of CD, just by changing the numbers of CDs. These supramolecular nano-sensitizers were used for the determination of phenanthrene and acenaphthene in water. 9, 9-Diflurofluorene produces a modest quenching of the fluorescence of CdTe in presence of γ-CD, when compared with acenaphthene. It was obsereved that relative FL intensities of CdTe QDs were found to decrease linearly with increasing concentration of phenanthrene and acenaphthene under optimal conditions with a detection limits of 0.53 μM (94.3 ng mL^{-1}) and 0.085 μM (13.1 ng mL^{-1}), respectively.

Bioinorganic conjugates made with highly luminescent semiconductor nanocrystals (CdSe–ZnS core–shell QDs) and antibodies were prepared by Goldman et al.[33], which can perform multiplexed fluoroimmunoassays. They performed sandwich immunoassays for the detection of cholera toxin, ricin, shiga-like toxin 1, and staphylococcal enterotoxin B simultaneously in single wells of a microtiter plate. Assay performance for the detection of each toxin was examined and then simultaneous detection of the four toxins from a single sample probed with a mixture of all four QD–antibody reagents was made. It was revealed that it was possible to deconvolute the signal from mixed toxin samples, using a simple linear equation-based algorithm, which allowed determination of all these four toxins simultaneously.

It is known that the biocompatible semiconductor quantum dots have some unique photophysical properties, which provided them an important place over organic dyes and lanthanide probes in fluorescence labeling applications. Ma et al.[34] prepared multicolor quantum QD-encoded microspheres via the layer-by-layer (LbL) assembly approach. Polystyrene microspheres (3 μm diameter) were used as templates for the deposition of different sized CdTeQDs/polyelectrolyte multilayers through electrostatic interactions. Two kinds of biofuntional multicolor microspheres were prepared by them with two different antibodies, anti-human IgG and anti-rabbit IgG. Human IgG and rabbit IgG can be detected as target antigens in the multiplexed fluoroimmunoassays. They also developed a novel microfluidic on-chip device for the detection of two kinds of antigen-conjugated multicolor QD-encoded microspheres. These microspheres can be easily distinguished from one another based on their respective fluorescence signals.

Water-soluble CdSe/ZnS (core–shell) semiconductor quantum dots were synthesized by Jin et al.[35] with surface-modification by tetrahexyl ether derivatives of p-sulfonatocalix[4]arene. These were then used for the optical detection of the neurotransmitter acetylcholine.

Yang et al.[36] developed a facile approach for highly sensitive and selective detection of tannic acid (TA) using nitrogen-doped fluorescent carbon dots (NCDs) as a novel fluorescent probe, sodium citrate and aminopyrazine as precursors. As-synthesized NCDs has multiple advantages such as high quantum yield (11.8%), good water solubility and satisfactory stability. The fluorescence quenching of NCDs was observed with the increase in TA concentration. The calibration curve displayed a wide linear region ranging from 0.40 to 9.0 µmol L^{-1} and a detection limit of 0.12 µmol L^{-1}. This fluorescent probe was also used in determining TA in beer samples and the average recoveries of TA was from 96.1 to 104.4%. The present method is a reliable, rapid and simple method for determination of TA in real samples.

Nitrogen and boron co-doped carbon quantum dots (NB-CQDs) were prepared by Liu et al.[37] via one-pot hydrothermal treatment of citric acid, borax, and p-phenylenediamine. It was reported that N and B species were efficiently doped into the carbon framework of the dots. N- and B-containing groups were formed on the surface of NB-CQDs during hydrothermal reaction. It was revealed that doping and surface functionalization of NB-CQDs efficiently modulated their physicochemical properties. NB-CQDs were found to be nearly mono-dispersed (average particle diameter 3.53 nm). The excitation and emission wavelengths were 360 and 490 nm, respectively. As-synthesized NB-CQDs showed regular photoluminescence (PL) emission and had pH and solvent polarity dependent PL properties. The NB-CQDs and a NB-CQD/poly(vinyl alcohol) composite film showed good PL responses in the presence of acetone solution and vapor. A PL sensor was then designed for the determination of acetone solution with a limit of detection as 0.54 µM. These NB-CQDs were also used for the determination of dopamine. The PL intensity of the NB-CQDs was reported to be inversely proportional to the concentration of dopamine in the range of 0.1–70 µM with LOD 11 nM. It was indicated that PL quenching occurred via a static quenching mechanism.

A simple and inexpensive mechanochemical method has been used by Siddique et al.[38] to prepare bulk quantities of self-passivated amorphous carbon dots. These carbon quantum dots are water soluble, like graphene quantum dosts and exhibited excitation-dependent photoluminescence with selectivity very high quantum yield (~40%). They used this photoluminescence property of carbon dots to detect trace amounts of the nitro-aromatic explosive, 2,4,6-trinitrophenol (TNP). It was revealed that benign nano-structures can selectively detect TNP over a wide range of concentrations (0.5–200 µM) simply by visual inspection. It has a detection limit of 0.2 µM, and therefore, better than almost all TNP sensor materials reported earlier.

3.3 CONCLUSION

Quantum dots have found various applications after their discovery. It is all because of their unique electronic structure, transport and optical properties. Quantum dot displays have a number of advantages over conventional LCD displays, and as a result, quantum dot-based screens have already gained a market. Quantum dots have find interesting sensing applications as chemical sensor for metal ions, anions, alcohol, acetone, antibiotics, plant growth regulators, herbicide, etc. and as biosensor for nucleic acids, certain pathogens, proteins and related substances, glucose, amino acids, ATP, dopamine, paraoxon, etc. Quantum dots will continue to revolutionize different fields in the near future, and will occupy a prominent position in the arena of sensors as quite sensitive chemical sensors and biosensors in future.

KEYWORDS

- **quantum dots**
- **chemical sensor**
- **biosensor**
- **antibiotics**
- **proteins**
- **metal ion**

REFERENCES

1. Ekimov, A. I.; Onushchenko, A. A. Quantum Size Effect in Three-Dimensional Microscopic Semiconductor Crystals. *JETP Lett.* **1981,** *34* (6), 345–349.
2. Brus, L. E. Electron-Electron and Electron-Hole Interactions in Small Semiconductor Crystallites: The Size Dependence of the Lowest Excited Electronic State. *J. Chem. Phys.* **1984,** *80*, 4403.
3. Reed, M. A.; Randall, J. N.; Aggarwal, R. J.; Matyi, R. J.; Moore, T. M.; Wetsel, A. E. Observation of Discrete Electronic States in a Zero-Dimensional Semiconductor Nanostructure. *Phys. Rev. Lett.* **1988,** *60* (6), 535–537.
4. Xie, H. Y.; Liang, J. G.; Zhang, Z. L.; Liu, Y.; He, Z. K.; Pang, D. W. Luminescent CdSe-ZnS quantum dots as selective Cu^{2+} probe. *Spectrochim. Acta. A* **2004,** *60* (11), 2527–2530.
5. Ruedas-Rama, M. J.; Hall, E. A. H. Azamacrocycle Activated Quantum Dot for Zinc Ion Detection. *Anal. Chem.* **2008,** *80* (21), 8260–8268.
6. Li, H.; Zhang, Y.; Wang, X.; Xiong, D.; Bai, Y. Calixarene Capped Quantum Dots as Luminescent Probes for Hg^{2+} ions. *Mater. Lett.* **2007,** *61* (7), 1474–1477.
7. Shang, Z. B.; Wang, Y.; Jin, W. J. Triethanolamine-Capped CdSe Quantum Dots as Fluorescent Sensors for Reciprocal Recognition of Mercury(II) and Iodide in Aqueous Solution. *Talanta* **2009,** *78* (2), 364–369.
8. Yang, Y.; Xiao, X.; Xing, X.; Wang, Z.; Zou, T.; Wang, Z.; Zhao, R; Wang, Y. Rhodamine B Assisted Graphene Quantum Dots Flourescent Sensor System for

Sensitive Recognition of Mercury Ions, *J. Lumin.* **2019**, *207*, 273–281, doi /10.1016/j. jlumin.2018.11.033.

9. Jin, W. J.; Fernández-Argüelles, M. T.; Costa-Fernández, J. M.; Pereiro, R.; Sanz-Medel, A. Photoactivated Luminescent CdSe Quantum Dots as Sensitive Cyanide Probes in Aqueous Solutions. *Chem. Commun.* **2005**, 883–885.

10. Qiao, G.; Lu, D.; Tang, Y.; Gao, J.; Wang, Q. Smart Choice of Carbon Dots as a Dual-Mode Onsite Nanoplatform for the Trace Level Detection of $Cr_2O_7^{2-}$. *Dyes Pigm.* **2019**,*163*, 102–110, doi:10.1016/j.dyepig.2018.11.049.

11. Lu, L. Q.; Ma, M. Y.; Tan, T.; Tian, X. K.; Zhou, Z. X.; Yang, C.; Li, Y. Novel Dual Ligands Capped Perovskite Quantum Dots for Fluoride Detection, *Sens. Actuators B.* 2018, *270*, 291–297.

12. Ji, X.; Zheng, J.; Xu, J.; Rastogi, V. K.; Cheng, T. C.; DeFrank, J. J.; Leblanc, R. M. (CdSe)ZnS Quantum Dots and Organophosphorus Hydrolase Bioconjugate as Biosensors for Detection of Paraoxon. *J. Phys. Chem. B.* **2005**, *109*, 3793–3799.

13. Vinayaka, A. C.; Basheer, S.; Thakur, M. S. Bioconjugation of CdTe Quantum Dot for the Detection of 2,4-dichlorophenoxyacetic Acid By Competitive Fluoroimmunoassay Based Biosensor. *Biosens. Bioelectron.* **2009**, *24* (60), 1615–1620.

14. Kashapov, R. R.; Bekmukhametova, A. M.; Petrov, K. A.; Nizameev, I. R.; Kadirov, M. K.; Zakharova, L. Y. Supramolecular Strategy to Construct Quantum Dot-Based Sensors for Detection of Paraoxon, *Sens. Actuators B,* **2018**, *273*, 592–599.

15. Yeh, H. C.; Ho, Y. P.; Wang, T. H. Quantum Dot-Mediated Biosensing Assays for Specific Nucleic Acid Detection. *Nanomed. Nanotechnol. Biol. Med.* **2005**, *1* (2), 115–121.

16. Zhao, D.; Jimei, Z.; Quanxi, D.; Ning, D.; Shichao, X.; Bo, S.; Yuehua, B. Adaption of Au Nanoparticles and CdTe Quantum Dots in DNA Detection. *Chin. J. Chem. Eng.* **2007**, *15* (6), 791–794.

17. Callan, J. F.; Mulrooney, R. C.; Kamila, S. Luminescent Detection of ATP in Aqueous Solution Using Positively Charged CdSe-ZnS Quantum Dots. *J. Fluoresc.* **2008**, *18*, 1157–1161.

18. Huang, S.; Xiao, Q.; Li, R.; Guan, H.L.; Liu, J.; Liu, X.R.; He, Z.K.; Liu, Y. A Simple And Sensitive Method for L-cysteine Detection Based on the Fluorescence Intensity Increment of Quantum Dots. *Anal. Chim. Acta.* **2009**, *645* (1–2), 73–78.

19. Han, C.; Li, H. Chiral Recognition of Amino Acids Based on Cyclodextrin-Capped Quantum Dots. *Small.* **2008**, *4* (9), 1344–1350.

20. Choi, J. H.; Chen, K. H.; Strano, M. S. Aptamer-Capped Nanocrystal Quantum Dots: a New Method for Label-Free Protein Detection. *J. Am. Chem. Soc.* **2006**, *128* (49), 15584–15585.

21. Gill, R.; Bahshi, L.; Freeman, R.; Willner, I. Optical Detection of Glucose and Acetylcholine Esterase Inhibitors by H_2O_2-Sensitive CdSe/ZnS Quantum Dots. *Angew. Chem. Int. Ed.* **2008**, *47*, 1676–1679.

22. Duan, X., Liu, Q., Wang, G., Su, X. WS_2 Quantum Dots as a Sensitive Fluorescence Probe for the Detection of Glucose. *J. Lumin.* **2019**, *207*, 491–496 doi:10.1016/j. jlumin.2018.11.034.

23. Ngo, Y. L. T.; Choi, W. M.; Chung, J. S.; Hur, S. H. Highly Biocompatible Phenylboronic Acid-Functionalized Graphitic Carbon Nitride Quantum Dots for the Selective Glucose Sensor. *Sens. Actuators B.* **2018**, doi:10.1016/j.snb.2018.11.031.

24. Wang, Y. Q.; Ye, C.; Zhu, Z. H.; Hu, Y. Z. Cadmium Telluride Quantum Dots as pH-Sensitive Probes for Tiopronin Determination. *Anal. Chim. Acta.* **2008,** *610* (1), 50–56.

25. Liang, J.; Huang, S.; Zeng, D.; He, Z.; Ji, Z.; Ai, X.; Yang, H. CdSe Quantum Dots as Luminescent Probes for Spironolactone Determination. *Talanta* **2006,** *69* (1), 126–130.

26. He, Y.; Wang, H.-F.; Yan, X.-P. Exploring Mn-Doped ZnS Quantum Dots for the Room Temperature Phosphorescence Detection of Enoxacin in Biological Fluids. *Anal. Chem.* **2008,** *80* (10), 3832–3837.

27. Zhong, Q.; Chen, Y.; Su, A.; Wang, Y. Synthesis of Catalytically Active Carbon Quantum Dots and its Application for Colorimetric Detection of Glutathione, *Sens. Actuators B.* **2018,** *273*, 1098–1102.

28. Ahamed, N. N. N.; Fan, W.; Schrlau, M.; Santhanam, K. S. V. A New Graphene Quantum Dot Sensor for Estimating an Antibiotic Concentration. *MRS Adv.* **2018,** *3* (15–16), 825–830.

29. Bhattacharyya, D.; Sarswat, P. K.; Free, M. L. Quantum Dots and Carbon Dots Based Fluorescent Sensors for TB Biomarkers Detection. *Vacuum* **2017,** *146*, 606–613.

30. Liu, Y. S.; Sun, Y.; Vernier, P. T.; Liang, C. H.; Chong, S. Y. C.; Gundersen, M. A. pH-Sensitive Photoluminescence of CdSe/ZnSe/ZnS Quantum Dots in Human Ovarian Cancer Cells. *J. Phys. Chem. C.* **2007,** *111*, 2872–2878.

31. Huang, C.-P.; Li, Y.-K.; Chen, T. M. A Highly Sensitive System for Urea Detection by Using CdSe/ZnS Core-Shell Quantum Dots. *Biosens. Bioelectron.* **2007,** *22* (8), 1835–1838.

32. Qu, F.; Li, H. Selective Molecular Recognition of Polycyclic Aromatic Hydrocarbons Using CdTe Quantum Dots with Cyclodextrin as Supramolecular Nano-Sensitizers in Water. *Sens. Actuators. B. Chem.* **2009,** *135* (2), 499–505.

33. Goldman, E. R.; Clapp, A. R.; Anderson, G. P.; Uyeda, H. T.; Mauro, J. M.; Medintz, I. L.; Mattoussi, H. Multiplexed Toxin Analysis Using Four Colors of Quantum Dot Fluororeagents. *Anal. Chem.* **2004,** *76* (3), 684–688.

34. Ma, Q.; Wang, X.; Li, Y.; Shi, Y.; Su, X. Multicolor Quantum Dot-Encoded Microspheres for the Detection of Biomolecules. *Talanta* **2007,** *72* (4), 1446–1452.

35. Jin, T.; Fujii, F.; Sakata, H.; Tamura, M.; Kinjo, M. Amphiphilic p-sulfonatocalix[4] arene-coated CdSe/ZnS Quantum Dots for the Optical Detection of the Neurotransmitter Acetylcholine. *Chem. Commun.* **2005,** 4300–4302.

36. Yang, H., He, L., Pan, S., Liu, H., Hu, X. Nitrogen-Doped Fluorescent Carbon Dots for Highly Sensitive and Selective Detection of Tannic Acid. *Spectrochim. Acta. Part A* **2019,** *210*, 111–119, doi:10.1016/j.saa.2018.11.029.

37. Liu, Y.; Li, W.; Wu, P.; Ma, C.; Wu, X.; Xu, M.; Luo, S.; Xu, Z.; Liu, S. Hydrothermal Synthesis of Nitrogen and Boron Co-Doped Carbon Quantum Dots for Application in Acetone and Dopamine Sensors and Multicolor Cellular Imaging, *Sens. Actuators B.* **2019,** *281*, 34–43.

38. Siddique, A. B.; Pramanick, A. K.; Chatterjee, S.; Ray, M. Amorphous Carbon Dots and Their Remarkable Ability to Detect 2,4,6-Trinitrophenol. *Sci. Rep.* **2018,** *8*, doi/10.1038/s41598-018-28021-9.

CHAPTER 4

Mn(II) AND Zn(II) CONTAINING LINSEED OIL-BASED POLY (ESTER URETHANE) AS PROTECTIVE COATINGS

ERAM SHARMIN[1,2*], MANAWWER ALAM[3], DEEWAN AKRAM[4], and FAHMINA ZAFAR[2]

[1]*Department of Pharmaceutical Chemistry, College of Pharmacy, Umm Al-Qura University, P.O. Box 715, 21955, Makkah Al-Mukarramah, Saudi Arabia*

[2]*Materials Research Laboratory, Department of Chemistry, Jamia Millia Islamia, New Delhi-110025, India*

[3]*Research Centre-College of Science, King Saud University, P.O. Box 2455, Riyadh 11451, Saudi Arabia*

[4]*Department of Chemistry, JRS College (Jamalpur), Munger, Bihar, India*

Corresponding author. E-mail: eramsharmin@gmail.com

ABSTRACT

Metal containing polymers bear synergistic properties of both the components, the polymer matrix as well as incorporated/dispersed metal. However, their cumbersome, multistep preparation methods at high temperatures and times prove disadvantageous. In this study, metal containing biobased poly(ester urethane) (M-PU) is prepared with Linseed oil polyol (Pol) as the bio-derived matrix, by simple "one-pot, two-step reaction." Pol was treated with phthalic anhydride, followed by metal acetates of Mn(II) and Zn(II) and toluylene-2,4-diisocyanate to produce M-PU. The reaction mechanism for their synthesis was also investigated. The effects of metal incorporation on the structural, physicochemical, and thermal behavior of M-PU were assessed by standard methods. M-PU produced scratch-resistant, impact-resistant, transparent and glossy coatings, at ambient temperature. M-PU may find applications as eco-friendly protective coatings up to 180°C.

4.1 INTRODUCTION

Biobased metal containing polymers consist of a bio-derived polymer as the main (organic constituent) matrix, with metals (inorganic constituent), either incorporated or dispersed in the polymer. These bear the advantages of both organic and inorganic materials, which include ease in processing, superior resistance to chemicals, excellent mechanical and optical properties.[1,2] However, the drawbacks associated with these metal containing polymers are their complex preparation methods that are expensive, time-consuming, occurring at elevated temperatures, and often involve hazardous solvents.[3]

Polyols constitute one of the important derivatives of plant oils. They occur in nature as ricinoleic or lesquerollic acids (in oils from seeds of Castor or Lesquerella) or are obtained in the laboratory from plant oils through various chemical reactions (hydroformylation, epoxidation, ozonolysis followed by hydrogenation). The research on polyols dates back to several decades. Interestingly, oil polyols are rich in hydroxyls, methylenes, double bonds, and often -OCH$_3$, -Cl, -Br, -OCOH, and oxirane rings, all of which can undergo various chemical modifications as green raw materials for polymers. Nonetheless, over the years, research has been focused mainly on the synthesis (involving –OH) and characterization of polyurethanes [PU] with applications as foams, lubricants, cosmetics, adhesives, plasticizers, and coatings.[4–10] It is presumed that the complicated structures of oil polyols and PU limit their other modifications. We have previously reported acrylation, boronation, silylation of chemically derived and natural polyols, with potential applications as protective coatings.[11–16] The research area holds wide scope for further studies.

In our earlier work, we have reported the synthesis, characterization, and coating properties of copper containing Linseed oil-based poly (ester urethane) coatings.[17]

The present work reports the synthesis and characterization of Mn(II), half-filled (d^5) and Zn(II), completely filled (d^{10}) d-orbitals containing Linseed oil-based poly(ester urethane) (M-PU, where M=Mn(II) and Zn(II)). The effect of metal incorporation on structural, physicochemical, thermal, and antibacterial behavior of these was investigated. The reaction mechanisms involved in the syntheses of M-PU were also elucidated.

The work aims (a) to overcome common disadvantages associated with preparation of metal containing polymers and (b) to investigate the prospective applications of these as protective coatings produced at ambient temperature.

4.2 EXPERIMENTAL

4.2.1 MATERIALS

Linseed polyol (Pol) was prepared according to our previously reported method.[11,14–18] Methyl isobutyl ketone (MIBK), xylene, manganese acetate, zinc acetate (Mn(OCOCH$_3$)$_2$, Zn(OCOCH$_3$)$_2$ (M-acetate)) (Merck, Mumbai, India), toluylene-2,4-diisocyanate (TDI) (Merck, Germany) and phthalic anhydride (PAN) (S.D. Fine Chem., India) were of analytical grade.

4.2.2 SYNTHESES OF M-PU (Mn-PU AND Zn-PU)

The synthesis was performed as described previously.[17,18] Pol was placed in a three-neck flat bottom flask equipped with a thermometer, condenser, and nitrogen inlet tube. Calculated amount of PAN was added in small pinches over a period of 15 min at 120±5°C with constant stirring, the latter was continued for additional 15 min. The temperature of the reaction mixture was lowered to 80±5°C. Next, slowly and continuously M-acetate (Mn(OCOCH$_3$)$_2$ or Zn(OCOCH$_3$)$_2$; 0.04, 0.05, 0.06 moles) was added for 15 min. The chemical reaction was carried out until the completion of the reaction. The progress of the reaction was monitored at regular intervals by FTIR. The reaction mixture was observed for clarity and changes in viscosity throughout the reaction. M-Pol (Mn-Pol/Zn-Pol) were obtained as clear, free-flowing, yellowish-brown liquids termed as 0.04Mn-Pol, 0.05Mn-Pol, 0.06Mn-Pol, 0.04Zn-Pol, 0.05Zn-Pol, and 0.06Zn-Pol, respectively (the prefix numbers are the amounts of M-acetate in moles). A similar reaction was also carried out without M-acetate to prepare Lpol.[17,18]

M-Pol (0.04Mn-Pol, 0.05Mn-Pol, 0.06Mn-Pol, 0.04Zn-Pol, 0.05Zn-Pol, and 0.06Zn-Pol) were dissolved in minimum amount of solvent blend (xylene:MIBK, 3:1v/v) and then treated with a predetermined amount of TDI (30 wt%), under constant stirring at 60±5°C.[17] The progress of reaction was monitored by hydroxyl value (HV) determination and thin layer chromatography. M-PU were obtained as clear, free-flowing, yellowish-brown products. M-PU were dissolved in solvent blend (xylene:MIBK (3:1v/v)) and filtered repetitively. The solvent blend was carefully removed from the filtrate by rotary vacuum evaporator. M-PU (labeled as 0.04Mn-PU, 0.05Mn-PU, 0.06Mn-PU, 0.04Zn-PU, 0.05Zn-PU, and 0.06Zn-PU, respectively) were further subjected to spectral analysis.

FTIR (Mn-PU/Zn-PU) (cm^{-1}): 3297–3276, 669 (–NH), 3049 (ArC=C–H), 3009 (–C=CH), 2927 (–CH$_2$ asym), 2856 (–CH$_2$ sym), 2264 (–NCO, residual), 1696 (urethane carbonyl), 1575 (metal carboxylate), 1730 (>C=O ester), 1575, 1508, 750, 702 (Ar–C= C), 1275 (C–C(=O)–O–C).

^{1}H NMR (Mn-PU/Zn-PU) (CDCl$_3$, ppm): 8.2 (–NH urethane), 7.6–7.0 (Ar–H), 5.38 (–OH), 5.27 (–CH=CH–), 3.6 (–CH$_2$–CHOOC–Ar), 3.40 (–CH–OH), 2.2 (–CH$_3$ of TDI), 2.1 (–CH$_2$–CHOOC–Ar), 1.36 (–CH$_2$–CH=CH–), 1.25–1.5 (chain –CH$_2$), 0.88 (–CH$_3$).

^{13}C NMR (Mn-PU/Zn-PU) (CDCl$_3$, ppm): 177.0 (>C=O metal carboxylate), 171.0 (>C=O Ar), 154.0 (–N=C=O), 134–120 (Ar ring carbons of TDI and PAN), 128.0–126.0 (–C=C–; olefinic carbon), 80.0 (–CH$_2$–CHOOC–Ar), 60.0 (–CH–OH), −29.0 (chain –CH$_2$), 24.0 (–CH$_3$ of TDI), 18.0 (chain -CH$_3$).

A similar reaction was also carried out with Lpol to prepare LPU.

4.2.3   CHARACTERIZATION

FTIR spectra were obtained with Perkin Elmer 1750 FTIR spectrometer (Perkin Elmer Cetus Instruments, Norwalk-CT) using NaCl cell. ^{1}H-NMR and ^{13}C-NMR spectra were recorded on a JEOL GSX 300 MHz FX-1000 spectrometer in deuterated chloroform (CDCl$_3$) with tetramethylsilane as an internal standard. Thermal analysis was carried out with a TGA51 (TA Instrument, USA) at a heating rate of 20°C/min under nitrogen.

To evaluate their coating properties, M-PU (Mn-PU/Zn-PU) were dissolved in solvent blend (60 wt% solution in xylene:MIBK; 3:1v/v), applied by brush on commercially available mild steel strips (size: 70×25×1 mm^3 and 30×10×1 mm^3) to evaluate scratch hardness (SH), impact resistance (IRt), bending ability (BT) and gloss of coatings by standard methods. The thickness of the coatings ranged from 90 to 100 μm, as measured by Elcometer model 345 (Elcometer Instruments, Manchester, UK). Chemical resistance test of M-PU coatings was carried out by dipping the coated panels in different media. The changes, such as loss in gloss, adhesion, weight of coated panels, were observed from time-to-time and recorded (Table 4.1).

TABLE 4.1 Physico-mechanical Properties of LPU, Mn-PU, and Zn-PU.

Sample code	SH (kg)	IRt (lb/inch)	BT (1/8inch)	Gloss (45°)
LPU	0.9	150	pass	64
0.04Mn-PU	1.1	200	pass	68
0.05Mn-PU	1.2	200	pass	69
0.06Mn-PU	1.3	200	pass	70
0.04Zn-PU	1.2	200	pass	68
0.05Zn-PU	1.3	200	pass	68
0.06Zn-PU	1.4	200	pass	69

4.3   RESULTS AND DISCUSSION

4.3.1   SYNTHESIS OF M-PU (Mn-PU/Zn-PU)

The preparation of Pol from Linseed oil is described in detail in previous reports.[11, 14-18] The reaction schemes for the preparation of Lpol, M-Pol, and M-PU are presented in Figure 4.1. The strategy in the preparation of Lpol, M-Pol, and M-PU comprises "single-pot multi-step" reactions, that is, condensation, followed by addition. The hydroxyl groups of Pol react with PAN forming Lpol, which then reacts with M-acetate forming M-Pol. The free hydroxyl groups of Lpol and M-Pol finally react with isocyanate groups of TDI forming LPU and M-PU. The synthesis of Lpol/M-Pol involves – OH and carboxylic end groups of polyol-polyester and carboxylic/acetate (phthalic anhydride/M-acetate) moieties (Fig. 4.1).[17,18]

The inherent fluidity of Pol, a typical characteristic of various plant oils and their derivatives, facilitates the dissolution of PAN and M-acetate powder. Thus, M-Pol was synthesized in the absence of solvent[13-16]. However, the synthesis of M-PU required a minimal amount of solvent because of the high viscosity and structural complexity of the product.[17] The overall approach eliminates the use of solvent in the first step of synthesis (M-Pol); the second step of synthesis involved only minimum possible solvent. Thus, the overall use of solvent in the synthesis strategy is cut off by almost 50%. Thus, overall, the use of solvent is reduced by 50%. The reaction time for M-PU syntheses is less (45–60 min) than LPU (90 min) and may be attributed to the catalytic efficiency of the interchain or intrachain metal in poly (ester urethane) metallohybrids.[19–23].

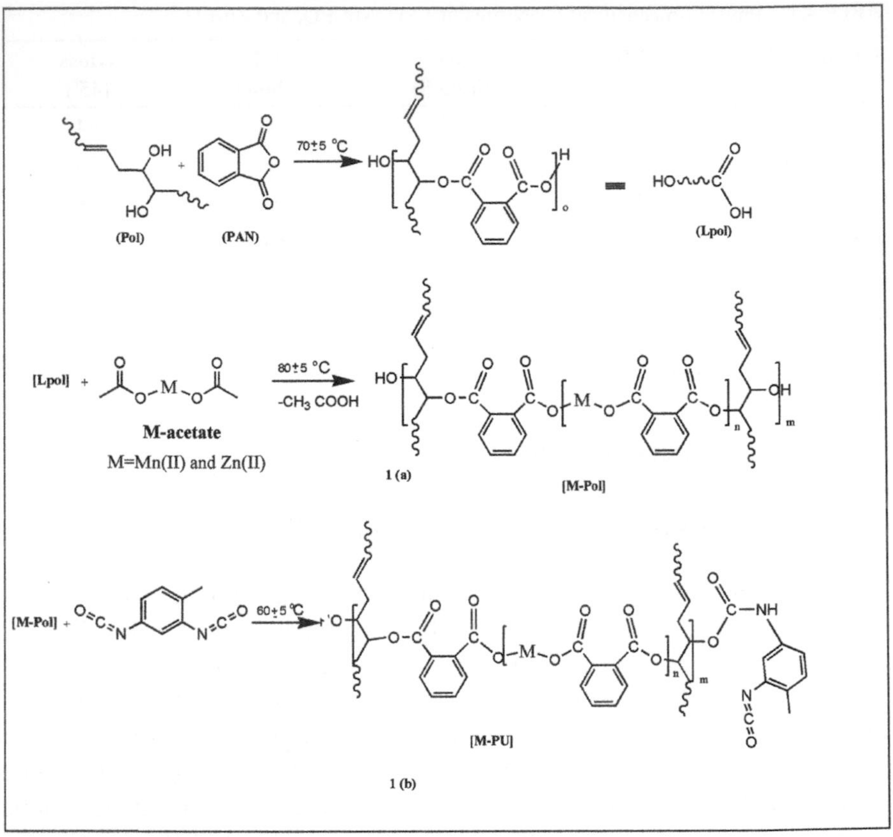

FIGURE 4.1   Reaction scheme for the synthesis of Lpol, M-Pol, and M-PU.

The reaction occurs both at hydroxylic and carboxylic ends of Lpol. The first step follows addition–elimination reaction at the carbonyl carbon of the carboxylic acid terminated end of Lpol and metal acetate. The reaction results in the incorporation of metal in Lpol backbone. The first step involves addition reaction at carbonyl carbon of Lpol leading to tetrahedral transition state developing partial negative charge on oxygen (from initial trigonal state), followed by the formation of tetrahedral intermediate. The latter then undergoes ejection reaction (of –OH) returning to initial trigonal state resulting in overall nucleophilic acyl substitution by addition-ejection/elimination reaction at the carbonyl carbon of carboxylic acid terminated end of Lpol. The reaction at hydroxylic end occurs by attack on carbonyl carbon of metal acetate, forming metal oxide.[17–25]

Urethanation appears to follow the mechanism proposed by Britain and Gemeinhardt for the metal-catalyzed polyaddition reaction between diol/polyol and isocyanate.[23] In this mechanism, interchain and/or intrachain metal interacts with -NCO and –OH resulting in an intermediate complex, which readily undergoes rearrangement to the urethane product.

4.3.2 COATING PROPERTIES OF M-PU (Mn-PU AND Zn-PU)

The curing process occurs in three stages, (a) by solvent evaporation: A physical process (the coatings become dry-to-touch), (b) by chemical reaction between free –NCO of LPU, Mn-PU, and Zn-PU, with moisture in the air, and (c) by auto-oxidation (the coatings become dry-to-hard).[11] The coatings were prepared at ambient temperature, curable by simple route, in contrast to their previously reported counterparts.[3] LPU coatings turn dry-to-touch in 120 min and dry-to-hard in 7–10 days, at ambient temperature (28–30°C). The coatings of Mn-PU and Zn-PU become dry-to-touch in 45–60 min and dry-to-hard within 7–8 days (at 28–30°C). We understand that interchain and intrachain incorporated metals facilitate drying of M-PU coatings. Metal complexes are also used industrially as "driers" in drying of coatings. These accelerate drying of paints by enhancing the rate of cross-link formation between binder molecules, that is, decomposition of and/or formation of hydroperoxides.[26–29] Thus, it is envisioned that in M-PU, the incorporated metal facilitates drying process through coordination of metals with donor groups of metallopolymer backbone, such as oxygen, nitrogen, and double bonds.[30]

Mn-PU and Zn-PU show higher SH values compared with LPU. On an increase in metal content (from 0.04Mn-PU, 0.05Mn-PU, 0.06Mn-PU, 0.04Zn-PU, 0.05Zn-PU, and 0.06Zn-PU), SH values increase (Table 4.1). LPU shows IRt value of 150 lb/inch. Mn-PU and Zn-PU pass 200 lb/inch IRt and 1/8-inch conical mandrel bend tests, which indicate that the introduction of metal provides additional flexibility, as well as adhesion, to Mn-PU and Zn-PU coatings relative to LPU. These exhibit good gloss as observed in oil-based coatings (Table 4.2).[13, 15]

TABLE 4.2　Chemical Resistance Performance of LPU, Mn-PU, and Zn-PU.

Sample code	NaOH (5wt%, 12 h)	HCl (5wt%, 36 h)	NaCl (3.5wt%, 36 h)	Water (36 h)	Xylene (36 h)
LPU	f	c	f	b	a
0.04Mn-PU	e	b	b	d	a
0.05Mn-PU	e	b	b	d	a
0.06Mn-PU	a	a	a	d	a
0.04Zn-PU	e	b	b	d	a
0.05Zn-PU	e	b	b	d	a
0.06Zn-PU	a	a	a	d	a

a = unaffected; b = loss in gloss; c = loss in gloss and weight; d = slight discoloration; e = slight loss in gloss, film unaffected; f = loss in gloss, weight and adhesion.

In alkaline (5wt%NaOH, 12 h) and saline media (3.5wt%NaCl, 36 h), LPU coatings showed loss in gloss, adhesion, and weight. The coatings of 0.06Mn-PU and 0.06Zn-PU remained unaffected in these media. 0.06Mn-PU and 0.06Zn-PU coatings remained unaffected in 5wt% HCl for 36 h compared with LPU, which showed dissolution of material leading to the loss in gloss and weight. Mn-PU and Zn-PU coatings exhibit improved performance than LPU in the acidic, alkaline, and salt environments.

Since the coatings of M-PU containing the highest amount of metal (0.06Mn-PU and 0.06Zn-PU) showed the best performance, only the thermal stability of 0.06Mn-PU and 0.06Zn-PU was investigated by TGA and DSC analysis.

4.3.3　THERMAL BEHAVIOR

TGA thermograms (Fig. 4.2) of Mn-PU and Zn-PU show 10wt% loss at 260°C and 265°C and 50wt% loss at 400°C and 410°C. LPU exhibit 10wt% and 50wt% losses at 250°C and 390ºC, respectively. DSC thermograms of Mn-PU and Zn-PU (Fig. 4.3) show broad endotherm from 50°C–100ᵛC eventually followed by an exotherm up to approximately 125°C. In TGA thermogram, thermal degradation starts at 180°C–190ºC. The initial degradation covering 2–5wt% loss may be attributed to evaporation of trapped solvent or moisture, followed by degradation of urethane bonds (>200°C), which results in formation of isocyanate, alcohol, CO_2, primary and secondary amine, and olefin[31]. These metal containing polymers may be safely used up to 180°C.

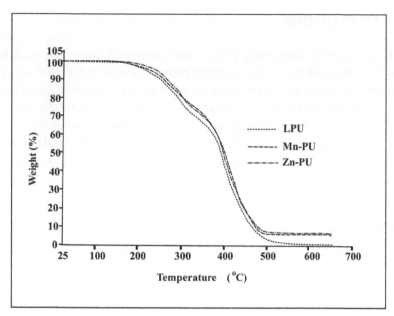

FIGURE 4.2   TGA thermograms of LPU, Mn-PU, and Zn-PU.

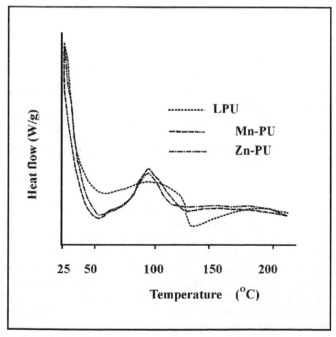

FIGURE 4.3   DSC thermograms of LPU, Mn-PU, and Zn-PU.

4.4   CONCLUSIONS

Mn(II) and Zn(II) containing poly (ester urethane) were prepared from Linseed oil derived polyol by a simple route. Linseed polyol served as the parent organic precursor and metal acetates as the inorganic constituents. A 50% reduction in the use of solvents was achieved during the syntheses processes. Poly (ester urethane) metallohybrids showed good physicomechanical and chemical resistance performance. The systems stand exemplary of "green materials" to be employed as protective coatings with safe usage up to 180°C.

KEYWORDS

- **linseed oil**
- **polyols**
- **polyurethanes**
- **coatings**
- **polymers**

REFERENCES

1. Bhushan, A.; Han, H.; Sutherland, A.; Boehme, S.; Yaghmaie, F.; Davis, C. E. *Appl. Organometal. Chem.* **2010,** *24,* 530.
2. Naoshima, Y.; Carraher, C. E.; Iwamoto, S.; Shudot, H. *Appl. Organometal. Chem.* **1987,** *1*, 245.
3. Kumar, A.; Vemula, P. K.; Ajayan, P. M.; John, G. *Nat. Mater.* **2008**; *7*, 236.
4. Petrovic, V. *Polym. Rev.* **2008,** *48,* 109.
5. Zlatanic, A.; Lava, C.; Zhang, W.; Petrovic, Z. S. *J. Polym. Sci.: Part B: Polym. Phys.* **2004,** *42*, 809.
6. Monteavaro, L. L.; Da Silva, E. O.; Cosia, A. P. O.; Samios, D.; Gerbase, A. E.; Petzhold, C. L. *J. Am. Oil Chem. Soc.* **2005,** *82*, 365.
7. Narine, S. S.; Yue, J.; Kong, X. *J. Am. Oil Chem. Soc.* **2007,** *84*, 173.
8. Kong, X.; Yue, J.; Narine, S. S.; *Biomacromolecules.* **2007,** *8*, 358.
9. Narine, S. S.; Kong, X.; Bouzidi, L.; Sporns, P. *J. Am. Oil Chem. Soc.* **2007,** *84*, 55.
10. Lligadas, G.; Ronda, J. C.; Galia, M.; Cadiz, V. *Biomacromolecules* **2010,** *11,* 2825.
11. Sharmin, E.; Ashraf, S. M.; Ahmad, S. *Int. J. Biol. Macromol.* **2007,** *40*, 407.
12. Sharmin, E.; Ashraf, S. M.; Ahmad, S. *Eur. J. Lipid Sci. Technol.* **2007,** *109*, 411.
13. Akram, D.; Sharmin, E.; Ahmad, S. *Prog. Org. Coat.* **2008,** *63*, 25.
14. Akram, D.; Sharmin, E.; Ahmad, S. *Macromol. Symp.* **2009,** *277*, 130.
15. Akram, D.; Sharmin, E.; Ahmad, S. *J. Appl. Polym. Sci.* **2009,** *116*, 499.

16. Akram, D.; Ahmad, S.; Sharmin, E.; Ahmad, S. *Macromol. Chem. Phys.* **2010**, *211*, 412.

17. Sharmin, E.; Akram, D.; Zafar, F.; Ashraf, S.M.; Ahmad, S. *Prog. Org. Coat.* **2012**, *73*, 118.

18. Sharmin, E.; Rahman, O.; Zafar, F.; Akram, D.; Alam, M.; Ahmad, S. *RSC Adv.* **2015**, *5*, 47928

19. Moroi, G. *React. Func. Polym.* **2008**, *68*, 268.

20. Moroi, G. *Macromol. Symp.* **2009**, *279*, 29.

21. Matsuda, H. *Polym. Adv. Technol.* **1997**, *8*, 616.

22. Saunders, K. J. *Organic Polymers Chemistry*; Chapman and Hall: New York,1988; 2nd ed; Chapter 16, pp 358–387.

23. Britain, J.W.; Gemeinhardt, P. G. *J. Appl. Polym. Sci.* **1960**, *4*, 207.

24. Zafar, F.; Ashraf, S. M.; Ahmad, S. *React. Funct. Polym.* **2007**, *67*, 928.

25. Morrison, R. T.; Boyd, R. N. *Organic Chemistry*; 6th ed; Prentice-Hall of India Pvt. Ltd, p 753, 2002.

26. Gorkum, R. V.; Bouwman, E. *Coordination Chem. Rev.* **2005**, *249*, 1709.

27. Erich, S. J. F.; van der Ven, L. G. J.; Huinink, H. P.; Pel, L.; Kopinga, K. *J. Phys. Chem. B.* **2006**, *110*, 8166.

28. Gorkum, R. V.; Bouwmann, E.; Reedijk, J.; *Inorg. Chem.* **2004**, *43*, 2456.

29. Oyman, Z. O.; Ming, W.; Linde, R. V. D. *Prog. Org. Coat.* **2005**, *54*, 198.

30. Zafar, F.; Zafar, H.; Sharmin, E.; Ashraf, S. M.; Ahmad, S. *J. In. Organomet. Polym. Mater.* **2011**, *21,* 646.

31. Javni, I.; Petrovic, Z. S.; Guo, A.; Fuller, R. *J. Appl. Polym. Sci.* **2000**, *77*, 1723.

relative to the engine. It was clearly observed that the nanodiesel fuels have significantly reduced CO, CO_2, NO, unburned HC, and enhanced the engine performance. According to the experimental results, the 100 ppm TiO_2 and 200 ppm CuO nanodiesel have shown almost the highest performance and lowest emissions comparable with neat diesel fuel and other nanodiesel samples. Owing to 100 ppm TiO_2 on hot start conditions, it is found that the CO, CO_2 NO, unburnt HC, exhaust temperature, and BSFC have been reduced by 41.4%, 37%, 38.3%, 81%, 4.9%, and 20.5%, respectively, at maximum load. Meanwhile, the brake power, RPM, and thermal efficiency have increased by 1.5%, 1%, and 2.65%, respectively. Regarding 200 ppm CuO on hot start conditions, it was found that CO, NO, unburned HC, exhaust temperature, fuel consumption, and brake-specific fuel consumption (BSFC) have been reduced by 42.6%, 22%, 33%, 7.4%, 2.7%, and 27.3%, respectively, meanwhile, CO_2, brake power, RPM, and thermal efficiency have increased by 9%, 1.5%, 0.09%, and 3.8%, respectively, at maximum load. One of the most important aspects of this experimental work is that the two samples of 100 ppm TiO_2 and 200 ppm of CuO nanodiesel fuel have reduced the CO emissions to an even lower value than that approved in Stage V for Euro standard emissions for nonroad mobile machinery. According to Euro standard stage (V), the CO emissions have been limited to 8 g/kW, whereas the 100 ppm TiO_2 and 200 ppm CuO nanodiesel fuel provide 7.112 g/kW h and 6.918 g/kW h, respectively. Another prospect of the experimental work is the reduction of fuel consumption provided by the nanodiesel fuel. It was observed that the 100 ppm TiO_2 sample has reduced fuel consumption by 20.5% compared with the neat diesel fuel, which is expected to positively influence the transportation sector.

5.1 INTRODUCTION

The industrial scale posesses a lot of operations and processes in which heat transfer occurs either in laminar or turbulent regime, which is permanently accompanied by high temperature and pressure. Many of these processes or applications would get benefits from reducing heat loss and decrease thermal resistance of the heat transfer fluid. Consequently, the heat transfer systems will get lower in terms of capital and operating cost and higher in thermal efficiency. Nanofluids surely have the ability to improve heat transfer process and reduce the thermal resistance as well. It is worth mentioning that the industrial applications that would get benefits are quite diverse like transportation, electronics, medical, food, and other manufacturing types.

"Nanofluid" is the term derived to describe a fluid in which nanometer-sized particles are suspended in a specific fluid. Nanofluids which consist of such nanometered particles suspended in the liquid have exhibited enhancement in terms of thermal conductivity and convective heat transfer comparable with the base liquids. Regarding the thermal conductivities of the nanomaterials, it is typically higher in value than base fluids like water, ethylene glycol, and light oils. Moreover, nanofluids with low concentrations have shown significant enhancement in thermal performance.

Nanotechnology concerns about nanomaterials and its wide applications especially in industrial scale, nanofluids are typically nanometerd particles in term of size that are stably suspending in liquid with a length range from 1 to 100 nm for heat transfer fluid applications. Many researches have pointed out the history, development, and field of applications for such nanomaterials. In addition, they have indicated that the thermal characteristics for these nanomaterials are completely different from conventional heat transfer liquids. A study has reported that such addition of nanomaterial by small amount, for example, 1% by volume, has led to increase in thermal conductivity of the fluid by approximately two times its conventional value.[1-5]

The evidence for that in 2006 the publications and research have extremely increased to be 100 published papers. The extensive increase in this area is directly related to establishment of research groups in well-known universities and institutes as well. In addition, the large international companies, as well as small ones, develop research plans in order to improve their products. On the other hand, the large-scale production for nanofluids to cope with industrial scale is still complicated and limited and requires a lot of studies to overcome the existing key barriers.[1,2]

Modern techniques and fabrication models have easily allowed the production of materials in their nanometer size. In general, nanomaterials show extremely unique physical and chemical characteristics that are considered completely different from the same material but in microscale or even larger. Not only the fabrication of these materials involves the main element itself, but also involves other materials as well. In addition, the preparation of these nanomaterials is mainly classified into two main categories that are physical and chemical processes. Oxides of ceramic has been widely used in nanofluid (Al_2O_3, CuO), nitride ceramics (AlN, SiN), carbide ceramics (SiC, TiC), metals (Ag, Au, Cu, Fe), and semiconductors (TiO_2), single, double, or multiwall carbon nanotubes and composite materials, for example, core polymer shell composites. Moreover, there are new structures and materials that are promising and very attractive for have shown a significant enhancement in the combustion characteristics which in turn reduce exhaust gas and soot emissions. Selvan et al. conducted his experiment investigations in two stages to study characteristics of emissions and the performance of CI engines using cerium oxide as nanoadditives to diesel fuel. In the first stage, he focuses on analyzing the stability of nanofluid, whereas in the second stage he has focused on the brake thermal power and ignition delay. He reported that cerium oxide has significant enhance over these two parameters.[1]

Richard A. Yetter has showed in his research the influence of nanoparticles on the combustion process. He reported that nanomaterials are characterized by their high catalytic activity as well as high tendency to store energy. Moreover, he reported that nanoparticles have low melting points and lower sintering temperatures comparable with larger parts of the same materials. Sadhikbash and Anand have examined the impact of carbon nanotubes as well as aluminum oxide on diesel on Jatopha biodiesel fuel.

They have observed significant improvement in the brake thermal efficiency , in addition, the exhaust emissions have been reduced for the neat diesel fuel. Because of enhancing combustion property, on the other hand, the alumina nanoparticles blended with biodiesel have reduced the peak pressure and heat released rate comparable with Jatopha biodiesel. Tyagi et al. have reported that addition of aluminum and aluminum oxide with various particle sizes and concentration have shortened the ignition delay in comparison with neat diesel fuel. Meanwhile, particle size and ignition delay have not shown any direct relation.[10–12]

Cerium oxide is mainly characterized by its ability to catalyze the combustion process by transferring oxygen atoms from its structure. Moreover, the catalytic activity is mainly based on other crucial properties, such as surface area, so using cerium oxide as nanoparticle additives will provide higher properties than bulky materials in normal size. The main role of adding nanoparticles, like cerium oxide, is to enhance combustion process by decomposing the unburnt hydrocarbon, in addition to soot emission, which resulted in reducing the amount of pollutants produced in the exhaust gas and fuel consumption. In addition, cerium oxide has shown that it can reduce the pressure in the combustion chamber which in turn, will disturb the formation of NOx as it requires high pressure to be formed and provide a smoothly efficient combustion process. Other advantage is that it can be used to treat particulate filters in diesel engines for a short term. Generally, the nanoparticles oxides enhance the combustion process and reduce the soot emissions that clog up the filter of the diesel engines, eventually, the exhaust emission will be lesser.[13–15]

   Aluminum nanoparticles, as well as microparticles, have been examined as a potential additive to diesel fuel. Aluminum has been known for its ability to increase output energy of diesel engines because of its high combustion energy. Recent advances in fabrication and characterization of nanoparticles have allowed more detailed research into the relationship of particle size and structure with performance benefit. A detailed study has been carried out by researchers at Purdue University and Indiana; they have found that aluminum's nanoparticle sizes have proved higher performance than microparticles. Moreover, many studies have proved that nanoparticle's suspension in ethanol-based fuels is much better than those in model hydrocarbons. So, it is greatly recommended that nanoaluminum will be an effective additive for bioethanol fuels.[16–18]

   Researchers from Anna University, India has published a study to investigate the potential usage of Cobalt oxide and magnesium, aluminum as nanoparticles additives to diesel and biodiesel fuels. Similar to cerium oxide, the presence of oxygen atoms in Cobalt oxide improves the combustion process. It was found that upon adding Cobalt oxide nanoparticles to diesel fuel, the combustion process has become efficient, clean. Moreover, the unburnt emissions, as well as soot emissions, have been significantly reduced.[19,20]

   The main concern, as with many nanomaterials, that limited their complete usage in the market is their environmental impact. Nanoparticles oxides have proved their capability to be a potential additive to enhance fuel efficiency and reduce exhaust emissions. On the other hand, nanoparticles can cause environmental problems if they escape with the exhaust gas. Many studies and researches have been conducted to discuss their environmental issues.

   Cerium oxide has been investigated; cerium addition of cerium oxide can result in escaping small amounts of nanoparticle with the exhaust emissions. The studies have shown that cerium oxide can accumulate in the environment especially in the roadside areas. The ability to explore the effect of additive particle size on performance has opened up a new range of potential to improve the performance of internal combustion engines, reduce fuel consumption, and counteract some of the performance compromises currently associated with using biofuels. The main challenge moving forward will be to fully assess the potential environmental impacts of releasing these nanoadditives into the environment and comparing the results with potential improvements to emissions.[13,14]

   The previous discussion, related to combustion aspects of metallic nanoparticles, provides the basis for utilizing their possible benefit when dispersed in liquid fuels as an alternative fuel for combustion in diesel engine.

Very limited experimental studies on combustion aspects of nanoparticles laden liquid fuels are available in the open literature, and almost no review work is available. In this work key findings of available work on combustion aspects of various nanoparticles doped liquid fuel are summarized.

Metal particles are known to have better thermal properties when they are suspended in micro- or nanoscale length in different concentrations with the base fluids. Several researches[110-113] in their experimental findings reported a 20% increase in thermal conductivity with a lower particle concentration (< 5%). The same incremental trend was observed in the heat transfer coefficient. Based on these above fascinating facts researches are attempting to use these energetic materials with liquid fuels in diesel engine for obtaining better combustion characteristics in terms of higher heating value. In order to gain fundamental understanding of combustion behavior, some experimental studies related to single droplet, and some in the bulk base fluid with micro- and a nanoscale additive, have been performed. Each study has its own significance. Mono droplet combustion studies provide an insight on general burning and vaporization characteristics and how different physical and chemical processes are involved in it. Tyagi et al. found that with a very small volume fraction (1–5% by weight) of n-Al and Al_2O_3 blended with diesel fuel, there has been a significant enhancement in the ignition probability compared to pure diesel fuel. This is due to better radiative heat and mass transfer properties that causes reduction in droplet ignition temperature.[21] Javed et al. investigated the evaporation characteristics of kerosene nanofluid droplets doped with dense concentration of n-Al (2.5–7%) at different elevated temperatures (400–800°C).

The authors concluded that vaporization does not occur according to D2 Law due to the inclusion of n-Al. However, at the early lifetime of droplets, micro-explosion phenomenon occurs at greater intensity either due to high loading rate of n-Al or an increase in ambient temperature. It was reported that at 800°C and at 2.5% n-Al concentration the evaporation rate was increased by 48.7%.[22] Kao et al. performed experiments on combustion of aqueous aluminum nanofluid with diesel fuel.[23] It was found that burning rates of hydrogen with diesel fuel in the presence of aqueous aluminum were quite high. The reason can be attributed due to small aluminum particle size (40–60 nm) and high oxidation rates that provides larger contact surface areas for decomposition of more hydrogen from water.

In addition, aluminum particles act as a catalyst that further accelerates the combustion process, which augments the total heat of combustion.[23] Xin et al. investigated the use of Cerium metal oxides (CeO_2) with diesel fuel to study the impact of nanosized additives of CeO_2 on combustion performance. The authors found an improvement in the combustion process that was attributed to high pressure rise and an extension in the combustion duration, which promotes full combustion of modified diesel fuel. Combustion and breakdown studies of fuel JP-10 with nano CeO_2 particles (120 nm) and iron oxide Fe_2O_3 (20 nm) were experimentally investigated by Devener et al.[139] The pyrolysis temperature of JP-10 was found 900°C in absence of any additives. It was also observed that the addition of CeO_2 and Fe_2O_3 with JP-10 causes a significant reduction in breakdown temperature that not only initiates the JP-10 breakdown, process but at the same time acts as oxidizer producing end products as CO_2, CO, and H_2O. This behavior is might be due to catalytic activity of CeO_2 and Fe_2O_3 that enhances combustion reaction rates.[24] Day by day the demand for energy has followed an increasing trend which in turn affects the quality of air and increases the air pollutants. It is worth mentioning that fossil fuels are not considered a sustainable source except for a limited source because of industrial revolution the demand all over the world the demand for fossil fuels has increased significantly. Apart from economic issues, fossil fuels have been widely used as a source of fuel for transportation sector which in turn contributes to long term environmental issues in terms of global warming and climate changes. As a result, for the continuous use of fossil fuels, the worldwide has suffered from fossil fuels depletion. Consequently, it has become first priority concern for the people who mainly depend on the fossil fuel as a source of energy. Compression ignition engines (CIEs) that are mainly gasoline and diesel engines have a significant contribution for the transportation sector. Recently many statistics have shown that lung diseases and cancer invasion have significantly increased because of wide use of diesel engines. Such that a comprehensive study on this issue becomes a necessity. Today fossil fuel contributes with 80% of the total energy required; 50% of it is directly linked to domestic transportation.

Diesel fuel mainly comprises aliphatic hydrocarbons C_{11}–C_{18} that is used mainly to power heavy vehicles as agriculture equipment. During combustion process emissions are produced from the combustion chamber, which is known as exhaust emissions.

These exhaust emissions are mainly THC, NO_x, PM, and carbon monoxide. All of these emissions are air pollutants. Moreover, the oxides of nitrogen and sulfur are the main contributors to the acid rains. The fossil fuels are considered the main source of greenhouse gases. Recently the whole world has raised concern over these air pollutants to find an alternative source of energy to replace fossil fuels.

The environmental problems associated with the continuous use of the fossil fuels have imposed to find a reasonable, clean, and efficient renewable source of energy. Many questions have been raised regarding the selection, distribution, and usage of the data to find appropriate answers to these questions. The answer to this is that researchers have to focus more on both the fuel efficiency and engine technology. One of the crucial issues to enhance the fuel properties and reduce the air pollutants is adding additives to the pure diesel fuel. The main additives that can be added to the neat diesel fuel are nanoparticles and oxygenated additives.

The main role of these additives is to enhance the physical properties, combustion process, and exhaust emissions.

5.3   EXPERIMENTAL MATERIALS AND METHODS

This section mainly describes the materials, as well as equipment, that were used in the experimental part. The experimental part is divided into two main sections. First, CuO and TiO_2 nanoparticles are added into 1 liter of diesel fuel and dispersed in the fuel by a means of ultrasonicator for a certain time range, from 15 to 40 min. The second section is the measurement's analysis where combustion of nanodiesel fuel is conducted in the diesel engine followed by passing the exhaust into gas analyzer to perform quantitative, as well as qualitative, analysis. During the experimental part of the combustion of nanodiesel fuel, the diesel engine was operated at full throttle to allow the maximum flow for the fuel and, in addition,the diesel engine was subjected to cumulative three equal loads to examine the performance of nanodiesel fuel (Fig. 5.1).

5.3.1   PREPARATION AND CHARACTERISTICS OF NANOPARTICLES

Copper oxide nanoparticles have been prepared via the sol–geltechnique. This technique offers many advantages in stabilizing the as-prepared nanoparticles (Table 5.1). In this method, CuO NPs involve the addition of an aqueous NaOH solution to the solution of $CuCl_2$ to form a precipitate.

TABLE 5.1 Characterization of Nanoparticles.

Properties	CuO	TiO$_2$
Appearance (color)	Dark brown	White
Appearance (form)	Powder	Powder
Solubility	Suspended	Suspended
Average size	35±6	34±6
Shape (Transmission electron microscopy (TEM))	Quasispherical shapes	Quasispherical shapes

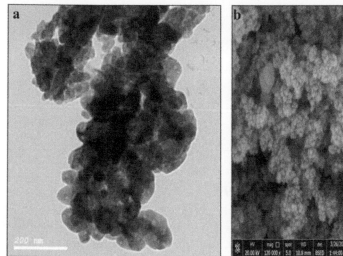

FIGURE 5.1 Transmission electron microscopy (TEM) for (a) CuO and (b) TiO$_2$ nanoparticles.

5.3.2 DIESEL FUEL SPECIFICATIONS

Diesel sample was obtained from Midor Refinery Plant located in Borg el Arab after the treatment process; however, there are no additives that have been added to the diesel sample. Table 5.2 shows the physic-chemical characteristics of the diesel sample.

TABLE 5.2   Midor Diesel Fuel Specifications.

Property	Unit	Method	Result
Density @15°C	$\dfrac{Kg}{m^3}$	ASTM D 4052	827.8
Color		ASTM D 1500	0.5
Distillation, IBP	°C	ASTM D 86	190.0
Distillation, 10%	°C	ASTM D 86	217.0
Distillation, 50%	°C	ASTM D 86	265.0
Distillation, 90 %	°C	ASTM D 86	329.0
Distillation, 95 %	°C	ASTM D 86	346.0
Distillation, FBP	°C	ASTM D 86	359
Distillation, 350 °C	Vol %	ASTM D 86	96
Copper corrosion (3 h @ 100°C)		ASTM D 130	1b
Pour point	°C	ASTM D 97	−6
Cetane index		ASTM D 4737	55.6
Total sulfur	Wt%	ASTM D 4294	0.06
Flash point	°C	ASTM D 93	71
Conradson crabon (on 10% residue)	Wt%	ASTM D 189	0.05

5.3.3 PREPARATION OF NANOFUEL SAMPLES

CuO and TiO_2 nanoparticles (Figs. 5.2 and 5.3) form a suspended solution when they are immersed in diesel fuel, to increase their solubility and form a homogenous phase with the diesel fuel, samples should be introduced to the sonicator followed by homogenizer in order to get effectively dispersed (nanoparticles) in diesel sample.

Sonicator is a device that supplies ultrasonic waves to the sample with a specified frequency usually above 20 Hz under controllable amplitude. Ultrasonic waves are generally supplied in a suspended solution to disperse the powder into the solvent (liquid).

FIGURE 5.2 Sonication stages for CuO nanodiesel samples. (a) Before sonication and after being sonicated for (b) 5 min, (c) 15 min, and (d) 30 min.

FIGURE 5.3 Sonication stages for TiO$_2$ nanodiesel samples. (a) Before sonication and after being sonicated for (b) 5 min, (c) 15 min, and (d) 30 min.

For the sake of better dispersion of nanoparticles in the diesel fuel, Sonicator UP200s is utilized to ensure higher energy output to the sample and full dispersion. The specifications of the used sonicator are shown in Table 5.3.

TABLE 5.3 Specifications of Sonicator Used in Experimental Work.

Technical data	Value
Power	200 W
Power control	Amplitude (20–100%)
Pulse range	0–100%
Operating frequency	24 Hz
Dimensions	(L×W×H) 257×157×130 mm
Weight	1.5 kg
Power supply	110–120 V, 4A, 50–60 Hz

5.3.4 ENGINE SPECIFICATIONS

The engine available in the BUE Applied Energy Lab is a diesel one (CIE) (Table 5.4). The specifications of diesel engine are as follow:

TABLE 5.4   Diesel Engine Specifications.

Engine model	Peter type
Fuel	Diesel ($C_{12}H_{23}$)
Heating values (kJ/kg)	42700
Number of cylinders	1
Compression ratio	17
Cylinder arrangement	Single Cylinder
Bore × stroke (mm ×mm)	85×110
Aspiration	Natural
Cycle	Four strokes
Displacement (L)	0.624
Cooling system	Water cooling
Speed range (rpm)	1200–1600
Rated power at 1350 rpm, (kw)	3.2

Gasoline and diesel engines are assumed to operate on Otto and Diesel cycles, respectively. The actual cycle shown in Figure 5.4 includes the main processes of intake, compression, combustion, expansion, and exhaust for both engines. In gasoline engines, air and fuel are mixed before inducted to the cylinder. In diesel engines, on the other hand, air only is inducted to the cylinder whereas fuel is injected inside the cylinder near the end of the compression process. Combustion occurs near the TDC via spark-generated flame propagation in gasoline engines and autoignition in diesel ones. The high-pressure combustion gases force the piston to move downward producing work.

In four-stroke engines (intake, compression, expansion and exhaust strokes), the crankshaft completes two revolutions for each thermodynamic cycle. In two-stroke engines (first stroke: intake, exhaust, and compression; second stroke: combustion, expansion, intake, and exhaust) only one revolution of the crankshaft is required for each cycle.

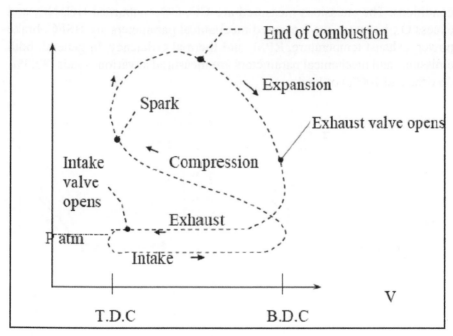

FIGURE 5.4   Combustion cycle for diesel engines.

The experimental part is divided into two main sections. First, CuO and TiO_2 nanoparticles are added into 1 liter of diesel fuel and dispersed in the fuel by a means of ultrasonicator for a certain time range from 15 to 40 min. The second section is the measurement's analysis where combustion of nanodiesel fuel is conducted in the diesel engine followed by passing the exhaust into gas analyzer to perform quantitative as well as qualitative analysis. During the combustion of nanodiesel fuel, the diesel engine was operated at full throttle to allow the maximum flow for the fuel and, in addition, the diesel engine was subjected to cumulative three equal loads to examine the performance of nanodiesel fuel.

5.4 RESULTS AND DISCUSSIONS

The following chapter represents the experimental results and discussions for the results obtained. The experimental work is mainly divided into two categories which are cold start and hot start. The cold start is mainly the stage at which the engine is operated for 30 min to be heated. The measured emissions and mechanical parameters are measured at cold and hot start conditions. The emissions measured are CO, CO_2, unburned HC, NO, and excess O_2; however, the measured mechanical parameters are BSFC, brake power, exhaust temperature, RPM, and thermal efficiency. In general, both emissions and mechanical parameters are measured at various loads (33.3%, 66.6 %, and 100%) of load.

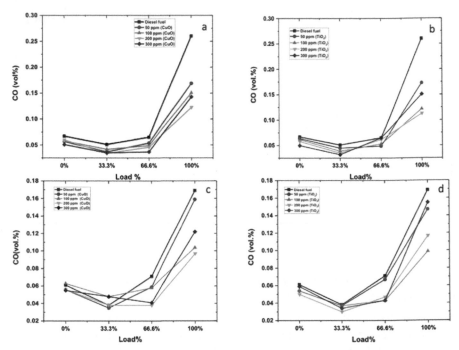

FIGURE 5.5   CO variations (a, c) with CuO nanodiesel fuel at various loads on cold and hot start conditions, respectively, (b, d) with TiO_2 nanodiesel fuel at various loads on cold and hot start conditions, respectively.

The following Figure 5.5a, c depict the relationship between CO emissions in exhaust and applied loads to the engine at the cold start conditions for base diesel fuel and nanodiesel fuels with CuO and TiO_2. The combustion process in the diesel engine is somehow random and complicated, consequently not all the fuel exhibits complete combustion that results in generating carbon monoxide and carbon dioxide. As the exerted load on the engine increases the fuel supply to the engine also increases and thus CO will increase too. Both figures illustrate that the base fuel (0 ppm) has the highest CO emissions all over the loads as compared with the blended diesel fuel with nanoparticles. Moreover, the blended nanoparticles diesel fuels show a great improvement in the CO emissions and thus they have sharply reduced among all the concentrations of nanoparticles. In Figure 5.5a nanodiesel with 200 and 300 ppm of CuO show the lowest CO emissions and these are 52.6 vol% and 45 vol%, respectively (cold start conditions). However, Figure 5.5b shows that all the nanodiesel samples produce lower CO emissions relative to base diesel fuel. In addition, both 100 and 200 ppm of TiO_2 represent a significant reduction in CO concentrations relative to other samples for cold start. The CO emissions have percentage reduction of 53 vol% and 56.5 vol% for 100 and 200 ppm TiO_2, respectively, in cold start conditions. The reduction in CO emissions is due to the catalytic activity provided by CuO and TiO_2 nanoparticles in the combustion process. Furthermore, CuO nanoparticles convert CO to CO_2 according to its reaction mechanism.

On the other hand, Figure 5.5b, d indicate the relationship between the CO emissions and the applied loads for the diesel engine at hot start conditions. The figures reveal that CO emissions have significantly reduced especially at the maximum load for all proportions that proves the positive influence of nanoparticles on the combustion of diesel fuel. Regarding Figure 5.5c, the CO emissions are 0.097 vol%, 0.122 vol% for 200 and 300 ppm CuO, respectively, at the maximum load, whereas it is 0.169 vol% for the base diesel fuel. However, in Figure 5.5d the CO emissions are 0.099 vol% and 0.117 vol% for 100 and 200 ppm TiO_2, respectively, at maximum load. It is worth mentioning that the emissions produced by diesel engine are completely different as compared with cold start conditions. Furthermore, in all experimental researches the hot start conditions simulate the real and practical conditions; consequently, it is taken more into consideration rather than cold start conditions.

Figure 5.6a, b indicate the CO_2 emissions with the applied loads to the diesel engine at cold start conditions for base diesel fuel and nanodiesel fuel with CuO and TiO_2. CO_2 is one of the main products that is produced from complete combustion of hydrocarbons; however, the combustion process always accompanied by CO due to incomplete combustion.

Figure 5.6a reveals that the base diesel fuel has the lowest CO_2 emissions as compared with CuO nanodiesel because of the incomplete combustion of diesel fuel at all loads which in turn produces lower CO_2 and higher CO relative to the nanodiesel fuel. On the other hand, for CuO nanodiesel samples, there is a significant increase in CO_2 emissions that reflects enhancement in the combustion process and converts CO to CO_2. Moreover, the 200 and 300 ppm samples of CuO increase CO_2 emissions by 10.2% and 25%, respectively. It is worth mentioning that 200 ppm sample shows the lowest CO_2 emissions among other nanodiesel samples, whereas as 300 ppm is the highest.

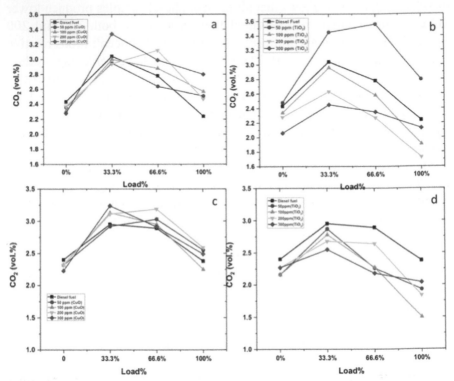

FIGURE 5.6 CO_2 variations (a, c) with CuO nanodiesel fuel at various loads on cold and hot start conditions, respectively. (b, d) with TiO_2 nanodiesel fuel at various loads on cold and hot start conditions, respectively.

Regarding Figure 5.6b, the TiO_2 nanodiesel samples follow a contradicting behavior to CuO samples at cold start conditions and thus the produced CO_2 in the combustion process has significantly reduced in comparison with the base diesel fuel except for 50 ppm sample. This can be explained as follows: The consumed fuel in the diesel engine in case of TiO_2 is much lower than the base diesel fuel so the increment of CO_2 produced is lower than the base diesel fuel. It is observed that the 100 and 200 ppm exhibits the lowest CO_2 emissions as compared with base diesel fuel. According to Figure 4.6, the values of CO_2 for 100 ppm and 200 ppm TiO_2 samples are 1.91 vol% and 1.73 vol% with a reduction percent to the base diesel fuel 14.7% and 22.7%, respectively. It is worth mentioning that the CO_2 emissions start increasing at small loads until they reach a maximum value, then it begins to reduce at high loads.

Figure 5.6c, d show the relationship between CO_2 emissions of the base diesel fuel and nanodiesel samples with TiO_2 and CuO under different loads at hot start conditions. In Figure 5.6c, all the CuO nanodiesel samples produce higher CO_2 emissions than base diesel fuel which indicates that combustion process proceeds towards completion. The 200 ppm CuO sample exhibits the highest CO_2 emissions relative to other concentrations with an increase of 9% relative to base diesel fuel. Owing to Figure 5.6d, the TiO_2 samples represent the same behavior as in cold start conditions, the CO_2 emissions reduce again with respect to base diesel fuel. The 100 ppm TiO_2 sample shows the lowest CO_2 relative to other concentrations, it is 1.84 vol% at the maximum load, whereas the base diesel fuel is 2.38 vol%, with a reduction of 22.6 vol%.

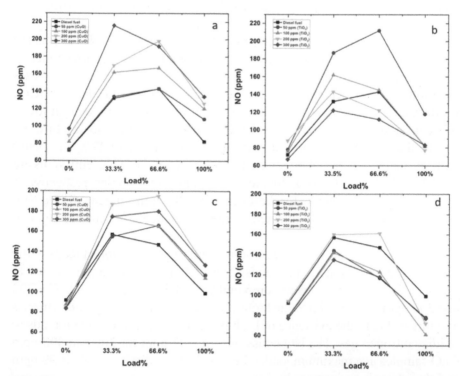

FIGURE 5.7 NO variations (a, c) with CuO nanodiesel fuel at various loads on cold and hot start conditions, respectively, (b, d) with TiO_2 nanodiesel fuel at various loads on cold and hot start conditions, respectively.

Figure 5.7a, c depict the NO emissions in the exhaust with the applied loads on diesel engine operated with the base diesel fuel and nanodiesel fuel of CuO. The formation of NO_x in diesel engine is directly related to three different chemical reactions including Thermal NO_x, Prompt NO_x, and Fuel NO_x.

The last type is related to the nitrogen content in the fuel itself and, since the used diesel fuel has nil nitrogen content, this type can be neglected and eliminated from the comparison. Thermal NO_x is the one formed from the oxidation of atmospheric nitrogen at high temperatures, whereas the prompt NO_x is formed from the reaction of atmospheric nitrogen with free radicals that are produced from the diesel fuel in the flaming zone. As shown in Figure 5.7a, the base diesel fuel produces the lowest NO emissions in exhaust relative to the nanodiesel fuel with CuO. At the maximum load, the 200 and 300 ppm provide the highest NO emissions among other samples and they are 126 and 134 ppm, respectively, whereas the base diesel fuel is 82 ppm.

This means that CuO increases NO emissions by 53% and 63 % for 200 and 300 ppm CuO samples, respectively. The nanodiesel fuels with CuO increase NO emissions to some extent due to the presence of high oxygen content in CuO nanoparticles that promote the formation of NO and as the concentration of CuO increases in the diesel fuel, the NO increases indeed.

With respect to Figure 5.7b, it is observed that as the load increases the produced NO decreases due to the mechanism of TiO_2 nanoparticles. Furthermore, it illustrates that the nanodiesel fuels with TiO_2 exhibit lower NO emissions than the base diesel fuel, especially at high loads. In other words, the main reason for that is the catalytic activity of TiO_2 and better ignition characteristics (shortening in ignition delay). The NO emissions are 82, 77, and 83 ppm for 100, 200, and 300 ppm TiO_2 samples, respectively, at maximum load while as for diesel fuel it is 82 ppm. In general, the NO_x emissions are significantly affected by the turbulence, efficient mixing, and lower equivalence ratio.

Figure 5.7c depicts the NO emissions with respect to applied loads to the diesel engine at hot start conditions. The NO emissions have increased at all loads for all proportions as compared with the base diesel fuel. The 200 and 300 ppm have exhibited the highest NO emissions among other samples and this is due to the existence of high oxygen content in CuO that tends to oxidize nitrogen gas. The NO emissions are 127 ppm for 200 and 300 ppm CuO samples at a maximum load, whereas the base diesel sample is 99 ppm and thus there is an increase by 28%.

As shown in Figure 5.7d, the NO emissions are affected with the applied loads of the diesel engine at hot start conditions. The NO emissions have significantly reduced at all loads almost for all samples with respect to the base diesel fuel. The NO emissions are 61, 72, and 77 ppm for 100, 200, and 300 ppm TiO_2 samples, respectively, at a maximum load, whereas for base diesel fuel it is 99 ppm. This means that the TiO_2 nanoparticles have reduced the NO emissions by 38 % at hot start conditions. The behavior of TiO_2 nanodiesel samples is completely different and contradicting with CuO nanodiesel samples. It is worth mentioning that NO_x emissions are hazardous gases and all engine researches and manufactures target to reduce them.

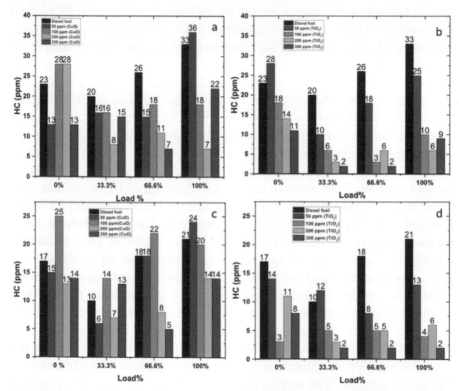

FIGURE 5.8 Unburned HC variations (a, c) with CuO nanodiesel fuel at various loads on cold and hot start conditions, respectively (b, d) with TiO_2 nanodiesel fuel at various loads on cold and hot start conditions, respectively.

Figure 5.8a, b relate the unburned hydrocarbons emitted in the exhaust at cold start conditions with respect to applied loads to the diesel engine operated by the base diesel and nanodiesel with CuO and TiO_2. It obviously appears that there are significant variations in HC emissions for the base diesel fuel and nanodiesel fuel with CuO and TiO_2; however, the base diesel fuel provides the highest HC emissions among all the applied loads comparable with nanodiesel fuel. It is previously known that diesel engines run at lean mixtures whenever small loads or no loads are applied to them, which means that excess amount of air is supplied, consequently, part of the mixture is overmixed leading to high HC emissions on exhaust. In addition, the evaporation process of fuel droplets becomes harder due to low wall temperature. On the other hand, the nanodiesel fuels produce lower HC in all the loads with respect to the base diesel fuel. According to Figure 5.8a, all CuO samples produce lower HC emissions in comparison with neat diesel fuel, in addition, as the load increases the HC emissions decrease that reflect the positive influence of CuO nanoparticles. The 200 and 100 ppm provide the lowest HC emissions for CuO samples relative to neat diesel sample. At the maximum load the HC emissions are reduced by 45% and 78% for 100 and 200 ppm CuO samples, respectively. The reason behind that is the more oxygen supplied by CuO which in turn combusts most of the HC supplied to diesel engine unlike to base diesel fuel, in addition, the ignition delay was shortened and the ignition characteristics were also improved.

Owing to Figure 5.8b the HC emissions provided by TiO_2 provide is much lower than neat diesel fuel and CuO samples at all loads. Almost all the TiO_2 samples show positive impact for reduction of HC emissions; it is found that HC emissions has been reduced by 69%, 81.8%, and 72 % for 100, 200, and 300 ppm TiO_2 samples, respectively. The reason behind that is the more oxygen supplied by TiO_2 which in turn combusts most of the HC supplied to diesel engine unlike to base diesel fuel, in addition, the ignition delay was shortened and the ignition characteristics were also improved. Figure 5.8c, d relate the HC emissions with the applied loads to the diesel engine at hot start conditions for base diesel fuel and nanodiesel fuel. The magnitude of HC emissions is totally different from the cold start conditions for both diesel fuel and nanodiesel samples. It is obviously observed that there is a significant reduction in the HC emissions either for CuO or TiO_2 nanodiesel samples. With respect to Figure 4.15, the 200 and 300 ppm CuO nanodiesel samples provide the lowest HC emissions among other samples; it is found to be 14 ppm at the maximum load that means there is a reduction by 33% in the magnitude HC emissions as compared with neat diesel fuel. Owing to Figure 5.8d, the 100, 200, and 300 ppm show the lowest emissions relative to neat diesel fuel. It is found to be 4, 6, 2 ppm, respectively. This means there is a reduction in HC emissions by 81%, 71%, and 90.4 % as compared with neat diesel fuel. It clearly appears that TiO_2 is more efficient than CuO due to existence of more oxygen atoms which in turn oxidize more HC.

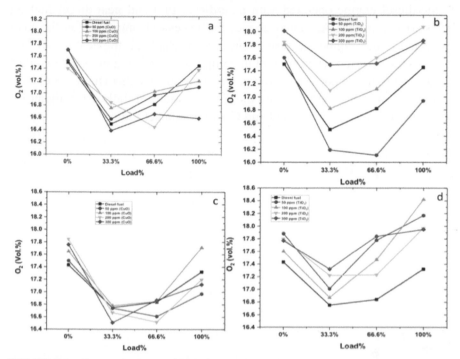

FIGURE 5.9 Excess O_2 (a, c) with CuO nanodiesel fuel at various loads on cold and hot start conditions, respectively, (b, d) with TiO_2 nanodiesel fuel at various loads on cold and hot start conditions, respectively.

Figure 5.9a, b represent the excess O_2 gas in exhaust with the applied loads to diesel engine operated with different fuels at cold start conditions. It is obviously observed that as the applied load increases the excess O_2 gas in exhaust increases too. It is worth mentioning that in order to have no O_2 gas in the exhaust, the air to fuel ratio must exist in its stoichiometric ratio; however, it is impossible to occur as combustion process does not stick to stoichiometric ratio rather than deviate from it.

As previously stated, the diesel engine runs at lean mixture at small loads, furthermore, the supplied air is greater than the stoichiometric value which in turn provides excess O_2 gas in the exhaust. It is clearly seen that there is randomness in the values of excess O_2 in the exhaust; moreover, it does not follow a specific pattern. Owing to Figure 4.18, the magnitude of excess O_2

gas in the exhaust is more than the neat diesel fuel and CuO nanoparticles as the oxygen content in one molecule of TiO_2 is double the oxygen content of CuO.

Figure 5.9c, d present the excess O_2 gas in exhaust with the applied loads in the diesel engine operated with nanodiesel fuel and neat diesel fuel on hot start conditions. As stated above, the excess O_2 increases as the load subjected to the engine increases; however, at 0% load, the neat diesel fuel exhibits the lowest oxygen content in comparison with nanodiesel fuel either CuO or TiO_2, this is due to high oxygen content that exists in nanodiesel fuel, in addition, at the maximum load the TiO_2 nanodiesel fuels offer higher O_2 content over the neat diesel fuel and CuO nanodiesel fuels.

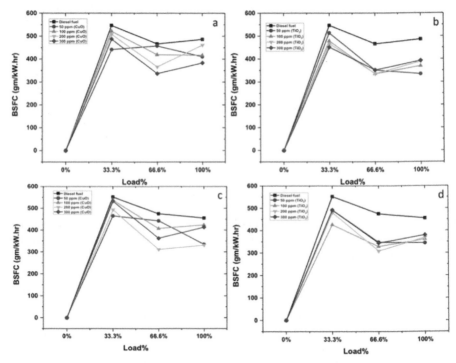

FIGURE 5.10   BSFC variations (a, c) with CuO nanodiesel fuel at various loads on cold and hot start conditions, respectively, (b, d) with TiO_2 nanodiesel fuel at various loads on cold and hot start conditions, respectively.

Figure 5.10a, b reveal the BSFC for the base diesel fuel, as well as nanodiesel fuel, with CuO and TiO_2 at different loads. BSFC is considered a crucial parameter to indicate both the efficiency and economy of engine to burn fuel and produce useful power (rotational power). According to figures, the base diesel fuel provides the highest BSFC in comparison with nanodiesel fuels at all loads on cold start conditions.

This means that the base diesel fuel consumes much fuel and provides low-brake power relative to nanodiesel fuel. Owing to Figure 4.22 the CuO nanodiesel fuel provides lower BSFC and brake power than neat diesel fuel on cold start conditions, which means the nanodiesel fuel consumes much lower fuel than base diesel fuel and produces higher brake power. Furthermore, the 300 ppm presents the highest BSFC among other nanodiesel fuel, it is 383.7817 $\frac{gm}{kw.hr}$ at the maximum load, meanwhile, the neat diesel fuel is 461.1041 $\frac{gm}{kw.hr}$, which means that CuO nanodiesel fuel reduces fuel consumption by approximately 16.7%

Regarding Figure 5.10b, all TiO_2 nanodiesel samples offer lower BSFC than neat diesel fuel on cold start conditions which in turn reflect the positive influence and the efficiency of TiO_2 nanoparticles as a potential additive to diesel fuel. Furthermore, the TiO_2 nanodiesel samples enhance the ignition characteristics and thermal conductivity of diesel fuel, consequently, it reduces fuel consumption. According to the figure, the 50 and 100 ppm TiO_2 samples provide the lowest BSFC relative to neat diesel sample, it is found to be 344.403 $\frac{gm}{kw.hr}$ and 370.117 $\frac{gm}{kw.hr}$, respectively, at maximum load, this means it reduces fuel consumption by 25% and 19.7% on cold start conditions.

Figure 5.10c, d depicts the BSFC for the base diesel fuel, as well as nanodiesel fuels, with CuO and TiO_2 at hot start conditions. The nanodiesel fuel improves thermal conductivity of diesel fuel and enhances the efficiency of the combustion process. Figure 4.23 shows that CuO nanodiesel samples reduce fuel consumption significantly as compared with base diesel fuel. The 200 ppm of CuO offer the lowest BSFC among other samples, it is found to be 330.83 $\frac{gm}{kw.hr}$ while as the neat diesel fuel is 455.2648 $\frac{gm}{kw.hr}$ at the maximum load. This means the 200 ppm CuO sample reduces fuel consumption by 27.3% on hot start conditions.

Owing to Figure 5.10d, it is clearly seen that the neat diesel fuel represents the highest BSFC among other samples that reflect the poor quality, as well as the uneconomic nature, of the combustion process. In addition, the TiO_2 samples provide lower BSFC over the neat diesel fuel. According to the figure, the 50 and 100 ppm TiO_2 samples present lowest BSFC among other samples and neat diesel fuel. It is found to be $344.403 \ \dfrac{gm}{kw.hr}$ and $361.71 \ \dfrac{gm}{kw.hr}$ at the maximum load that reduces about 24% and 20.5%, respectively, at maximum load.

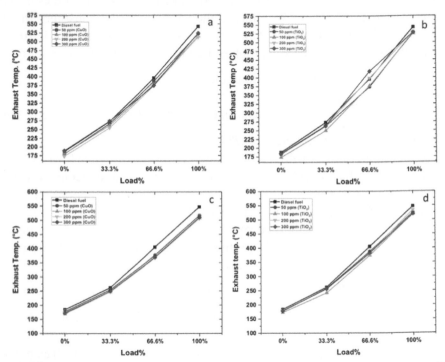

FIGURE 5.11 Exhaust temperature (a, c) with CuO nanodiesel fuel at various loads on cold and hot start conditions, respectively, (b, d) with TiO_2 nanodiesel fuel at various loads on cold and hot start conditions, respectively.

Figure 5.11a, b relate the exhaust temperature with the applied loads to the diesel engine operated with base diesel fuel, as well as nanodiesel, with CuO and TiO_2 at cold start conditions. It is clearly seen that the exhaust temperature increases as the applied load to the engine increases. The base diesel fuel shows the highest exhaust temperature under all loads that indicates much heat loss from the combustion process, meanwhile the nanodiesel fuel with CuO and TiO_2 enhances the combustion process and thermal conductivity of diesel fuel, thus, the exhaust temperature has signifi-cantly reduced under all loads in both CuO and TiO_2. According to Figure 4.25, it is obviously seen that all the nanodiesel samples reduce the exhaust temperature that extensively appeared at the maximum load. The 200 ppm of CuO sample represents the lowest exhaust temperature over the other concentrations, it is reduced by 5% from 543 to 513°C at the maximum load comparable with the neat diesel fuel.

Owing to Figure 4.26, the exhaust temperature reduces significantly at all loads except for 300 ppm sample that is slightly higher than neat diesel fuel at small loads, According to the figure, the 100 and 200 ppm TiO_2 present almost the same exhaust temperature in cold start conditions at maximum load, it is found to be 525°C and 530°C, respectively, which means a reduction in exhaust temperature by 3.3% and 2.3%, respectively, at the maximum load.

Figure 5.11c, d relate the exhaust temperature with respect to applied loads to the diesel engine operated with diesel fuel and nanodiesel fuel (CuO, TiO_2) on hot start conditions. As stated above, the exhaust temperature increases as the applied load to the engine increases. According to Figure 5.11c, it is clearly seen that the magnitude of exhaust temperature is much smaller than the cold conditions as appeared in two figures. Furthermore, the CuO nanodiesel samples exhibit lower exhaust temperatures than neat diesel fuel. The 200 ppm CuO sample provides the lowest exhaust temperature among other samples with a reduction by 7.4% at the maximum load. Owing to Figure 5.11d, it follows the same behavior as the cold start but with a smaller magnitude than the cold one. All TiO_2 nanodiesel samples present lower exhaust temperatures with respect to the neat diesel fuel, in addition, the 100 TiO_2 sample has the lowest exhaust temperature among other samples with a reduction by 5.1% at the maximum load.

Figure 5.12a, b illustrate the brake power delivered to the generator under various loads applied to the engine operated with neat diesel fuel and nanodiesel fuel with CuO and TiO_2 on cold start conditions. Brake power is defined as the produced power by the engine at the output shaft, in other words, the useful power produced to do work.

Owing to Figure 5.12a, it is clearly seen that almost the base diesel fuel, 50 and 100 ppm of nanodiesel fuel with CuO, have the same brake power at all loads; however, for 200 and 300 ppm nanodiesel fuel with CuO there are minor increase in the transmitted brake power especially at high loads as the engine efficiency increases to overcome the loads. At the maximum load, the brake power of 200 and 300 ppm is 2.587 Kw, whereas the base diesel fuel is 2.548 Kw that means that the CuO nanoparticles increase brake power by 1.53%. With regard to Figure 4.30, it follows the same trend as CuO nanoparticles, it is obviously seen that neat diesel fuel, 50 and 100 ppm have the same brake power, meanwhile the higher concentrations as 200 and

300 ppm present higher brake power as compared with neat diesel fuel, it is to found to be 2.587 and 2.574 Kw with an improvement by 1.53 % and 1%, respectively.

Figure 5.12c, d represent the brake power delivered to the generator for an engine operated with neat diesel fuel and nanodiesel fuel with CuO and TiO$_2$ under various loads on hot start conditions. Owing to Figure 5.12d, the CuO nanoparticles follow the same behavior as in cold start conditions; 200 ppm and 300 ppm exhibit the highest brake power over the other samples and neat diesel fuel. It is as same as in cold start conditions with an increase of 1.53% and 1%, respectively.

According to Figure 5.12d, the TiO$_2$ nanodiesel samples on hot start conditions follow the same trend as in cold start conditions; in addition, the 200 ppm and 300 ppm TiO$_2$ samples provide the highest brake power over the other samples. It is slightly higher than the cold start conditions with an improvement of 2% and 1% over the neat diesel fuel.

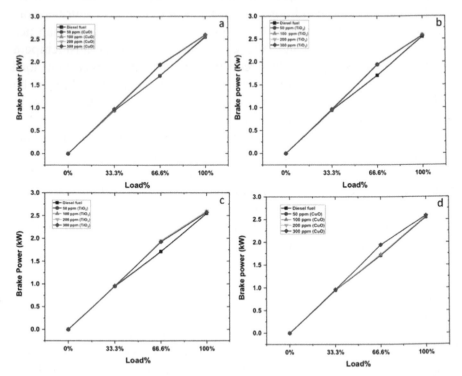

FIGURE 5.12 Brake power produced (a, c) with CuO nanodiesel fuel at various loads on cold and hot start conditions respectively, (b, d) with TiO$_2$ nanodiesel fuel at various loads on cold and hot start conditions, respectively.

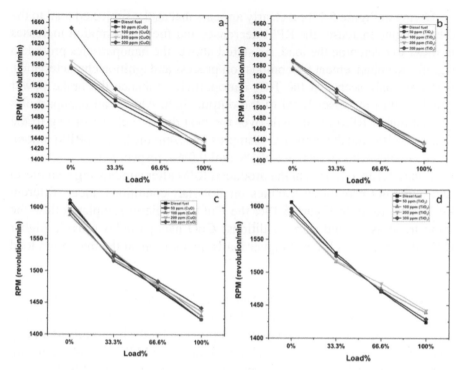

FIGURE 5.13 RPM produced (a, c) with CuO nanodiesel fuel at various loads on cold and hot start conditions, respectively, (b, d) with TiO$_2$ nanodiesel fuel at various loads on cold and hot start conditions, respectively.

Figure 5.13a presents the produced RPM after combusting the base diesel fuel as well as the nanodiesel fuel with CuO nanoparticles under various loads on cold start conditions. It is worth mentioning that as the applied loads to diesel engine increase, the RPM decrease, whereas the fuel consumption increases in order to overcome the load. The results indicate that the samples higher than 50 ppm provide the highest RPM with respect to base diesel fuel at all loads, whereas the nanodiesel fuel with 50 ppm CuO exhibits different trends at small loads.

The variation in the produced RPM is an indication for the performance of the nanodiesel fuel; furthermore, it means that addition of CuO nanoparticles obviously enhances the combustion process and reduces the loss in energy due to its catalytic activity. The 200 and 300 ppm samples offer the highest RPM in all loads with an increase of 1% and 1.4%, respectively, relative to neat diesel fuel.

Owing to Figure 5.13b, it clearly appears that as the subjected load to the diesel engine increases the RPM decreases and fuel consumption increases in order to overcome the load. As stated above, the nanoparticles provide a significant enhancement for combustion process and ignition characteristics, it is obviously seen that the TiO_2 nanoparticles enhance the brake power which in turn increases RPM of the output shaft. Almost all nanoparticles present higher RPM at all loads over the neat diesel fuel. According to the figure, the 100 and 200 ppm TiO_2 samples represent the highest RPM among other samples with an increase of 1% and 0.08%, respectively.

Figure 5.13c, d represent the produced RPM after combusting neat diesel fuel and CuO nanodiesel samples on hot start conditions under different loads. According to Figure 5.13c, the CuO nanodiesel samples follow the same trend as in cold start conditions; CuO nanoparticles provide higher RPM than neat diesel fuel at higher loads, in addition, at the maximum load the 300 ppm CuO presents the highest RPM (1442) with an increase by 1.2% over the neat diesel fuel on hot start conditions.

According to Figure 5.13d, the TiO_2 nanodiesel samples show higher RPM at the higher loads in comparison with neat diesel fuel. Owing to the figure, it clearly appears that 100 and 200 ppm samples exhibit the highest RPM among other samples at the maximum load. It is found to be 1439 and 1442 rpm with an increase of 1.05% and 1.2%, respectively, over the neat diesel fuel.

Figure 5.14a, b relate the thermal efficiency of the diesel engine operated with base diesel fuel, as well as the nanodiesel fuel, with CuO and TiO_2 nanoparticles at different loads on cold start conditions. The thermal efficiency is defined as the amount of useful energy (brake power) transmitted to the generator from the input power supplied to the engine.

Figure 5.14a indicates that the thermal efficiency increases as the applied load to the engine increases to overcome the load. The nanodiesel fuels enhance the combustion process; in addition, the nanodiesel fuels increase the brake power delivered to the generator and reduce the fuel consumption due to its high catalytic activity as previously mentioned. The 200 and 300 ppm CuO samples provide the highest efficiency among other concentrations, they are 24% and 20.61%, respectively, at the maximum load, whereas for the base diesel fuel it is 16.27%. Regarding Figure 4.38, as stated above the TiO_2 nanoparticles improve thermal efficiency and enhance the combustion process. It is clearly seen that TiO_2 samples provide higher thermal efficiency over the neat diesel fuel in all loads. Moreover, the 50 and 100 ppm present the highest thermal efficiency among other concentrations at maximum load; they are 23.6% and 21.3%, respectively, whereas the neat diesel fuel is 16.27%.

Figure 5.14c, d relate the thermal efficiency of the combustion process with the subjected loads to the diesel engine operated with neat diesel fuel and nanodiesel fuel with CuO and TiO_2 on hot start conditions. As mentioned above, as the load increases the thermal efficiency increases too. Owing to Figure 5.14c, the neat diesel fuel shows the poorest efficiency relative to CuO nanodiesel samples. Moreover, the 200 ppm CuO sample provides the highest efficiency over the neat diesel fuel and other samples, it is found to be 24%, whereas for the neat diesel fuel it is 17.3%.

According to Figure 5.14d, the TiO_2 nanodiesel samples provide higher efficiency than neat diesel fuel over all the loads. As mentioned above, the 50 and 100 ppm TiO_2 exhibit the highest efficiency among other samples and neat diesel fuel, they are 22.9% and 22%, respectively, whereas the neat diesel fuel is 17.3%.

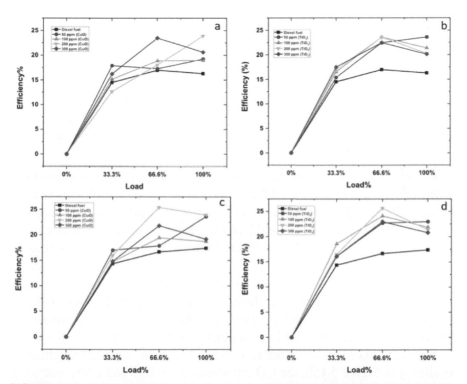

FIGURE 5.14 Efficiency of diesel engine (a, c) with CuO nanodiesel fuel at various loads on cold and hot start conditions, respectively, (b, d) with TiO_2 nanodiesel fuel at various loads on cold and hot start conditions, respectively.

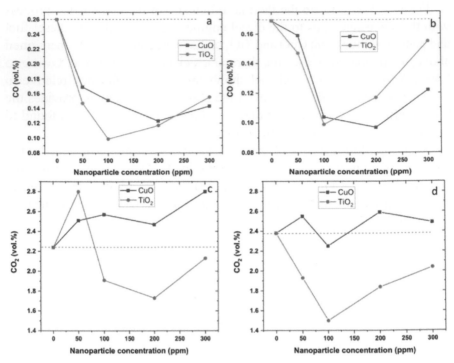

FIGURE 5.15   Influence of nanoparticles concentrations on CO variations at maximum load (a, c) on cold start conditions, respectively. (b, d) on hot start conditions, respectively.

Figure 5.15a, b represent the produced CO emissions at maximum load during combustion of neat diesel fuel and nanodiesel fuel at different concentrations. The dotted line represents CO emissions of neat diesel fuel, they are 0.26 vol% and 0.169 vol% on cold and hot start, respectively. It is clearly shown that the neat diesel fuel exhibits highest emissions either in cold or hot start conditions. On the other hand, addition of nanoparticles enhances the combustion process and reduces emissions significantly; however, at a certain point, the emissions pick their way upward again. The results have shown that 100 ppm TiO_2 and 200 ppm CuO have the lowest CO emissions at maximum load either in cold or hot start conditions. As previously stated, the hot start conditions are more reliable and realistic to actual conditions. As shown in Figure 5.15b, the CO emissions are 0.099 and 0.097 vol% for 100 ppm TiO_2 and 200 ppm CuO, respectively, whereas for neat diesel fuel it is 0.169 vol%. In other words, it means that 100 ppm TiO_2 and 200 ppm CuO nanodiesel fuels have reduced the CO emissions by 41.4% and 42.6%, respectively.

Figure 5.15c,d relate the produced CO_2 at maximum load during combustion process of neat diesel fuel and nanodiesel fuel at various concentrations. CO_2 is one of the main products that resulted from complete combustion of any hydrocarbons, unfortunately, the combustion process is always accompanied by CO due to incomplete combustion. Regarding the neat diesel fuel, the magnitude of CO_2 produced is lower than TiO_2 nanodiesel fuel and higher than CuO nanodiesel fuel either in cold or hot start conditions. The results have shown that the behavior of CuO and TiO_2 nanodiesel fuels are contradicting with each other either cold or hot start condition. The CuO nanoparticles provide higher CO_2 emissions almost in all samples either in cold or start conditions, whereas the TiO_2 nanodiesel fuel exhibits lower CO_2 than neat diesel fuel either in hot or cold start conditions. Regarding CuO nanodiesel fuel, the increase for CO_2 indicates that the combustion process proceeds toward completion rather than incomplete one and this, in turn, enhances combustion process and its efficiency. It is found that 200 ppm CuO produces the highest CO_2 with a value of 2.59 vol% and represents an increase of 9% on hot start conditions. On the other side hand, TiO_2 nanodiesel fuels show lower CO_2 emissions than neat diesel fuel. This is mainly due to their high efficiency that delivers more power and reduces the fuel consumption, consequently, the CO_2 is very low as compared with neat diesel fuel and CuO nanodiesel fuels. Moreover, the 100 ppm TiO_2 provides the lowest CO_2 emissions among other concentrations and neat diesel fuel at maximum load. It is 1.5 vol% with a reduction of 37% on hot start conditions.

Figure 5.16a, b represent the produced NO emissions at maximum load during combustion of diesel fuel and nanodiesel fuels at various concentrations on cold and hot start conditions. Regarding Figure 5.16a, the results indicate that the neat diesel fuel has lower magnitude for NO emissions than CuO nanodiesel fuels and almost the same as TiO_2 nanodiesel fuel. It is worth mentioning that as concentration of CuO nanoparticles increases, the NO emissions increase, indeed, due to high supply of O_2 atoms from nanoparticles. As shown in Figure 5.16b, the NO emissions have increased by 22% comparable with neat diesel fuel for 200 ppm CuO. On the other hand, the TiO_2 nanodiesel fuels provide contradicting results to CuO nanodiesel fuels, furthermore, the NO emissions follow a declining trend over the neat diesel fuel on hot start conditions then it picks its way upward again upon increasing concentration more than 100 ppm. The results show that the 100 ppm TiO_2 nanodiesel fuel has the lowest NO emissions with a reduction by 38.3% as shown in Figure 5.16b. This is mainly due to its high reactivity that delivers more power and reduces the fuel consumption, consequently the NO emissions are very low as compared with neat diesel fuel and CuO nanodiesel fuels.

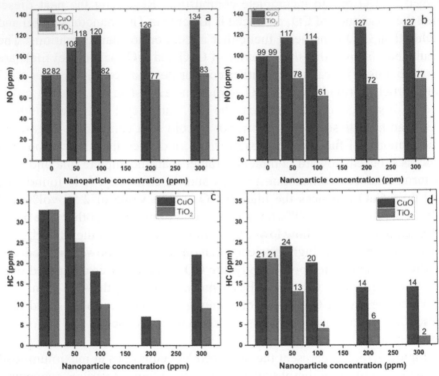

FIGURE 5.16　Influence of nanoparticles concentrations on NO variations at maximum load (a, c) on cold start conditions, respectively, (b, d) on hot start conditions, respectively.

Figure 5.16c, d provide the unburned HC emissions during combustion process of neat diesel fuel and nanodiesel fuel at various concentrations under the maximum loads on both cold and hot start conditions. The results indicate that the neat diesel fuel has almost the highest emissions for unburned HC over the TiO_2 and CuO nanodiesel fuels on both hot and cold start conditions. In general, the nanodiesel fuels enhance the combustion process and reduce the unburned HC emissions, significantly, especially for the TiO_2 nanodiesel fuels that provide promising results. Regarding the CuO nanodiesel fuel, it is found that the 200 ppm CuO nanodiesel fuel shows the lowest unburned HC emissions among other samples and neat diesel fuel on both cold and hot start conditions.

According to Figure 5.16d, it is found to be 14 ppm with a reduction of 33% relative to neat diesel fuel. Owing to TiO_2 nanodiesel fuels, they exhibit higher performance than CuO nanodiesel fuel either in cold or hot start conditions, this is mainly due to high reactivity of TiO_2 comparable with CuO, which significantly influenced the combustion process in term of HC emission. As shown in Figure 5.16d, the 100 ppm TiO_2 shows the lowest emissions for unburned HC with a reduction of 81% relative to neat diesel fuel.

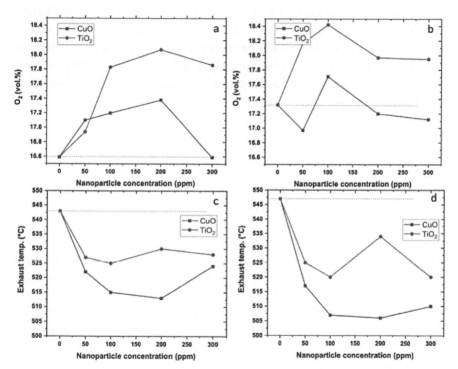

FIGURE 5.17 Influence of nanoparticles concentrations on excess O_2 at maximum load (a, c) on cold start conditions, respectively, (b, d) on hot start conditions, respectively.

Figure 5.17a presents the excess O_2 in the exhaust after the combustion process comes to an end at which is the engine is fueled by neat diesel fuel and nanodiesel fuel with different concentrations at the maximum load on cold start conditions. As stated above, the diesel engine runs at lean mixture in which an excess air supplied to ensure complete combustion, therefore, at the end of the process, excess air appears in exhaust in terms of O_2 and nitrogen. The figure shows that the neat diesel fuel has the lowest O_2 content comparable with nanodiesel fuel either CuO or TiO_2. It is also seen that almost all TiO_2 nanodiesel fuels exhibit the highest oxygen content comparable with neat diesel fuel and CuO nanodiesel fuel, in addition, as the dosage of TiO_2 increases the O_2 gas increases too till it picks its way downward at 300 ppm dosage. The reason behind that is high oxygen content per one molecule of O_2 with respect to CuO nanodiesel fuel.

Figure 5.17b presents the relationship between excess O_2 in the exhaust and various concentrations of nanodiesel fuels on hot start conditions. The dashed line represents the excess O_2 in exhaust at the maximum load. The behavior of nanoparticles in hot start conditions is somehow contradicting, the CuO nanodiesel samples provide lower oxygen content that can be explained as follows: Most of the oxygen present in CuO is involved in the combustion process and the remaining O_2 is mainly excess and the evidence for that is low emissions for unburned HC, NO, and high CO_2 emissions.

On the other hand, the TiO_2 nanodiesel samples exhibit the same behavior as in cold start, owing to the high oxygen content for TiO_2 nanodiesel samples, all TiO_2 nanodiesel fuels show higher O_2 content than neat diesel fuel and CuO nanodiesel fuel. It is worth mentioning that although the O_2 content in exhaust is high, the unburned HC and NO emissions are very low compared with CuO nanodiesel fuel that reflects the role and the high catalytic activity of TiO_2 in combustion process. In addition, the 100 ppm TiO_2 has the highest O_2 content among other samples and above this concentration the excess O_2 picks its way downward.

Figure 5.17c, d reveal the exhaust temperature for an engine operated with various concentrations of nanodiesel fuels on both cold and hot start conditions. The dashed line represents the exhaust temperature for the neat diesel fuel at the maximum load. It is clearly observed that the nanodiesel fuels provide lower exhaust temperature than neat diesel fuel on all concentrations, either cold or hot start conditions. For CuO nanodiesel, as the dosage of nanoparticles increases, the exhaust temperature decreases till 200 ppm then it picks its way upward again either in hot or cold start conditions. As shown in Figure 5.17d, the 200-ppm sample provides the lowest exhaust temperature of 506°C comparable with neat diesel fuel with a reduction of 7.4%. Owing to TiO_2 nanodiesel samples, they provide lower exhaust temperature over the neat diesel fuel; however, their magnitude is slightly higher than CuO nanodiesel fuel, this is mainly due to higher content of O_2 gas in exhaust that acts as a good carrier for temperature. According to Figure 4.52, the exhaust temperature of 100 ppm TiO_2 sample is 520°C with a reduction of 4.9% in comparison with neat diesel fuel at maximum loads. The role and influence of nanodiesel fuel have been significantly observed in the reduction of exhaust temperature, furthermore, it is an indication for efficient combustion process and limitation of energy loss as well. This mainly occurs due to high catalytic activity of nanoparticles and shortening ignition delay of fuel.

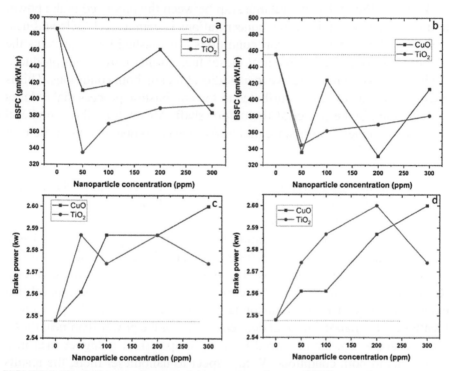

FIGURE 5.18   Influence of nanoparticles concentrations on BSFC at maximum load (a, c) on cold start conditions, respectively, (b, d) on hot start conditions respectively.

Figure 5.18a, b provide a relationship between BSFC of diesel engine and various concentrations of nanodiesel fuels and neat diesel fuel on hot and cold start conditions. BSFC is considered a crucial parameter to indicate the efficiency as well as the economy of engine to burn fuel and produce useful power (rotational power). The dashed line represents the BSFC of the neat diesel fuel at the maximum load. It is clearly observed that neat diesel fuel has the highest BSFC that reflects the poor quality of the combustion process in terms of economy and efficiency. As seen in Figure 5.18b, the BSFC is 455.26 $\frac{gm}{kw.hr}$ for the neat diesel fuel meanwhile the 200 ppm CuO nanodiesel fuel shows the lowest BSFC, that is, 330.83 $\frac{gm}{kw.hr}$ among all nanodiesel samples with a reduction of 27.3% over the neat diesel fuel. Furthermore, the 50 and 100 ppm of TiO_2 shows the lowest BSFC, 344.403 $\frac{gm}{kw.hr}$ and 361.71 $\frac{gm}{kw.hr}$, among TiO_2 samples with a reduction of 24.3% and 20.5%, respectively.

Figure 5.18c, d depict a relationship between the produced brake power in a diesel engine and various concentrations of nanodiesel fuels and neat diesel fuel on cold and hot start conditions. The dashed line represents the produced brake power for the neat diesel fuel, furthermore, it has produced the lowest brake power relative to nanodiesel fuel on both conditions. This is mainly due to many losses attributed to the combustion process that reduces the delivered brake power at the output shaft. Meanwhile, the nanodiesel fuel shows higher performance and brake power comparable with the neat diesel fuel. With respect to Figure 5.18d, the TiO_2 samples show higher brake power than CuO samples in most of the samples because of high content of O_2 in its molecule relative to CuO. Moreover, the 200 ppm TiO_2 sample shows the highest brake power, 2.6 kW, among all samples with an increase of 2% over the neat diesel fuel.

Figure 5.19, b provide a relationship between the produced RPM in a diesel engine and various concentrations of nanodiesel fuels and diesel fuel on cold and hot start conditions. The dashed line represents the delivered RPM for the neat diesel fuel; it is clearly observed that the neat diesel fuel exhibits the lowest RPM comparable with nanodiesel fuels. As previously mentioned, the nanodiesel fuel delivers higher brake power than neat diesel fuel; consequently, the produced RPM by output shaft is higher than neat diesel fuel in both conditions. With respect to nanodiesel fuels, the results indicate that as the concentration of nanoparticles increases the produced RPM increases too. As seen in Figure 4.58 the 300 ppm CuO and 200 ppm TiO_2 samples show the highest RPM 1442 among other samples with an increase of 1.2% over the neat diesel fuel. Owing to TiO_2 nanodiesel samples, the results indicate that the concentration of TiO_2 sample the RPM increase until a certain value (200 ppm) and then it picks its way downward as shown in Figure 5.19b.

Figure 5.19, d represent the thermal efficiency of a diesel engine operated by neat diesel fuel and nanodiesel with various concentrations at maximum load. The dashed line indicates the thermal efficiency of the combustion process for the neat diesel fuel. With respect to Figure 5.19c, d, the neat diesel fuel shows the lowest efficiency comparable with the nanodiesel samples on both conditions. This is mainly due to many losses accompanied by combustion process consequently low brake power delivered to the output shaft; meanwhile, the nanodiesel fuels exhibits higher efficiency than neat diesel in all samples due to its high catalytic activity. With respect to Figure 5.19c, the results indicate that as the concentration of the CuO increases, the efficiency of the combustion process also increases, except for 200 ppm, whereas its contradicting with TiO_2 samples. Regarding Figure 5.19d, the CuO nanodiesel fuels do not follow a uniform pattern in contrast with the TiO_2 nanodiesel samples that follow a declining pattern. In addition, the 200 ppm CuO and 50 ppm TiO_2 samples show the highest efficiency 24% and 22.9%, respectively, whereas for neat diesel fuel it is 17%. The nanodiesel fuels have significantly enhanced the combustion process and ignition characteristics of the diesel that is extensively observed in terms of emissions and mechanical parameters.

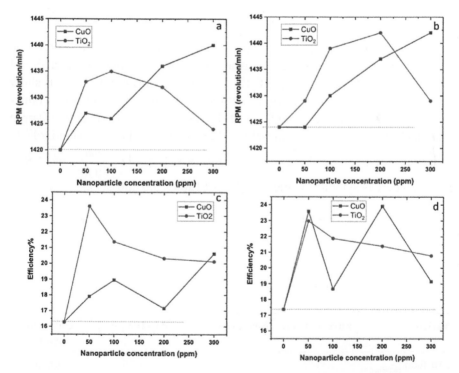

FIGURE 5.19 Influence of nanoparticles concentrations on RPM at maximum load (a, c) on cold start conditions, respectively, (b, d) on hot start conditions, respectively.

TABLE 5.5 Output Variables for Optimum Samples and Diesel Fuel.

Variable	CuO (200 ppm)	TiO₂ (100 ppm)	Diesel fuel
Sonication time	40 min	30 min	–
Mixing time	15 min	10 min	–
CO	$6.918\ \dfrac{gm}{kw.hr}$	$7.112\ \dfrac{gm}{kw.hr}$	$14.579\ \dfrac{gm}{kw.hr}$
CO₂	$290\ \dfrac{gm}{kw.hr}$	$169.369\ \dfrac{gm}{kw.hr}$	$322.637\ \dfrac{gm}{kw.hr}$
NO	127 ppm	61 ppm	99 ppm
Unburned HC	14 ppm	4 ppm	21 ppm
Excess O₂	17.2 vol%	18.42 vol%	17.32 vol%
Fuel consumption	0.3308 Kg/kW.h	0.36171 Kg/kW.h	0.45526 Kg/kW.h
Brake power	2.587 kW	2.587 kW	2.548 kW
Exhaust temperature	506°C	520°C	547°C

Table 5.5 summarizes the values of the measured variables related to the combustion process and makes a reasonable comparison between neat diesel fuel, 100 ppm TiO_2 and 200 ppm CuO. In addition, the obtained results show that 100 ppm TiO_2 is more efficient than 200 ppm CuO in terms of emissions, ease of preparation, and the cost. In addition, it is found that 100 ppm TiO_2 and 200 ppm CuO nanodiesel fuel have matched the Euro standard CO emissions for nonroad diesel engines with a value $7.112\dfrac{gm}{kw.hr}$ and 6.918 $\dfrac{gm}{kw.hr}$, whereas standard value is $8\dfrac{gm}{kw.hr}$ (Tables 5.6 and 5.7).

$$4C_{12}H_{23+}(17.75O_2 +66.74\ N_2)\longrightarrow CO_2 + CO + O_2 + 66.74N_2 + H_2O \quad (5.1)$$

$$N_2 + O_2 \longrightarrow 2NO \quad (5.2)$$

TABLE 5.6 Emissions of Diesel Fuel According to Euro V Specifications for Nonroad Diesel Engines.

Variables	Euro V specs.	Neat diesel fuel	TiO₂ (100 ppm)	CuO (200 ppm)
(Air/fuel) Theoretical	–	17.75	17.75	17.75
(Air/fuel) actual	–	24.14	25.205	27.5125
ʎ	–	1.36	1.42	1.55

TABLE 5.6 *(Continued)*

Variables	Euro V specs.	Neat diesel fuel	TiO_2 (100 ppm)	CuO (200 ppm)
CO_2	–	$322 \frac{gm}{kw.hr}$	$169 \frac{gm}{kw.hr}$	$290 \frac{gm}{kw.hr}$
CO	$8 \frac{gm}{kw.hr}$	$14.579 \frac{gm}{kw.hr}$	$7.112 \frac{gm}{kw.hr}$	$6.918 \frac{gm}{kw.hr}$
O_2	–	$1707.558 \frac{gm}{kw.hr}$	$1512.65 \frac{gm}{kw.hr}$	$1402.03 \frac{gm}{kw.hr}$
NO	–	$0.831 \frac{gm}{kw.hr}$	$0.468 \frac{gm}{kw.hr}$	$0.971 \frac{gm}{kw.hr}$
HC	–	$1.08 \frac{gm}{kw.hr}$	$0.171 \frac{gm}{kw.hr}$	$0.595 \frac{gm}{kw.hr}$
HC+NO	$7.5 \frac{gm}{kw.hr}$	$1.911 \frac{gm}{kw.hr}$	$0.639 \frac{gm}{kw.hr}$	$1.566 \frac{gm}{kw.hr}$

TABLE 5.7   Advantage of Nanodiesel Fuel in the Transportation Sector.

Type of vehicle	Number of licensed vehicles in Egypt	Average daily consumption	Neat diesel consumption	TiO_2 nanodiesel consumption	Money saved/ month
Trucks	120,000	$107 \frac{km}{day}$	$15.12 \frac{Liter}{day}$	$12.1 \frac{Liter}{day}$	$15,826,823 \frac{L.E}{day}$
Buses	142,800				
Heavy trucks	93,000				

In general, the metal nanoparticles and their oxides are characterized by their high catalytic activity as they belong to transition metals. Transitions are widely used as oxidizing agents due to their high reactivity in chemical reactions. Most of nanoparticles are employed as catalysts in heat transfer applications for many reasons.

- High thermal conductivity
- High surface area to volume ratio
- Ease of availability and low cost

As previously mentioned in this chapter CuO and TiO_2 nanoparticles have extensively enhanced the combustion process and reduced its emissions. This is mainly due to enhancing ignition characteristics and heat transfer in the combustion process. However, the exact mechanism of nanoparticle metal oxides and how it affects the combustion process is unclear, it requires more scrutinization in order to come up with the exact mechanism that is expected to be dependent upon the metal oxide nanoparticles. D. Fino et al. have suggested a simple mechanism for hydrocarbon combustion, oxidation of soots, and reduction of NO emissions.[24]

$$C_xH_y + (2x+y)\ TiO_2 \longrightarrow (\tfrac{2x+y}{2})\ Ti_2O_3 + \tfrac{x}{2}\ CO_2 + \tfrac{y}{2}\ H_2O \qquad (5.3)$$

$$C_{soots} + 4TiO_2 \longrightarrow 2\ Ti_2O_3 + CO_2 \qquad (5.4)$$

$$Ti_2O_3 + NO \longrightarrow 2TiO_2 + N_2 \qquad (5.5)$$

In general, particles in dispersion may adhere to each other or repel from each other based on their charges and physical attraction force, like Van der Wall. In case of attracting each other, this will lead to forming aggregates with higher size that may settle out due to gravity. On the other hand, if the repulsive force is higher than adhering force, this, in turn, will enhance the properties of nanofluids as it will be stable for a long time. The stability of nanodiesel fuel is a key issue to characterize the potential use of nanoparticles as an additive to enhance the thermophysical properties of the base fluid. The agglomeration phenomena are directly related to the attraction force of the nanoparticles to each other if a high Van der Wall attraction exists between particles. There are many ways to evaluate the stability of nanofluids like sedimentation process, zeta potential, and centrifugation method. Sedimentation is considered the most common and economic type to evaluate the stability of nanodiesel fuel. Sedimentation is an indication to the weight or the volume of aggregates formed in the sample. The lower the sedimentation, the higher is the stability of nanodiesel fuel.

5.5 CONCLUSIONS

This study provides an overview of the research progress in the use of dispersion of various nanoadditives to diesel fuel for engine performance enhancement and emissions reduction. The main contributions and find-ings of this research are summarized as follows: CuO and TiO_2 provide high catalytic activity in the combustion process. Various nanoadditives' concentrations showed that increasing concentrations of nanoadditives do not increase performance or reduce emissions proportionality and, in some cases, it is intensive beyond a certain concentration level. Consequently, the determination of the optimum dosage for every nanoparticle is a very crucial issue. It is understood from the emissions characteristics that, there is a significant reduction in CO, NO, and unburned HC. However, for CO_2, it is reduced in case of TiO_2 and increased in case of CuO. It is understood from the performance of diesel engine that there is a reduction in exhaust temperature, fuel consumption, and on the other hand there is an increase in brake power, thermal efficiency, and RPM. According to experimental results, 100 ppm TiO_2 and 200 ppm CuO are the optimum dosage each for its type. In addition, the 100 ppm provides the lowest emissions and highest improvement in mechanical parameters in comparison with 200 ppm CuO.

KEYWORDS

- nanoparticles
- diesel engines
- exhaust
- emissions
- metal oxides

REFERENCES

1. Wenhua, Y. U.; David, M. F.; Jules, L. R.; Stephen, U. S. Review and Comparsion of Nanofluid Thermal Conductivity and Heat Transfer Enhancements. Taylor and Francis group, 2015, Vol 29, no. 5, pp 1–30.
2. Eastman, J. A.; Phillpot, S. R.; Choi, S. U. S.; Keblinski, P. Thermal Transport in Nanofluids. Annual Review in Material Research, 2011, Vol 34, pp 219–246.
3. Lee, S.; Choi, S. U. S.; Li, S.; Eastman, J. A. Measuring Thermal Conductivity of Fluids Containing Oxide Nanoparticles, Transactions of the ASME. *J. Heat Trans.* **2009,** *121,* 280–289.
4. Choi, S. U. S.; Zhang, Z. G.; Keblinski, P. *Nanofluids, in Encyclopedia of Nanoscience and Nanotechnology.* Nalwa, H. S., Ed.; Vol 6, pp 757–737.
5. Keblinski, P.; Eastman, J. A.; Cahill, D. G. Nanofluids for Thermal Transport. *Mater. Today,* **2005,** 36–44.

producing humidity indicators,[1,19] catalysts,[18] treatment of waste water for removal of harmful substances .[26,27]

Silica hydrogel can be prepared by different methods using water glass. Water glass is acidified by either sulphuric acid,[28] hydrochloric acid, [9, 14,17] carbonic acid, phosphoric acid[29]or by ion exchange.[5] For example, colloidal silicas were treated with ion exchanger Dowex Marathon C to lower their pH to 7.0–7.8. After separation by filtration and addition of $NaCl_{(aq)}$ they were gelled to give silica hydrogels with 9.0 % silica.[5] The overall reaction of silica formation from sodium silicate can be represented by eq 6.1 and 6.2.

$$Na_2O.xSiO_2.yH2O +H_2SO_4 \rightarrow Si(OH)_4 + Na_2SO_4 \text{ hydrolysis} \qquad (6.1)$$

$$Si(OH)_4 + Si(OH)_4 \rightarrow 2SiO_2 + 4H_2O \text{ condensation} \qquad (6.2)$$

The complexity of the silica growth can be gauged from its three consecutive stages, which dominate at different times but also run concurrently to some extent: (1) polymerization to form the initial particles, (2) growth of these particles, and (3) aggregation to form networks, which eventually span the containing vessel, defining the gel time.[8] Primary particles of mean hydrodynamic diameter of 1.5 nm are found to be present within 20 minutes of mixing sodium silicate solution and sulphuric acid.[4] Clustering then occurs during siloxane polymerization to produce secondary particles with a mean diameter up to 4.5 nm after 30 h. The growth rate depended on silicate concentration and time to micro gelation. Subsequent condensation to 4 nm diameter particles occurs within 1 week as particle syneresis dominates.[4] On the other hand silica precipitation under very acidic conditions, at hydrochloric acid concentration of 2–8 mol dm^{-3} proceeds through two distinct steps. First, the monomeric form of silica is quickly depleted from solution as it polymerizes to form primary particles of ~5 nm in diameter. Second, the primary particles formed then flocculate.[9]

The structure of silica hydrogel was investigated by using three dimensional electron tomography.[5] All the gels provided an accessible pore volume fraction of 30–40% when taking the third dimension into account.[5]

6.1.1 DIFFUSION OF IONS IN WATER

Diffusion of Cobaltous chloride in water was concentration dependent and it decreased as the concentration of $CoCl_2$ was increased.[22] It was 1.29×10^{-9} and 1.08×10^{-9} m^{2}s^{-1} for 0.008 and 0.3 mol dm^{-3} concentrations respectively.[22]

6.1.2   DIFFUSION OF IONS IN SILICA HYDROGEL

Diffusion coefficient of $Ni(NO_3)_2$ in silica hydrogel was determined by using a flow cell by Takahashi et al.[24] The diffusion coefficient was found as 60% of the diffusion coefficient in water. The data suggested that the silica gel network allowed relatively fast transport of the liquid phase.[24] The diffusion coefficient of nickel nitrate dissolved in water in porous silica plates with different pore size was measured to investigate the effect of pore size on the liquid-phase diffusion resistance. The pore-diffusion coefficient of nickel nitrate varied with the pore size of the silica gel. In pores of diameter > 5 nm, the diffusion resistance can be ignored, and the pore-diffusion coefficient varies proportionally to the porosity of the silica plates. In pores of diameter < 5 nm, however, the diffusion resistance increased rapidly with decreasing pore diameter. Microporous silica showed the largest diffusion resistance: the pore-diffusion coefficient in micro pores was less than 1% of the diffusion coefficient in an unbounded system.[24]

Diffusion of sodium chloride in silica hydrogel was investigated by single reservoir test by Kaczmarek, and Kazimierska-Drobny.[13,14] The diffusivity of NaCl in silica hydrogel was 1.57×10^{-9} m^2s^{-1} which was slightly lower than the value in pure water, 1.6×10^{-9} m^2s^{-1}.

The diffusion of chromate ion in colloidal silica gel set to hydrogel by calcium chloride was investigated by Tantemsapya and Meegoda .[25] To measure diffusion coefficients of chromium in the colloidal silica gel, a measurement method based on digital photography was used. The adsorption isotherm of chromate ions to colloidal silica gel was found to be linear at pH 7; the partition coefficient was calculated to be 0.549. The apparent diffusion coefficient of chromium in colloidal silica gel ranged from 1.76×10^{-10}–8.48×10^{-10} $m^2 s^{-1}$ depending mainly on the concentration of silica in the gel with chromium concentration less than 10^{-2} mol dm^{-3}. Higher silica concentrations yielded lower diffusion coefficients due to the obstruction to the free movement of chromium. The gel behaved as a porous material with silica network forming continuous solid phase and its pore space saturated with water. The chromium ions diffuse in porous silica gel on a tortuous path. Therefore, the bulk diffusion dominates. Thus, the silica can be represented as a fixed and impenetrable immersion in the solution. The presence of these motionless silica chains leads to an increase in the mean path of the diffusing molecules between two points in the system. However chromium was not adsorbed on silica under the experimental conditions.

6.1.3 PRECIPITATION IN SILICA HYDROGEL

Calcium phosphate crystals were synthesized by diffusing calcium ions into silica hydrogels containing phosphate ions. Hydroxyapatite (HAp) and octacalcium phosphate(OCP) with different types of crystal morphology were formed in the gel. The changes in crystal morphology of the HAp and OCP were ascribed to the degree of supersaturation of the reaction environment and the rate-determining step in the HAp and OCP crystal growth mechanism. The diffusion rate of ions in the gels relates to the rate of crystal growth.[31]

Controlled calcium phosphates precipitation has been obtained by calcium chloride diffusion through metasilicate solutions polymerized at different densities and pH values with phosphoric acid.[10] At the metasilicate solution density of 1.03 g cm^{-3} periodical precipitation calcium phosphates, Liesegang rings were observed. The number of these rings increased with the time and the distance between two consecutive rings increased toward going from the top to the bottom of the vial.[10]

6.1.4 CoCl$_2$ IMPREGNATION FOR HUMIDITY INDICATING SILICA GEL PRODUCTION

Well-defined CoCl$_2$-containing silica gels were prepared by impregnation of the aqueous solution of the salt to silica hydrogel, drying and aging methods. Silica gels having 392–437 m^2 g^{-1} surface area and 0.21–0.37 cm^3 g^{-1} pore volume and having an average particle size of 3 mm were obtained. The CoCl$_2$-containing gels were successfully used in dynamic column experiments, with linear relation between velocities of inlet air and movement of blue to pink boundary. The color change also made possible the detection of the defects in column filling which causes air channeling.[1] Indicating silica gel was obtained from silica gel prepared from rice husk ash waste material and CoCl$_2$.[19] The silica hydrogel was prepared by acid neutralization of sodium silicate solution which was extracted from the ash through sodium hydroxide leaching. The final pH was kept at pH 6. The silica hydrogel was impregnated with cobalt chloride solution of 0.5 mol cm^{-3} concentration at pH values 2-10 and dried at 150°C , kept in 90% relative humidity for 1 day and then re-dried at 150°C for even distribution of CoCl$_2$. The CoCl$_2$ containing silica gel prepared at pH 2 was nearly colorless, since it contained very little amount of cobalt compound. This may be due to low adsorption of cobalt on the gel since the isoelectric point of silica is near pH 2. The most prominent blue to pink color changing behavior was observed for pH 6 gel. The Co(II) has tetrahedral $[CoCl_4]^{2-}$ complex structure in dry state, which is deep blue in color. Upon exposure to moisture the structure changes to the octahedral $[Co(H_2O)_6]^{2+}$ complex, which is pink in color.[19]

Kesareva et al[15] investigated preparation of indicating silica gels. They prepared spherical silicahydrogel particles from silica hydrosol at pH 8–8.5 by adding it to hot turbine oil drop by drop. They showed that the diffusivity of cobalt salts in silica hydrogel changes with type of the salt. The diffusivity of $CoSO_4$, $CoCl_2$ and $Co(NO_3)_2$ was found as 1.9×10^{-10}, 2.7×10^{-10} and 3.6×10^{-10} in silica hydrogel spheres, respectively. The minimum salt concentration at which blue to pink color change on moisture adsorption and the relative humidity of air for observation of color change changed with the anion of the cobalt salt. For CNS^-, $H_2PO_4^-$ Cl^- cobalt compounds the color change was observed at 45–50 %, 35–40% and 25–30 % relative humidity respectively. The minimum Co concentration for these salts was 0.2, 0.4 and 0.5%, repetively.[15]

6.1.5 CO(II) SORPTION ON SILICA

Sorption of Co(II) on SiO_2 xH_2O (hydrous silica gel) has been investigated as a function of time by Pathac and Choppin.[20] Hydrous silicon dioxide $(SiO_2.xH_2O)$ was used for this purpose. Using the sorption kinetics data, the diffusion coefficient of Co(II) was calculated to be $(6.86 \pm 0.44) \times 10^{-12}$ m^2s^{-1} under particle diffusion-controlled conditions. The sorption data followed the Freundlich, Langmuir, and Dubinin-Radushkevich (D-R) isotherms. Cobalt sorption decreased with increased ionic strength. A gradual decrease in pH with increased ionic strength supported the sorption of Co(II) by an ion exchange mechanism.[20]

6.1.6 ADSORPTION AND DIFFUSION OF DYES IN SILICA HYDROGEL

Methylene blue adsorption behavior of a silica hydrogel was investigated by Top et al.[26] Linear and non-linear forms of Langmuir, Freundlich and Temkin models were tested to fit equilibrium data. Methylene blue adsorption capacity was obtained as ~13 mg g^{-1} hydrogel. Non-linear forms of pseudo first order, pseudo second order and Elovich models agreed with the kinetic data of the hydrogel in slab form with R^2 values greater than 0.99.

Silica hydrogel which had a porous structure and might be formed in any shape by gelation of silica hydro sol was employed for the adsorption of Malachite Green (MG) from water by Top et al .[27] A transparent silica hydrogel in slab form was obtained from aqueous sodium silicate and sulphuric acid and purified by water. The equilibrium and kinetics of MG adsorption was investigated by conducting in situ measurements with a conventional and fiber optic spectrophotometer, respectively. It was shown that nonlinear forms of kinetic models fitted to the experimental data more thoroughly than their linear forms by giving evenly distributed errors.

Effective diffusion coefficient of Malachite Green and Xylenol Orange in silica hydrogel was determined to be $(1.4 \pm 0.1) \times 10^{-10}$ and $(0.27 \pm 0.2) \times 10^{-10}$ respectively by Perrulini et al.[21]

The effective diffusion coefficient of protons in silica hydrogel was found to be dependent on the pore structure which depended on preparation pH of silica hydrogel.[2]

The effective diffusivities of various ions and substances in water, silica hydrogel prepared by different methods or silica gel are summarized in Table 6.1. The order of the diffusion coefficient of ions in water is 10^{-9} m² s⁻¹.[2,13,22]Depending on the pore structure and porosity of the silica hydrogels the effective diffusivities of the ions are lowered. The effective diffusivity of strongly adsorbed ions such as malachite green and methylene blue or Co(II) ions at pH 8 are at the order of 10^{-10} m² s⁻¹ since they were diffused in silica hydrogel after fast adsorption on the surface of silica particles.[21, 15]In silica gels dispersed in water adsorption process is much slower due to their smaller pore volume and pore size of hydrous gels, the effective diffusivity of Co(II) ions in hydrous silica gel is at the order of 10^{-12} m² s⁻¹ [20]

TABLE 6.1   Effective Diffusivities of Various Ions and Substances in Water, Silica Hydrogel, and Silica Gel.

Species	pH	Medium	$D \times 10^9$, m²s⁻¹	Reference
Protons	7	Water	9. 9.19.9.1	2
NaCl	7	Water	1.6	13
CoCl$_2$	7	Water	1.29–1.08	22
Protons	1–4	Silica hydrogel	$-17 + 21$ pH-3.7 pH ^{2}	2
NaCl	7	Silica hydrogel	1.57	14
Ni(NO)$_3$		Silica hydrogel	0.6	23
Ni(II)		TEOS derived silica gel	0.41–0.7	24

TABLE 6.1   *(Continued)*

Species	pH	Medium	$D \times 10^9$, m^{2}s^{-1}	Reference
CoSO$_4$	8–8.5	Silica hydrogel	0.16	15
CoCl$_2$	8–8.5	Silica hydrogel	0.27	15
Co(NO$_3$)$_2$	8–8.5	Silica hydrogel	3.6	15
Chromate	9	Silica hydrogel from colloidal silica	0.176–0.848	25
Malachite Green	4.5–9	Silica hydrogel from colloidal silica	0.14±01	21
Xylenol Orange	4.5–9	Silica hydrogel from colloidal silica	0.27±0.2	21
CoCl$_2$	6.8–10.8	Hydrous silica gel	0.00686	20

6.1.7   CATALYTIC APPLICATIONS

Supported cobalt is one of the common catalysts used in Fischer-Tropsch synthesis (FTS). Cobalt nano particles supported on silica was synthesized by strong electrostatic adsorption (SEA) by Ling et al[18] and Chee et al.[7] Cobalt nitrate was used as the catalyst precursor and non-porous silica spheres, which were synthesized using the modified Stöber method, were used as a catalyst support. High cobalt uptake at basic pH and low cobalt uptake at acidic pH indicated electrostatic interaction between the cobalt complexes in the precursor solution and the hydroxyl group on the support's surface. The optimum pH was determined as 8.76–9.93.[7–18]

Cobalt-containing SiO$_2$-based nanospheres with multi-nanochambers were prepared via a micro emulsion method by Wang et al[29]. Silica spheres with nanochambers containing cobalthydroxide were obtained. This novel material was found to be an efficient catalyst for the decomposition of cyclohexyl hydroperoxide to K–A oil (cyclohexanone and cyclohexanol) which is a basic intermediate of Nylon 6 and 66.[29]

The aim of this study is to investigate the diffusion of protons in silica hydrogel during washing process and the preparation of Co(II) containing silica that can be used as either humidity indicator or catalyst. The silica hydrogel slabs prepared from water glass and sulfuric acid were contacted with CoCl$_2$ solutions for this purpose. The change of the cobalt concentration the solutions with time was monitored either by visible spectroscopy or atomic absorption spectroscopy. Inductively coupled plasma (ICP) was used to analyze the Co(II) solution in equilibrium with silica hydrogel. The Co(II)

silica gel were analyzed by EDX using Philips XL30S model scanning electron microscope. The SEM micrographs of the ground silica hydrogel and silica gel were also obtained in the same instrument. The FTIR spectrum of silica gel was obtained by KBr disc method using Shimadzu FTIR-8201 model Fourier transformed infra-red spectrometer.

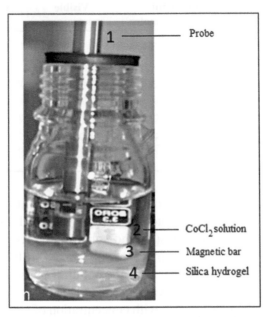

FIGURE 6.1   In-situ monitoring Co(II) concentration by fiber optic spectrophotometer.

6.3   RESULTS AND DISCUSSION

6.3.1   SILICA HYDROGEL PREPARATION

Silica hydrogel in the present study was prepared at pH 1 since particles with suitable size (3 mm) and sufficient mechanical strength for using as a packing material in a drying column thus obtained.

6.3.2   WASHING OF SILICA HYDROGEL

The by-product sodium sulfate and the excess acid were leached from silicahydrogel by washing it with water at least 15 times for 30 min. The acid leaching process for silica hydrogels with different layer thickness was monitored for the first washing step. As shown in Figure 6.2 the fraction of the acid transferred to water phase at any time (M_t) to at equilibrium(M_∞) was linear with the square root of time. However the lines did not pass through origin indicating a step change in pH of the wash solution at the start of the washing process.

The simplified equation for diffusion from a slab in one direction

$$\frac{M_t}{M_\infty} = \frac{4}{l}\sqrt{D_e \frac{t}{\pi}} \tag{6.3}$$

Where l is the thickness of the slab, D_e is the effective diffusion coefficient and t is the time. Thus from the slopes of the lines and the eq 6.3 the D_e values for protons in silica hydrogel was calculated and are reported in Table 6.2.

The pH of the solutions was monitored for a sufficient period of time till the equilibrium pHs of the solutions reported in Table 6.3 were measured.

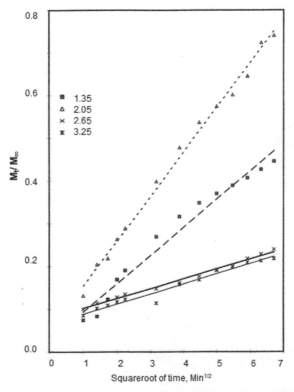

FIGURE 6.2 Fraction of protons removed from silica hydrogel with different layer thickness. The layer thickness in cm is the parameter.

TABLE 6.3 The Effective Diffusion Coefficient of Protons in Silica Hydrogel with Different
Layer Thickness for Washing with 200 cm³ Water Initially at pH 6.

Layer thickness, cm	Equilibrium pH of wash water	$D_e \times 10^9$, m²s⁻¹
0.75	1.96	2.40
1.35	1.50	1.38
2.05	1.73	1.35
2.65	1.56	2.58

The average value of the effective diffusion coefficient of protons in silica hydrogel was $1.92\pm0.56\times10^{-9}$ m² s⁻¹ at 25°C. This value was lower than that of the diffusion coefficient, D, of protons in water 9.1×10^{-9} m²s⁻¹.[2] This showed that the proton movement in silica hydrogel is obstructed by the silica network in silica hydrogel. The tortuosity factor, k_t of silica hydrogel was found as 2.07 from Eq 6.4 taking void volume fraction, ε, as 95.2 % which was the volume occupied by water in silica hydrogel.

$$D_e = D\left(\varepsilon / kt^2\right) \tag{6.4}$$

The k_t value has been obtained in the present study is close to the value which has been found previously in 1.78 cm layer thickness silica hydrogel prepared at pH 1 and washed with 200 cm³ water.[3]

The efficiency of the washing was controlled by FTIR spectroscopy. The FTIR spectrum of the ten times washed at 25°C and dried silica hydrogel at 150°C seen in Figure 6.3 indicated the presence of hydrogen bonded O-H streching vibrations at 3645 cm⁻¹, H₂O bending vibrations

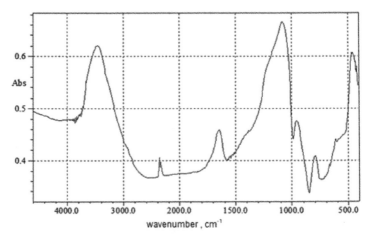

FIGURE 6.3 FTIR spectrum of silica gel obtained by drying of 15 times washed silica hydrogel at 150°C.

at 1645 cm^{-1}, Si-OH streching vibration at 962 cm^{-1}. There was a small shoulder for SO$_4^{-2}$ vibrations at 620 cm^{-1} indicating the removal of SO$_4^{-2}$ ions was not complete.

6.3.3 PH CHANGE AND Co(II) TRANSPORT

The inital pH of silica hydrogel used for Co(II) sorption experiments was 1.77–1.80 which was the pH of the water phase in 10th washing step of the hydro gels as reported in Table 6.4. When the silicahydrogel was equilibriated with Co(II) solutions with initial pH 6.44–6.60 protons diffused from the silica hydrogel phase to solution phase and the pH of the Co(II) solution decreased to 2.16–2.18 as reported in Table 6.4.

TABLE 6.4   The pH and Mass Change for Co(II) Sorption Process and for Drying of Silica Hydrogel.

Co(II) concentration, %	2.43	1.97	1.44	0.91
pH of silica hydrogel	1.77	1.80	1.78	1.77
Initial pH of Co(II) solution	6.60	6.60	6.51	6.44
Equilibrium pH of Co(II) solution	2.16	2.18	2.18	2.16
Mass decrease during Co(II) sorption, %	4.70	2.70	4.40	2.60
Mass change during drying, %	90.7	90.5	90.8	90.3
Co(II) % in dry gels	1.3	0.76	0.45	0.24

6.3.4 DRYING OF Co(II) CONTAINING SILICA HYDROGEL

The mass loss due to evaporation of water during drying at 150°C for 2 hours was 90.3–90.8 % as reported in Table 6.4. TG analysis of the wet gels seen in Figure 6.4 showed the gels reached constant mass up to 220°C for gels in equilibrium with 0.24 g dm^{-3} Co(II) solution and 235°C for 0.51 gdm^{-3} Co(II) solution on heating at 10°C min^{-1} rate.

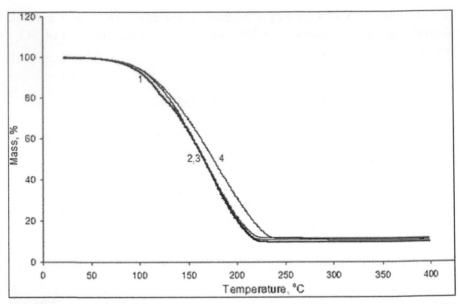

FIGURE 6.4   TG analysis curves for silica hydrogel equilibriated with 1. 0.24, 2.0.37, 3.0.42, 4. 0.41 g dm⁻³ Co(II) solutions.

The SEM micrographs of silica hydrogel prepared form 0.53 g dm⁻³ Co(II) solution and its dried form are seen in Figure 6.5. The silica hydrogel contracts to small granular particles during drying due to surface tension of water.

The Co(II) content in dry gels that was determined from the material balance of the Co(II) sorption and drying changed with the initial Co(II) concentration. It was in the range of 0.24–1.3% from sorption from 0.91–2.43 % Co(II) solution as reported in Table 6.4.

EDX analysis of the Co(II) containing gel equilibrated with 0.52 g Co(II) solution is reported in Table 6.5. In EDX analysis the samples are examined in vacuum conditions. Thus a fraction of water was removed from the surface of the silica hydrogel. The EDX analysis indicated the presence of O, Na, Si, S and Cl and Co elements at the surface of the particles shown in SEM micrographs in Figure 6.5. In addition, it indicated that not all of the Na_2SO_4 formed as by product was leached from silica hydrogel in 10 times washing process. It was expected that the Cl⁻ ions should diffuse in silica hydrogel together with Co(II) ions for keeping the charge neutrality. However the Cl⁻ content both in silica hydrogel and silica gel was very low. Neverthless, it can be concluded that Co(II) ions exchanged with Na⁺ ions and protons in silica hydrogel. The transport of

Na$^+$ ions into Co(II) solutions was confirmed with ICP analysis results shown in Table 6.6. The solution contains $1.147 - 1.527$ g dm^{-3} Na element and 0.1-0.205 g dm^{-3} Si element (corresponding to 0.21–0.44 g dm^{-3} silica). The solubility of silica at pH 2 at 25°C is reported to be 0.150 g dm^{-3}. The dissolved silica is reported to be in monomeric form.[11] Consequently, the solution phase in the present study contained silica particles besides dissolved monomeric silica. This could be due to attrition of silica hydrogel phase by the magnetic bar revolving over silica hydrogel phase in order to mix the solution phase.

TABLE 6.5 Elemental Composition of Silica Hydrogel Equilibrited with 0.52 g dm^{-3} Co(II) and Silica Gel Obtained by Drying it by EDX Analysis.

Element	Silica hydrogel		Silica gel	
	Wt %	Std dev %	Wt %	Std dev %
O	51.57	9.47	53.30	6.11
Na	1.22	0.46	5.91	3.12
Si	45.44	9.16	31.25	6.48
S	0.88	0.32	4.89	2.16
Cl	0.22	0.20	0.26	0.21
Co	0.67	0.57	4.39	8.47

TABLE 6.6 ICP Analysis for Initial and Equilibrium Concentration of Co(II), Si, and Na in Solution.

Initial concentration of Co(II), g dm^{-3}	Equilibrium concentration of Co(II), g dm^{-3}	Equilibrium concentration of Si, g dm^{-3}	Equilibrium concentration of Na, g dm^{-3}
0.244	0.207	0.119	1.269
0.310	0.249	0.126	1.147
0.317	0.275	0.100	1.165
0.465	0.402	0.131	1.498
0.511	0.467	0.205	1.527

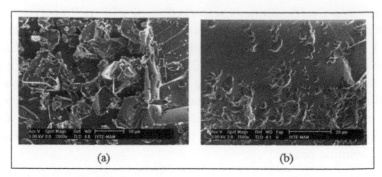

FIGURE 6.5   SEM micrographs of Co(II) containing (a) silica hydrogel and (b) dry silica gel prepared from 0.53 g dm^{-3} Co(II) solution.

6.3.5   COLOR OF CO(II) CONTAINING SILICA HYDROGEL AND SILICA GEL

The silica hydrogel with Co(II) appears pink in color and the color becomes blue on drying at 150°C for 2 h. As seen in Figure 6.6 the blue color intensity changes with the concentration of Co(II) in silica gel.

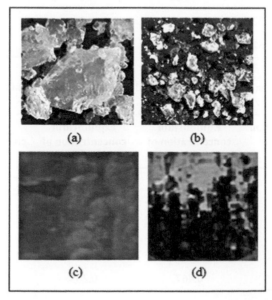

FIGURE 6.6   The silica hydrogel in equilibrium with initial Co(II) concentration of (a) 0.52 g dm^{-3}, (c) 2.43 g dm^{-3}; silicagel from the silica hydrogel in equilibrium with initial (b) 0.52 g dm^{-3}, (d) 2.43 g dm^{-3}.

6.3.6 EQUILIBRIUM OF CO(II) SORPTION

UV-visible spectra of the initial and equilibrited Co(II) solutions with silica hydrogel are shown in Figure 6.7. The absorbance of the absorbtion maximum at 560 nm was used in Co(II) concentration measurement since it was linear with the Co(II) concentration up to 0.52 g dm^{-3}. During Co(II) transport to silica hydrogel protons, Na$^+$ and SO$_4^{2-}$ ions are transferred in opposite direction to the aqueous phase. The transport of these ions changed the spectrum of Co(II) ions in the aqueous phase. ICP analysis of the aqueous phase indicated the presence of Si. There are two new absorption peaks at 230 nm and 320 nm in addition to 560 nm peak in the aqueous phase in equilibrium with silica hydrogel. Wang et al reported a broad band in the UV region of Co-SiO$_2$ spectrum centered at 224 nm assigned to a low energy charge transfer between the oxygen ligands and central Co(II) ion in tetrahedral symmetry and another broad absorption was centered at 356 nm for Co(III) species formed via the automatic oxidation of the Co(II) by dissolved oxygen.[29] Zola et al reported that broad band located at 250 nm was due to absorption by small particles.[31]

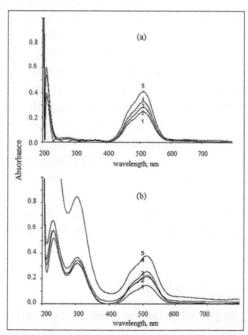

FIGURE 6.7 UV–visible spectrum obtained by fiber optic spectrophotometer (a) initial Co(II) solution (b) equilibrium Co(II) solution for 1.0.25 , 2. 0.32, 3.0.37, 4. 0.42, 5.0.52 g dm^{-3} initial Co(II) concentration.

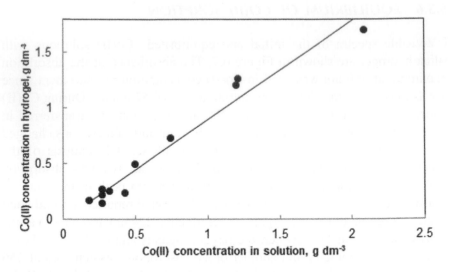

FIGURE 6.8   Adsorption isotherm of Co(II) in silica hydrogel.

The amount of Co(II) ions in silica hydrogel equilibrium with Co(II) solutions were calculated using the data obtained by fiber optic spectrometer, visible spectrometer and atomic absorption. The adsorption isotherm of Co(II) on silicahydrogel is as shown in Figure 6.8. The adsorption of Co(II) from aqueous solution on silica depends on the double layer potential. The farther the separation of the solution pH from the zero charge of the silica particles, the higher the adsorption of Co(II) on silica surface is.[12] The isoelectric point of silica is reported as pH 1.7 or 2.02.[1,11] Reaction of Co(II) with the silicilic acid formed from silicates was also reported.[6] Since the difference in solution pH at equilibrium (around 2) and isoelectric point of silica pH(1.7–2.0) is very small, the amount of Co(II) adsorbed on the silica particles should be very small[12] and Co(II) should remain in aqueous phase of silica hydrogel.The adsorption isotherm shown in Figure 6.8 can be represented by a straight line having slope 0.89 with R^2 value of 0.96.

$$q = 0.89c \tag{6.5}$$

where q is g Co(II) in dm^3 silica hydrogel and c is g Co(II) in dm^3 solution. If there were no adsorption on silica surfaces, the Co(II) concentration in both aqueous phases would be the same and the equilibrium constant showing the concentration in silica hydrogel and solution phase would be 0.95 since the 95% of the silicahydrogel volume was water. Considering some pores of the silica hydrogel was not accessable as indicated by Blomqvist et al[5] the equilibrium constant 0.89 in eq 6.5 is reasonable.

6.3.7   KINETICS OF Co(II) SORPTION

The Co(II) concentration of the aqueous solution in contact with silica hydrogel decreased in an oscillating manner as seen in Figures 6.9 and 6.10. In two weeks asdorption time the amount adsorbed was found by integrating the concentration change of solution versus time data. Co(II) distribution in silica hydrogel was homogeneous since the same pink color intensity was observed throughout their vertical crossection as seen in Figure 6.11. The higher degree of oscillations observed at high liquid to gel ratio made the kinetic analysis impossible. At low liquid to gel ratio the oscillations were relatively damped. The simultaneous processes occuring during Co(II) transport were the cause of the oscillations in solution concentrations at high solution to gel ratio.

The effective diffusion coefficient of Co(II) in silica hydrogel could only be calculated for 1:1 solution to gel ratio using the slopes of the M_t/M_∞ versus $t^{1/2}$ linear relations up to 60 min time period in Figure 6.12. The data was linear with a correlation coefficient in the range of 0.58 –0.92. The low correlation coefficient should be due to oscillating nature of the process. The oscillations were due to water transport from the silica hydrogel phase to aqueous phase and the transport other ions such as Na^{+1}, Cl^{-1}, $SO4^{2-}$, H^+, SiO_4^{2-} between the phases. Thus a simplified approach to describe quantitatively the transport of Co(II) from the aqueous phase to silica hydrogel phase can be made.

The diffusion of electrolytes in a solid is composed of two processes, that is a simple mass diffusion process and an ionic diffusion process. Oscillatory permeation of electrolytes through composite membranes was observed due to these processes.[16] There are studies about modelling surface diffusion of liquids in porous solids. The addition of surface diffusion to pore diffusion has a large effect on the overall rate of adsorption for a slightly adsorbing solid.[17]

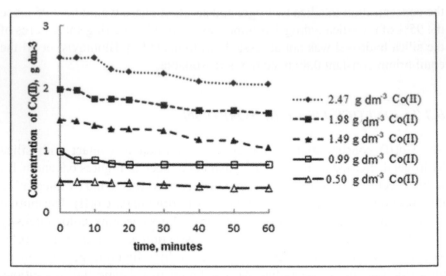

FIGURE 6.9   Change of Co(II) concentration with time for silica hydrogel/liquid ratio of 1: 1.

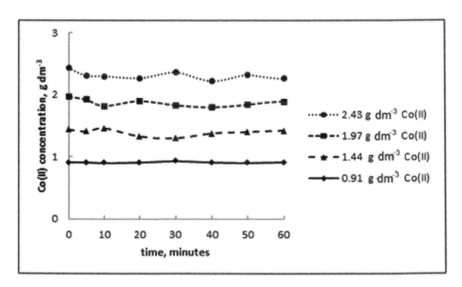

FIGURE 6.10   Change of Co(II) concentration with time for silica hydrogel/liquid ratio of 1:5.

FIGURE 6.11 The colored picture of silica hydrogel in equilibrium with Co(II) for 1:5 silica hydrogel/solution ratio for 1. 2.43, 2. 1.97, 3.1.44, 4. 0.99 g dm^{-3} Co(II) initial solutions.

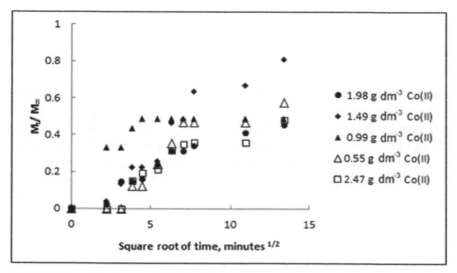

FIGURE 6.12 The change of M_t/M_∞ with squareroot of time for Co(II) diffusion from 200 cm^3 Co(II) solution to 200 cm^3 silica hydrogel with layer thickness of 3.5 cm.

The effective diffusion coefficient of Co(II) for different initial Co(II) concentrations are listed in Table 6.7. The average effective diffusion coefficient of Co(II) in silica hydrogel $(13.5 \pm 8.1) \times 10^{-9}$ m^2 s^{-1} is substantially higher than the diffusivity of Co(II) in water, 1.29×10^{-9} m^2 s^{-1}.[22] The surface diffusion of Co(II) on silica network outweighed the hindered diffusion due to tortuous structure of the gels. The effective diffusivity of $CoCl_2$ in silicahydrogel spheres prepared at pH 8–8.5 was reported to be 2.7×10^{-10} $m^2 s^{-1}$ indicating the effect of silica network prepared at higher pH value hindered the diffusion of $CoCl_2$.[15]. However in that study it was assumed that only free $CoCl_2$ remaining diffused after its fast chemisorption process to silica surface. The Co(II) ions were strongly adsorbed to silica at this pH value. [12]. The diffusion coefficient of Co(II) in hydrated silica, $(6.86 \pm 0.44) \times 10^{-12}$ $m^2 s^{-1}$ reported by Pathak and Robin had also indicated particle diffusion controlled conditions.[20]

TABLE 6.7 Effective Diffussion Coefficient of Co(II) in Silica Hydrogel for Gel Thickness 3.5, Solution Volume: Silica Hydrogel Volume 1:1 for Different İntial Concentrations.

c, g dm^{-3} Co(II)	D×10^9, m^2 s^{-1}	R^2
2.47	7.41	0.81
1.98	7.41	0.92
1.49	16.9	0.85
0.99	26.3	0.58
0.55	9.62	0.74

Sorption and diffusion of Co(II) ions in silica hydrogel showed a different behavior than the sorption and diffusion of basic dyes such as methylene blue[26] and malachite green[27]. While the dyes were adsorbed at the interface of silica hydrogel and aqueous solution phase at a fast rate they diffused in silica hydrogel phase very slowly in time, Co(II) was not adsorbed on silica and diffused in the pores of silica hydrogel which had a large pore size and on the surface of silica particles.

6.4 CONCLUSION

Silica gel loaded with Co(II) ions which could be used as humidity indicators or catalyst was prepared from washed silica hydrogel at pH 1.77–1.8 and $CoCl_2$ solution. The washing kinetics of silica hydrogel with water was a solid diffusion controlled process. Co(II) was present in the aqueous phase of silica hydrogel and cobalt sorption was an oscillating process due to simultaneous water transport from the silica hydrogel phase to aqeous phase and transport of Na^+, Cl^-, SO_4^{2-}, H^+, SiO_4^{2-} ions between the phases. The average proton and Co(II) diffusion coefficients in silica hydrogel were determined as $(1.92 \pm 0.56) \times 10^{-9}$ m^2 s^{-1} and $(13.5 \pm 8.1) \times 10^{-9}$ m^2 s^{-1} respectively in the present study. While protons moved slower in silica hydrogel than in water due to turtous structure, Co(II) moved faster since surface diffusion exist.

ACKNOWLEDGMENTS

The authors thank Gül Özpinar, Ismail Dizdar, Filiz Ozmihci Omurlu, Aynur Küçük and Gülendam Yilmaz for their contribution to experimental work.

KEYWORDS

- silica hydrogel
- cobaltous chloride
- diffusion
- adsorption
- absorption
- syneresis

REFERENCES

1. Balkose, D.; Ulutan, S.; Ozkan, F. C.; Celebi S.; Ulku S. Dynamics of Water Vapor Adsorption on Humidity-Indicating Silica Gel. *Appl. Surf. Sci.* **1998,** *134*, 39–46
2. Baltacioglu, H.; Balkose D. Diffusion of Protons In Silica Hydrogel. *Colloid Polym Sci.,* **1989,** *267* (5), 460–464.
3. Baltacioglu, H. Nem Çekici Silikajel Üretimi Prosesinin Geliştirilmesi Ve Pilot Tesis Tasarımı, MS thesis, Ege University İzmir Turkey, 1985
4. Birch, D. J. S.; Geddes C. D. Cluster Dynamics, Growth and Syneresis During Silica Hydrogel Polymerisation. *Chem. Phys. Letters* **2000,** *320* (3-4): 229–236.
5. Blomqvist, C. H.; Geback, T.; A. Altskär A.; Hermansson, A. M.; Gustafsson, S.; Lorén, N.; Olsson, E. Interconnectivity Imaged in Three Dimensions: Nano-Particulate Silica-Hydrogel Structure Revealed Using Electron Tomography. *Micron* **2017,** *100*, 91–105.
6. Charlet, L.; Manceau, A. Evidence for the Neoformation of Clays Upon Sorption of Co(II)) and Ni(II) on Silicates. *Geochimica Et Cosmochimica Acta.* **1994,** *58* (11), 2577–2582.
7. Chee K. L.; Asmawati, N.; Zabidi, M.; Chandra , M. S. Synthesis of Cobalt Nano Particles on Silica Support Using the Strong Electrostatic Adsorption (SEA) Method. *Defect Diffus. Forum,* **2011,** 312-315, 370–375.

8. Cleary, A.; J. Karolin J.; Birch D. J. S. pH Tracking of Silica Hydrogel Nanoparticle Growth. *Appl. Phys. Lett.* **2006,** *89* (11), 113–125

9. Gorrepati, E. A., Wongthahan, P.; Raha, S.; Fogler H. S. Silica Precipitation in Acidic Solutions: Mechanism, pH Effect, and Salt Effect. *Langmuir* **2010,** *26* (13), 10467–10474.

10. Iafisco, M., Marchetti, M.; Morales J. M.; Hernández-Hernández,M. A.; Ruiz J. M. G.;Roveri N. Silica Gel Template for Calcium Phosphates Crystallization. *Crystal Growth Design.* **2009,** *9* (11), 4912–4921.

11. Iler, R. K. *Chemsitry of Silica*, Wiley; New York ,1979

12. James R. O.; Healy, T. W. Adsorption of Hydrolyzable Metal-İons at Oxide Water İnterface. 1. Co(II) Adsorption on SiO_2 and TiO_2 as Model System, *J. Colloid Interface Sci.* **1977,** *40*, 65.

13. Kaczmarek, M.; Kazimierska-Drobny, K. Estimation-Identification Problem for Diffusive Transport in Porous Materials Based on Single Reservoir Test: Results for Silica Hydrogel. *J. Colloid Interface Sci.* **2007,** *311*(1): 262–275.

14. Kaczmarek, M.; Kazimierska-Drobny, K. Identification Problem of Interface Boundary Conditions for Diffusive Transport Between Water and Silica Hydrogel. *Mater. Sci. Poland* **2007,** *25* (3), 851–859.

15. Keserova, G. M.; Shamrikov, V. M.; Malkimon, V. I.; Belotserskovskoya; N. G., *USSR Appl. Chem.* **1987,** *4*, 882–885

16. Lee, K. B.; Lee Y. H. Periodic Characteristics of Composite Membrane-Permeability. *J. Microencapsul.* **1989,** *6* (1), 59–70.

17. Leyvaramos, R.; Geankoplis. C.J. Model Simulation and Analysis of Surface-Diffusion of Liquids in Porous Solids *Chem. Eng. Sci.* **1985,** *40* (5), 799–807.

18. Ling, C. K.; Zabidi, N. M.; Chandra, M. Synthesis and Characterization of Silica-Supported Cobalt Nano catalysts Using Strong Electrostatic Adsorption. *J. Appl. Sci.* **2011,** *11* (7), 1436–1440

19. Nayak, J. P. ; Bera, J. Preparation of an Efficient Humidity Indicating Silica Gel from Rice Husk Ash, *Bull. Mater. Sci.* **2011,** *34* (7) 1683–1687.

20. Pathak, P. N.; Choppin G. R. Effects of pH, Ionic Strength, Temperature, and Complexing Anions on the Sorption Behavior of Cobalt on Hydrous Silica. *Soil Sediment Contam.* **2009,** *18* (5), 590–602.

21. Perullini, M.; Jobbagy, M.; Japas M. L.; Nilmes S. S. New Method for the Simultaneous Determination of Diffusion and Adsorption of Dyes in Silica Hydrogels. *J. Colloid Interface Sci.* **2014,** *425*, 91–95.

22. Ribeiro, C. F.; Lobo, V. M. M.; Natividade J. J. S. Diffusion Coefficients in Aqueous Solutions of Cobalt Chloride at 298.15 K *J. Chem. Eng. Data* **2002,** *47,* 539–541

23. Takahashi, R.; Sato, S.; Sodesawa, T.; Kamomae,Y. Measurement of the Diffusion Coefficient of Nickel Nitrate in Wet Silica Gel Using UV/VIS Spectroscope Equipped with a Flow Cell. *Phys. Chem. Chem. Phys.* **2000,** *2* (6), 1199–1204

24. Takahashi, R.; Sato, S.; Sodesawa, T.; Nishida H. Effect of Pore Size on the Liquid-Phase Pore Diffusion of Nickel Nitrate. *Phys. Chem. Chem. Phys.* **2002,** *4* (15), 3800–3805

25. Tantemskaya N.; Meegoda, J. N. Estimation of Diffusion Coefficient of Chromium in Colloidal Silica Using Digital Photography, *Environ. Sci. Technol.* **2004,** *38*, 3950–3967

26. Top, A.; Akdeniz,Y.; Sevgi Ulutan, S.; Balköse, D. An Investigation of Kinetic and Equilibrium Behavior of Adsorption of Methylene Blue on a Silica Hydrogel. In *Physical Chemistry for Engineering and Applied Sciences Theoretical and Methodological*

Implication Haghi, A. K.; Aguilar, C. N.; Thomas, S.; Praveen K. M., Eds.; Apple Academic Press, pp. 3–24.

27. Top, A.; Kaplan, H.; Savrık, S. A.; Balkose, D. Adsorption of Malachite Green to Silica Hydrogel. In *Applied Chemistry and Chemical Engineering, Volume 5; Research Methodologies in Modern Chemistry and Applied Science*, Haghi A. K.; Ribeiro A. C. F.; Balkose D.; Torrens F.; Mukbaniani O. V., Eds: Apple Academic Press; Toronto, 2018; pp. 3–20

28. Ülkü, S.; Balköse, D.; Baltacioglu, H. Effect of Preparation pH on Pore Structure of Silica Gels. *Colloid Polym. Sci.* **1993,** *271,* 709–713.

29. Wang, M.; Chen, C.; Ma, J.; Zheng, X.; Li, Q.; Jinb Y.; Xu, J. Cobalt Ammonia Complex Mediated Preparation Of Hollow Silica Nanospheres With Multi-Nanochambers *J. Mater. Chem.* **2012,** *22,* 11904–11907

30. Yokoi, T.; Kawashita, M.; Ohtsuli C. Synthesis of Calcium Phosphate Crystals in a Silica Hydrogel Containing Phosphate Ions. *J. Mater. Res.* **2009,** *24* (6), 2154–2160.

31. Zola, A. S.; da Silva, L. S.; Moretti, A. L.; Fraga, A. C.; Sousa-Aguiar, E. F.; Arroyo. P. A. Effect of Silylation and Support Porosity of Co/MCM-41 and Co/SiO$_2$ Catalysts in Fischer-Tropsch Synthesis. *Topics Catalysis* **2016,** *59* (2-4): 219–229.

CHAPTER 7

FEATURES OF SELECTIVE SORPTION OF LANTHANUM FROM SOLUTION, WHICH CONTAINS IONS OF LANTHANUM AND CERIUM, BY INTERGEL SYSTEM HYDROGEL OF POLYMETHACRYLIC ACID: HYDROGEL OF POLY-2-METHYL-5-VINYLPYRIDINE

T. K. JUMADILOV* and R. G. KONDAUROV

JSC "Institute of Chemical Sciences after A.B. Bekturov,"
Almaty, the Republic of Kazakhstan

Corresponding author. E-mail: jumadilov@mail.ru

ABSTRACT

The study is a short review devoted to features of selective sorption of lanthanum ions from the solution, which contains lanthanum and cerium ions. It was found that specific electric conductivity increases, the concentration of hydrogen ions decreases during mutual activation of polymethacrylic acid (PMAA) and poly-2-methyl-5-vinylpyridine (P2M5VP) hydrogels in intergel system on their basis in an aqueous medium. Also, it should be noted that a significant increase of the swelling degree of PMAA and P2M5VP is observed in water. Sorption of lanthanum ions by intergel system hPMAA–hP2M5VP is accompanied by radically different changes of electrochemical and conformational properties of polymer hydrogels. There is a decrease of specific electric conductivity, pH, swelling degree due to folding of polymer globe in process of lanthanum ions sorption. Mutual activation with further transition into highly ionized state provides much higher values of lanthanum ions extraction degree of the intergel system comparatively with initial hydrogels of PMAA and P2M5VP. Maximum sorption of lanthanum is observed at 50%hPMAA–50%hP2M5VP ratio, sorption degree is 89.65%. Polymer-chain-binding degree (in relation to lanthanum ions) has maximum values at hydrogels ratio 50%hPMAA–50%hP2M5VP, the value of binding degree is 74.67%. Intergel system hPMAA–hP2M5VP has maximum values of cerium ions extraction degree at 67%hPMAA–33%hP2M5VP (sorption degree is 87.67%). Polymer-chain-binding degree at this ratio also has the highest values, it is 72.72%. Selective sorption of lanthanum ions by intergel system 50%hPMAA–50%hP2M5VP indicates that selectivity of the intergel system is manifested to lanthanum ions. Extraction degree of lanthanum ions is 65.40%, of cerium ions 24.38%. Polymer-chain-binding degree is in relation to lanthanum ions—54.25%, in relation to cerium ions—20.22%.

7.1   INTRODUCTION

Main feature of intergel system is the absence of direct contact between polymer hydrogels in solution.[1–3] In other words, interaction of polymer hydrogels of acid and basic nature occurs remotely. Intergel system is presented in Figure 7.1. Polyacids and polybases are put in a special glass filter, pores of which are permeable for low-molecular ions but nonpermeable for hydrogels dispersion.

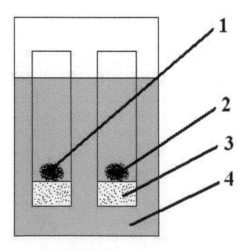

FIGURE 7.1   Intergel system: (1) polyacid, (2) polybase, (3) glass filter, and (4) solution.

During remote interaction of hydrogels the following chemical reactions occur:

(1) Dissociation of –COOH groups of internode links

$$-COOH \rightarrow COO^- \cdots H^+ \rightarrow -COO^- + H^+$$

It should be mentioned that first there is an occurrence of ionization with ionic pairs formation after that ionic pairs partially dissociate on separate ions.

(2) Nitrogen atom in pyridine ring is ionized and partially dissociates

$$\equiv N + H_2O \rightarrow \equiv NH^+ \cdots OH^- \rightarrow \equiv NH^+ + OH^-$$

(3) After that nitrogen atom also interacts with proton, which is cleaved from carboxyl group:
$$\equiv N + H^+ \rightarrow \equiv NH^+$$

(4) H^+ and OH^- ions, formed in result of the interaction of functional groups with water molecules, form water molecules (it is valid for equimolar concentrations of protons and hydroxyl ions):

$$H^+ + OH^- \rightarrow H_2O$$

As seen from above mentioned processes, in the result of the remote interaction effect of polymer hydrogels their mutual activation occurs. Mutual activation proposes the transfer of hydrogels into a highly ionized state. The result of this phenomenon is significant change of electrochemical properties (specific electric conductivity, pH) of solutions and changes of conformational and sorption properties of macromolecules.

According to Equation 7.1 dissociation of carboxyl groups on carboxylate anions and protons occurs, and it depends on the degree of dissociation. Due to the association of protons by heteroatoms total amount of hydrogen ions in solution decreases, which, in turn, provides additional dissociation of other (nonreacted) functional carboxyl groups. It occurs according to Le Chatelier principle—due to shift of the equilibrium to right (to the side of protons formation).[4]

These interactions provide the formation of same-charged groups on internode links of both (acid and basic) hydrogels.[5] Due to the laws of electrostatics, these groups repulse from each other and provide unfolding of macromolecular globe. The end result of such electrostatic interactions is a significant increase in the swelling of polymeric macromolecules.[6,7]

7.2 EXPERIMENTAL PART

7.2.1 EQUIPMENT

For measurement of solutions specific electric conductivity conductometer "MARK-603" (Russia) was used, hydrogen ions concentration was measured on Metrohm 827 pH meter (Switzerland). Sample's weight was measured on analytic electronic scales Shimadzu AY220 (Japan). La^{3+} ions concentration in solutions was determined on spectrophotometers SF-46 (Russia) and Jenway-6305 (UK).

7.2.2 MATERIALS

Studies were carried out in 0.005 M 6-water lanthanum nitrate solution. Polymethacrylic acid hydrogel (hPMAA) was synthesized in the presence of crosslinking agent N,N-methylene-bis-acrylamide and redox system $K_2S_2O_8$–$Na_2S_2O_3$ in the water medium. Synthesized hydrogels were crushed into small dispersions and washed with distilled water until constant conductivity value of aqueous solutions was reached. Poly-2-methyl-5-vinylpyridine (hP2M5VP) hydrogel of Sigma-Aldrich Company (linear polymer cross-linked by divinylbenzene) was used as a polybase.

For investigation task from synthesized hydrogels an intergel pair polymethacrylic acid hydrogel–poly-2-methyl-5-vinylpyridine hydrogel (hPMAA–hP2M5VP) was created. Swelling degrees of hydrogels are $\alpha_{(hPMAA)} = 20.65$ g/g; $\alpha_{(hP2M5VP)} = 3.20$ g/g.

7.2.3 ELECTROCHEMICAL INVESTIGATIONS

Experiments were carried out at room temperature. Studies of intergel system were made in the following order: each hydrogel was put in separate glass filters, pores of which are permeable for low molecular ions and molecules, but impermeable for hydrogels dispersion. After that filters with hydrogels were put in glasses with lanthanum nitrate solution. Electric conductivity and pH of overgel liquid were determined in the presence of hydrogels in solutions.

7.2.4 DETERMINATION OF HYDROGELS SWELLING

The swelling degree was calculated according to the equation:

$$K_{sw} = \frac{m_2 - m_1}{m_1}$$

where m_1 is the weight of dry hydrogel and m_2 is the weight of swollen hydrogel.

7.2.5 METHODOLOGY OF LANTHANUM IONS DETERMINATION

Methodology of lanthanum ions determination in solution is based on formation of colored complex compound of organic analytic reagent Arsenazo III with lanthanum ions.[8]

Extraction (sorption) degree was calculated by the equation:

$$\eta = \frac{C_{initial} - C_{residual}}{C_{initial}} \times 100\%$$

where $C_{initial}$ is the initial concentration of lanthanum in solution (g/L); $C_{residue}$ is the residual concentration of lanthanum in solution (g/L).

Polymer-chain-binding degree was determined by calculations in accordance with the equation:

$$\theta = \frac{v_{sorb}}{v} \times 100\%$$

where v_{sorb} is the quantity of polymer links with sorbed lanthanum (mol) and v is the total quantity of polymer links (if there are two hydrogels in solution, it is calculated as the sum of each polymer hydrogel links) (mol).

Effective dynamic exchange capacity was calculated by the formula:

$$Q = \frac{v_{\text{sorbed}}}{m_{\text{sorbent}}}$$

where v_{sorbed} is the amount of sorbed metal (mol); m_{sorbent} is the mass of the sorbent (if there are two hydrogels in solution, it is calculated as the sum of the mass of each of them) (g).

7.2.6 MUTUAL ACTIVATION OF HYDROGELS IN INTERGEL SYSTEM HPMAA–HP2M5VP IN AN AQUEOUS MEDIUM

Dependence of specific electric conductivity from molar ratios in time is shown in Figure 7.2. Conductivity increases with time for all ratios of hPMAA:hP2M5VP. The absence of the second hydrogel in the presence of polyacid or polybase in an aqueous medium provides impossibility of transfer into highly ionized state, due to which these areas are areas of minimum conductivity. Maximum values of specific electric conductivity are observed at 33%hPMAA–67%hP2M5VP ratio at 48 h.

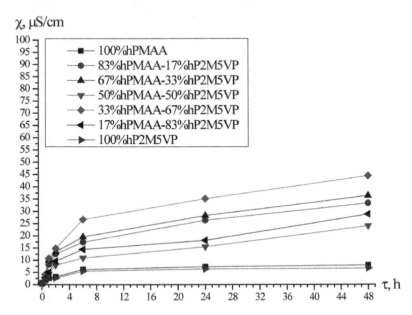

FIGURE 7.2 Dependence of specific electric conductivity of aqueous solution from time in presence of intergel system hPMAA–hP2M5VP.

However, one cannot judge about degree of mutual activation solely from the specific electric conductivity, because high values may indicate to the dominance of the dissociation of carboxyl groups over the proton association by the heteroatom of the P2M5VP hydrogel.

Figure 7.3 represents the dependence of concentration of hydrogen ions in an aqueous medium from time in presence of intergel system hPMAA–hP2M5VP. Minimum values of pH are observed in presence of only acid hydrogel during all time of polymer interaction with the solution. Such phenomenon is due to the release of protons in the solution as a result of carboxyl groups electrolytical dissociation. Obtained results show that high cleavage rate of H^+ is observed before 2 h. Maximum values of pH are observed at 67%hPMAA–33%hP2M5VP at 48 h of hydrogels remote interaction in an aqueous medium.

So, the concentration of H^+ ions determines equilibrium in two processes:

(1) Swelling rate, at which H^+ ions are released in solution as a result of –COOH groups dissociation and

(2) binding rate of H^+ ions by links of polybases.

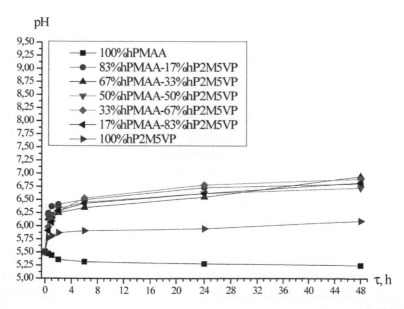

FIGURE 7.3   Dependence of pH of the aqueous solution from time in presence of intergel system hPMAA–hP2M5VP.

Figure 7.4 shows that there is swelling degree increase due to long-range effect of the polymer hydrogels in intergel system hPMAA–hP2M5VP.

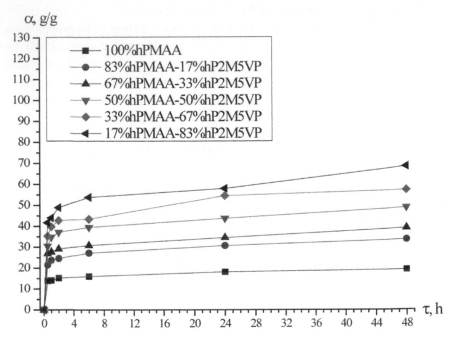

FIGURE 7.4 Dependence of hPMAA swelling degree in presence of hP2M5VP from time in an aqueous medium.

Minimum swelling of hPMAA is observed in presence of individual hydrogel of polymethacrylic acid in aqueous solution during 48 h. Swelling of polyacid increases proportionally to increase of share of polybases in the solution. Maximum values of swelling degree of polymethacrylic acid are seen at 48 h of remote interaction of the hydrogels at ratio 17%hPMAA:83%hP2M5VP. Significant increase of swelling of polyacid is true for all ratios during 6 h. Further swelling occurs more slightly.

Figure 7.5 characterizes the change of swelling degree of hP2M5VP in an aqueous medium in the presence of hPMAA. With polyacid share increase in solution, there is additional swelling of basic hydrogel of hP2M5VP.

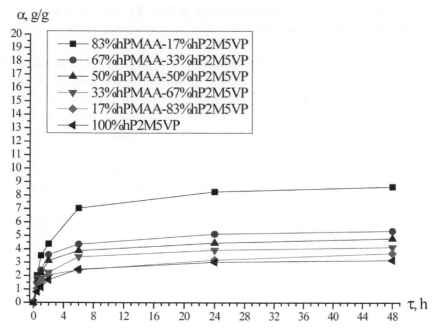

FIGURE 7.5   Dependence of hP2M5VP swelling degree in presence of hPMAA from time in an aqueous medium.

This phenomenon points to additional ionization of links of the poly-bases in result of mutual activation of hydrogels of PMAA and P2M5VP. It should be noted that polybases share increase provides increase of protons association degree of heteroatom of polyvinylpyridine, which provides shift of equilibrium to the right in reaction of dissociation of carboxyl groups. Minimum values of swelling degree of polybases are observed in presence of only hP2M5VP during the time of interaction of the polymer with water. Maximum values of the parameter are observed at ratio 83%hPMAA–17%hP2M5VP at 48 h of hydrogels remote interaction.

Mutual activation of polymer hydrogels provides significant increase of swelling degree of hP2M5VP. As known, hP2M5VP is weak polybase, consequently has low values of swelling degree in an aqueous medium. Significant increase of swelling of hP2M5VP points to high ionization degree of vinylpyridines links. This may be due to binding of proton, which was cleaved from carboxyl group, by nitrogen atoms of the polybase.

Obtained data shows that hydrogels of PMAA and P2M5VP undergo remote interaction in an aqueous medium. Result of such interaction is the formation of same-charged groups with uncompensated counter ions.

7.2.7 SORPTION OF LANTHANUM IONS BY INTERGEL SYSTEM HPMAA–HP2M5VP

Lanthanum nitrate presents in solution in dissociated state. Dissociation of lanthanum nitrate occurs in three stages. Dissociation constant of 1st stage is much higher than constant of 2nd and 3rd. Due to this fact sorption of dissociated ions in intergel system occurs according to different mechanisms. Nitrate of lanthanum formed in 1st stage of dissociation is binded by ionic mechanism. In result of dissociation of lanthanum nitrate by 2nd and 3rd stages, there is an increase of ionic pairs and molecular forms concentration in solution, which, in turn, provides their binding by coordination mechanism.

In the presence of the intergel system in the salt solution, there is an occurrence of the following reaction:

(1) dissociation of lanthanum nitrate along with dissociation of carboxyl groups;
(2) mutual activation of the hydrogels due to protons association by polybase; and
(3) sorption of lanthanum ions.

These reactions impact electrochemical equilibrium and dominance of any of them will provide changes of specific electric conductivity and pH.

In the solution of lanthanum nitrate in presence of abovementioned intergel system in the beginning, there is ionization of the polymer hydrogels similarly to mutual activation in an aqueous medium, further ionization occurs due to formation of coordination bonds with lanthanum ions. Due to sorption of lanthanum by coordination mechanism the polymers do not have the same charged on internode links, which provides folding of the macromolecules and swelling decrease.

Extraction of lanthanum ions from the solution by the intergel system hPMAA–hP2M5VP is accompanied with decrease of specific electric conductivity (Fig. 7.6). Significant decrease in electric conductivity occurs during 30 min after intergel system comes in contact of with the lanthanum nitrate solution. Character of decrease of electric conductivity at all ratios of hydrogels is different. At ratio 83%hPMAA–17%hP2M5VP after 6 h, decrease of the parameter is not very intensive. This indicates to a not high degree of ionization of the polymer structures during their remote interaction. Minimum values of electric conductivity in the solution of lanthanum nitrate are observed at ratio 50%hPMAA–50%hP2M5VP at 48 h. During mutual activation of the polymer hydrogels, there is a release of protons in the solution; however, electric conductivity decreases, which points to the fact of the metal sorption by the polymers.

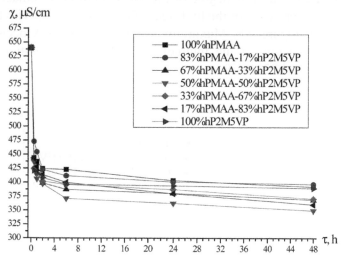

FIGURE 7.6 Dependence of specific electric conductivity of lanthanum nitrate from time in presence of intergel system hPMAA–hP2M5VP.

Dependence of pH of lanthanum nitrate solution from time in presence of the intergel system hPMAA–hP2M5VP is presented in Figure 7.7.

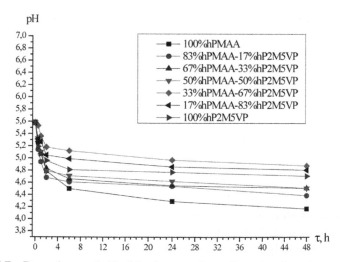

FIGURE 7.7 Dependence of pH of lanthanum nitrate from time in presence of intergel system hPMAA–hP2M5VP.

Extraction degree of lanthanum of individual hydrogels of PMAA and P2M5VP is not high. Hydrogel of polymethacrylic acid sorbes 66.28% of lanthanum, hydrogel of poly-2-methyl-5-vinylpyridine—63.65%. The increase of sorption ability of the initial hydrogels in the intergel system is due to their transition into highly ionized state in the result of hydrogels mutual activation during their remote interaction in the intergel system. The highest values of sorption degree are reached at 67%hPMAA–33%hP2M5VP and 50%hPMAA–50%hP2M5VP ratios. Maximum amount of lanthanum is extracted at 50%hPMAA–50%hP2M5VP hydrogels ratio, extraction degree of lanthanum ions is 89.65%.

Dependence of polymer-chain-binding degree (in relation to lanthanum ions) of the intergel system hPMAA–hP2M5VP from time is presented in Figure 7.11.

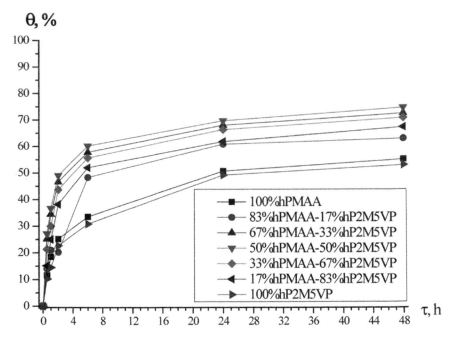

FIGURE 7.11   Dependence of polymer-chain-binding degree of intergel system hPMAA–hP2M5VP from hydrogels molar ratios in time.

Individual hydrogels of PMAA and P2M5VP do not have a very high binding degree: 55.17% for hPMAA and 53.00% for hP2M5VP. In result of mutual activation of polymer hydrogels of polymethacrylic acid and

poly-2-methyl-5-vinylpyridine significant changes of their sorption properties occur. The highest values of polymer-chain-binding degree are observed
at 50%hPMAA–50%hP2M5VP, binding degree is 74.67%.

Dependence of effective dynamic exchange capacity of the intergel
system hPMAA–hP2M5VP from hydrogels molar ratios in time is presented
in Figure 7.12. As seen from obtained results, the intergel system based
on rare-cross-linked polymer hydrogels of polymethacrylic acid and poly-
2-methyl-5-vinylpyridine has significantly higher values of exchange
capacity comparatively with individual hydrogels. It should be noted that
two polymers have bulk methyl substituents and their ionization occur slow.
High values of the parameter are seen at ratios 67%hPMAA–33%hP2M5VP
and 50%hPMAA:50%hP2M5VP. Sufficiently high values of exchange
capacity are not observed at ratios 83%hPMAA–17%hP2M5VP and
17%hPMAA–83%hP2M5VP due to a not very high degree of ionization of
the polymers.

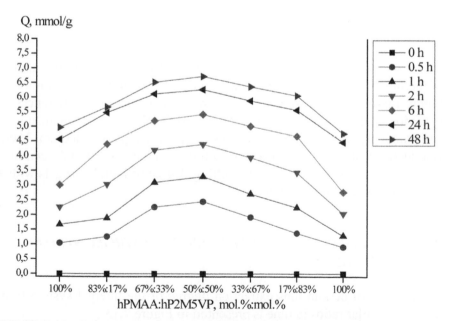

FIGURE 7.12 Dependence of effective dynamic exchange capacity of intergel system
hPMAA–hP2M5VP from hydrogels molar ratios in time.

Obtained results point to the fact that transition of polymer hydrogels in
intergel systems into highly ionized state is characterized by a significant

increase (up to 30%) of effective dynamic exchange capacity comparatively with initial hydrogels.

In Table 7.1, sorption parameters (for lanthanum sorption) of individual hydrogels of PMAA and P2M5VP and intergel system on their basis are presented.

TABLE 7.1 Sorption Parameters of Individual Hydrogels and Intergel System During Lanthanum Sorption.

Individual hydrogel/ Intergel system	hPMAA	hP2M5VP	50%hPMAA–50%hP2M5VP
Lanthanum extraction degree (%)	66.28	63.65	89.65
Polymer-chain-binding degree (%)	55.17	53.00	74.67
Effective dynamic exchange capacity (mmol/g)	4.97	4.77	6.72

As seen from the table, there is a significant increase of binding ability of the initial polymers in intergel pairs, which is due to their transition into highly ionized state. It should be also noted that there are differences in behavior of the intergel system hPMAA–hP2M5VP in lanthanum nitrate comparatively with activation of the polymers in an aqueous medium. There is a decrease of specific electric conductivity, an increase of hydrogen ions concentration, and an increase of swelling degree of both hydrogels with a further decrease of this parameter.

7.2.8 SORPTION OF CERIUM IONS BY INTERGEL SYSTEM HPMAA–HP2M5VP

Dependence of cerium ions extraction degree of the intergel system from hydrogels molar ratios in time is presented in Figure 7.13.

As seen from the figure, individual hydrogels of PMAA and P2M5VP do not have a very high degree of cerium ions extraction (60.33% for hPMAA and 50.00% for hP2M5VP). Maximum values of cerium sorption are observed at hydrogels ratio 67%hPMAA–33%hP2M5VP, sorption degree is 87.67%.

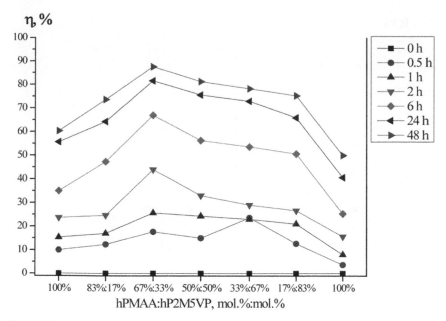

FIGURE 7.13 Dependence of cerium ions extraction degree of intergel system hPMAA–hP2M5VP from hydrogels molar ratios in time.

Dependencies of polymer-chain-binding degree (in relation to cerium ions) of the intergel system hPMAA–hP2M5VP from the time are shown in Figure 7.14. Similarly to other parameters, the polymer-chain-binding degree is in great influence from the state in which polymer hydrogel presents itself in the solution. In other words, hydrogels in intergel pairs are in highly ionized state due to their mutual activation in result of their remote interaction. As seen from the obtained data, the maximum values of binding degree are observed at hydrogels ratio 67%hPMAA–33%hP2M5VP, it is 72.72%. Also, high values are in intergel pair 50%hPMAA–50%hP2M5VP, which, in turn, points to high ionization degree of the hydrogels in the intergel pair. Not sufficiently high values of polymer-chain-binding degree at hydrogels ratios 83%hPMAA–17%hP2M5VP and 17%hPMAA–83%hP2M5VP are the consequence of not very high ionization degree of the polymer macromolecules. Interaction of the individual hydrogels of PMAA and P2M5VP with cerium nitrate shows that equilibrium in the solution is reached rather fast, subsequence of what is not high values of polymer-chain-binding degree. Binding degree is 50.05% for hPMAA and 41.47 for hP2M5VP.

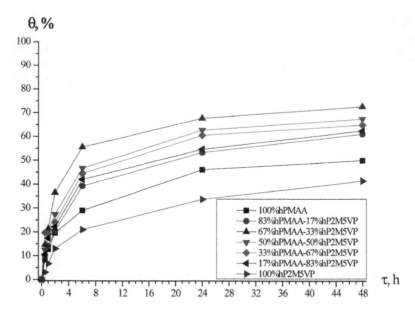

FIGURE 7.14   Dependence of polymer-chain-binding degree of intergel system hPMAA–hP2M5VP from hydrogels molar ratios in time.

Figure 7.15 reflects the dependence of effective dynamic exchange capacity (in relation to cerium ions) of the intergel system

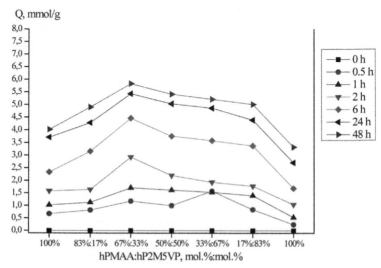

FIGURE 7.15   Dependence of effective dynamic exchange capacity of intergel system hPMAA–hP2M5VP from hydrogels molar ratios in time.

hPMAA–hP2M5VP from hydrogels molar ratios in time. The results show that maximum values of the parameter are observed at hydrogels ratio 67%hPMAA–33%hP2M5VP at 48 h. High values of exchange capacity are also observed at 50%hPMAA–50%hP2M5VP and 33%hPMAA–67%hP2M5VP ratios. The minimum values of effective dynamic exchange capacity are observed in the presence of individual polymer hydrogels of PMAA and P2M5VP. This is the result of the absence of the phenomenon of mutual activation.

In Table 7.2, sorption parameters (for cerium sorption) of individual hydrogels of PMAA and P2M5VP and intergel system on their basis are presented.

TABLE 7.2 Sorption Parameters of Individual Hydrogels and Intergel System During Cerium Sorption.

Individual hydrogel/ Intergel system	hPMAA	hP2M5VP	67%hPMAA–33%hP2M5VP
Cerium extraction degree (%)	60.33	50.00	87.67
Polymer-chain-binding degree (%)	50.05	41.47	72.72
Effective dynamic exchange capacity (mmol/g)	4.02	3.33	5.84

As seen from Table 7.2, the sorption parameters of the intergel system hPMAA–hP2M5VP is almost 25–30% higher in comparison with individual hydrogels of PMAA and P2M5VP. Remote interaction of the hydrogels in the intergel system provides a significant increase of cerium ions extraction degree, polymer-chain-binding degree, and effective dynamic exchange capacity.

7.2.9 SELECTIVE SORPTION OF LANTHANUM IONS BY INTERGEL SYSTEM HPMAA–HP2M5VP

For selective extraction of lanthanum ions from solution, which contains ions of lanthanum and cerium, the ratio 50%hPMAA–50%hP2M5VP was taken due to the fact that maximum sorption of lanthanum occurs at this ratio.

Figure 7.16 represents dependencies of lanthanum and cerium ions extraction degrees of the intergel system 50%hPMAA–50%hP2M5VP from time. High level of mutual activation in the intergel pair during 2 h provides a significant increase of lanthanum and cerium ions extraction. At this time of hydrogels, remote interaction in the solution 29.25% of lanthanum and 7.56% of cerium is extracted from the solution. Furthermore, there is an occurrence of extraction degree of both the metals. At 24 h of the hydrogels interaction, lanthanum ions sorption degree is 52.97% and 18.46% for cerium ions. It should be noted that an increase up to 48 h occurs very slightly, which indicates that system is reaching the equilibrium state. At 48 h, 65.40% of lanthanum is sorbed, at 24.38% of cerium is sorbed.

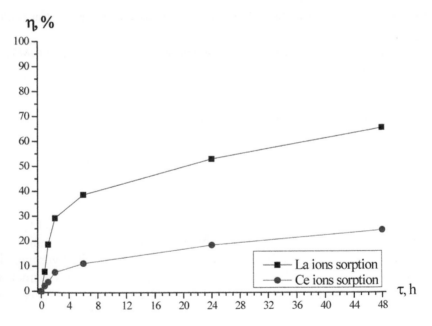

FIGURE 7.16   Dependence of lanthanum and cerium ions extraction degree of intergel system 50%hPMAA–50%hP2M5VP from time.

Dependence of polymer-chain-binding degree (in relation to lanthanum and cerium ions) of the intergel system 50%hPMAA–50%hP2M5VP from time is presented in Figure 7.17. As seen from the obtained data, a sharp increase of the polymer-chain-binding degree of the intergel system is observed during first 2 h, binding degree of lanthanum is 24.26%, of cerium, is 6.27%. Further interaction of the polymer hydrogels in the intergel pair provides a subsequent increase of binding degree. Maximum values of binding degree are observed at 48 h of remote interaction, wherein 54.25% of lanthanum and 20.22% of cerium is a bind.

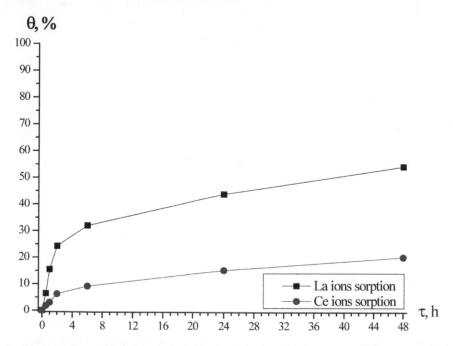

FIGURE 7.17 Dependence of polymer-chain-binding degree of intergel system 50%hPMAA–50%hP2M5VP from time.

Figure 7.18 presents the dependence of effective dynamic exchange capacity of the intergel system 50%hPMAA–50%hP2M5VP from time. There is an increase in the parameter with time.

Effective dynamic exchange capacity in relation to lanthanum is 2.5 times higher compared with cerium. These data evidence that intergel system 50%hPMAA–50%hP2M5VP shows selectivity to lanthanum ions during its sorption from solution, which contains lanthanum and cerium ions.

In Table 7.3, the values of the sorption parameters (extraction degree, polymer-chain-binding degree, effective dynamic exchange capacity) of the intergel system 50%hPMAA–50%hP2M5VP at selective sorption of lanthanum ions are presented. It should be noted that these values are the results of measurements at 48 h.

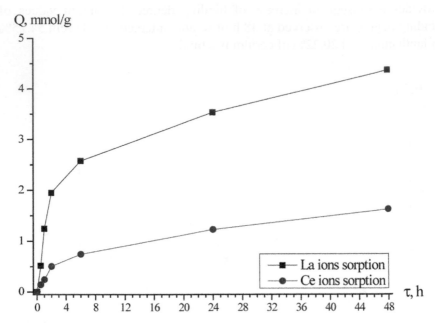

FIGURE 7.18 Dependence of effective dynamic exchange capacity of intergel system 50%hPMAA–50%hP2M5VP from time.

TABLE 7.3 Sorption Parameters of Intergel System 50%hPMAA–50%hP2M5VP at Lanthanum Selective Sorption.

Intergel system	50%hPMAA–50%hP2M5VP
Lanthanum extraction degree (%)	65.40
Cerium extraction degree (%)	24.38
Polymer-chain-binding degree (in relation to lanthanum) (%)	54.25
Polymer-chain-binding degree (in relation to lanthanum) (%)	20.22
Effective dynamic exchange capacity (in relation to lanthanum) (mol/g)	4.36
Effective dynamic exchange capacity (in relation to lanthanum) (mmol/g)	1.63

7.3   CONCLUSION

(1) Mutual activation of PMAA and P2M5VP hydrogels in intergel system on their basis in aqueous medium provides significant increases in electrochemical properties of polymer macromolecules. Specific electric conductivity increases, the concentration of hydrogen ions decreases.

(2) Result of remote interaction of polymer hydrogels of PMAA and P2M5VP in water medium is a sharp increase of swelling, which is due to the formation of the same charges of internode links of polymer chains, what, in turn, provides unfolding of macromolecular globe and increase of swelling degree.

(3) Sorption of lanthanum ions by intergel system hPMAA–hP2M5VP is accompanied by radically different changes of electrochemical properties of polymer hydrogels. There is a decrease of specific electric conductivity and an increase of protons concentration in the solution. Also, it should be noted that in the process of lanthanum sorption amount of same-charged groups on internode links of polymer chains, which provides folding of polymer globe and decrease of swelling degree.

(4) All intergel pairs in the intergel system hPMAA–hP2M5VP have much higher values of lanthanum ions extraction degree comparatively with initial hydrogels of PMAA and P2M5VP. Wherein maximum sorption of lanthanum is observed at 50%hPMAA–50%hP2M5VP ratio, sorption degree is 89.65%. Polymer-chain-binding degree (in relation to lanthanum ions) has maximum values at hydrogels ratio 50%hPMAA–50%hP2M5VP, the value of the binding degree is 74.67%.

(5) Intergel system hPMAA–hP2M5VP have maximum values of cerium ions extraction degree at 67%hPMAA–33%hP2M5VP (sorption degree is 87.67%). Polymer-chain-binding degree at this ratio also has the highest values, it is 72.72%.

(6) Selective sorption of lanthanum ions by intergel system 50%hPMAA–50%hP2M5VP indicates that selectivity of the intergel system is manifested to lanthanum ions. Extraction degree of lanthanum ions is 65.40%, of cerium ions 24.38%. Polymer-chain-binding degree in relation to lanthanum ions is 54.25%, in relation to cerium ions is 20.22%.

ACKNOWLEDGMENTS

The work was financially supported by two projects AP05131302 and AP05131451 by the Committee of Science of Ministry of Education and Science of the Republic of Kazakhstan.

KEYWORDS

- polymethacrylic acid
- poly-2-methyl-5-vinylpyridine
- remote interaction
- La^{3+} ions
- Ce^{3+} ions
- sorption

REFERENCES

1. Alimbekova, B. T.; Korganbayeva, Zh. K.; Himersen, H.; Kondaurov, R. G.; Jumadilov, T. K. Features of Polymethacrylic Acid and Poly-2-Methyl-5-Vinylpyridine Hydrogels Remote Interaction in an Aqueous Medium. *J. Chem. Chem. Eng.* **2014,** *3,* 265–269.
2. Alimbekova, B.; Erzhet, B.; Korganbayeva Zh.; Himersen, H.; Kaldaeva, S.; Kondaurov, R.; Jumadilov, T. Electrochemical and Conformational Properties of Intergel Systems Based on the Crosslinked Polyacrylic Acid and Vinylpyridines. In *Proceedings of VII International Scientific-Technical Conference "Advance in Petroleum and Gas Industry and Petrochemistry"* (APGIP-7), Lviv, Ukraine, May 2014; p 64.
3. Jumadilov, T. K.; Himersen, H.; Kaldayeva, S. S.; Kondaurov, R. G. Features of Electrochemical and Conformational Behavior of Intergel System Based on Polyacrylic Acid and Poly-4-Vinylpyridine Hydrogels in an Aqueous Medium. *J. Mater. Sci. Eng., B* **2014,** *4,* 147–151.
4. Jumadilov, T. K.; Abilov Zh. A.; Kaldayeva, S. S.; Himersen, H.; Kondaurov, R. G. Ionic Equilibrium and Conformational State in Intergel System Based on Polyacrylic Acid and Poly-4-Vinylpyridine Hydrogels. *J. Chem. Eng. Chem. Res.* **2014,** *1,* 253–261.
5. Jumadilov, T. K.; Abilov Zh. A.; Kondaurov, R. G.; Eskalieva, G. K. Mutual Activation of Hydrogels of Polymethacrylic Acid and Poly-2-Methyl-5-Vinylpyridine. *Chem. J. Kazakhstan* **2015,** *2,* 75–79.
6. Jumadilov, T.; Abilov Zh.; Kondaurov, R.; Himersen, H.; Yeskalieva, G.; Akylbekova, M.; Akimov, A. Influence of Hydrogels Initial State on Their Electrochemical and Volume-Gravimetric Properties in Intergel System Polyacrylic Acid Hydrogel and Poly-4-Vinylpyridine Hydrogel. *J. Chem. Chem. Technol.* **2015,** *4,* 459–462.

7. Jumadilov, T.; Kaldayeva, S.; Kondaurov, R.; Erzhan, B.; Erzhet, B. Mutual Activation and High Selectivity of Polymeric Structures in Intergel Systems. In *High Performance Polymers for Engineering Based Composites*; Jumadilov, T., Omari, V., Mukbaniani, M., Abadie, J. M., Tatrishvilli T., Eds.; CRC Press: Boca Raton, FL, 2015; pp 111–119.

8. Petruhin, O. M. Methodology of Physico-chemical Methods of Analysis. *M.: Chem.* **1987,** 77–80.

CHAPTER 8

MEMBRANE TECHNOLOGY FOR WATER TREATMENT: DESIGN, DEVELOPMENT, AND APPLICATIONS

AMJAD MUMTAZ KHAN* and YAHIYA KADAF MANEA

Department of Chemistry, Faculty of Science, Aligarh Muslim University, Aligarh, India

*Corresponding author. E-mail: amjad.mt.khan@gmail.com

ABSTRACT

Membrane technology for water treatment has gained a huge interest of scientists and researchers in the academic and industrial fields. This chapter highlights recent advance preparation methods, fabrication strategies, and water purification applications of membranes. For decades, removal of pollutants from water, such as hardness, heavy metal ions, arsenic anions, organic pollutants, carcinogenic dyes, and biological substrates are considered the biggest challenge. This chapter will be valuable for readers to understand the recent advances in membrane design and fabrication that includes high porosity, exchangeability, photocatalytic activity, and their potential purification mechanisms toward different water pollutants. In addition, it will be helpful for developing new kind of membrane for quick, economically viable, and high-performance water purification system.

8.1 INTRODUCTION

Recently, membrane technique is broadly utilized in numerous regions of water filtration and treatment. Membrane technology has been utilized to deliver consumable water from surface water, groundwater, seawater, and wastewaters before discharged or reused. Membrane filtration systems have many features over conventional systems including clarification, coagulation, aerobic, and anaerobic treatments. Separation systems by using membrane filtration are easy to operate and the performance is more reliable. Purifying water using today's technology is expensive and energy based, therefore there is an urgent need to understand and identify the approaches of purifying water without addition of chemicals with relatively small energy consumption and easy and well-arranged process conductions in a compact module design and reducing the harmful effects in environment. Membrane technique has proven viable for clearification and filitration of water for decades. Membrane technique has distinct advantages, such as high-water quality with easy repairing, fixed parts with compact build units, anti-chemical sludge effluent, and excellent efficiency for separation and treatment of water. Recent innovations in the field related to the fabrication of analytical tools and more recent advanced membrane technologies have emerged showing number of applications for water purification. Currently, polymeric membrane is most widely used for water purification due to its straightforward pore forming mechanism, higher flexibility, smaller footprints required for installation, and relatively low costs compared to inorganic membrane equivalents.[1] In this chapter, the fabrication techniques for pressure-driven membrane processes have been discussed. The features of the fabricated membranes and performance in water filtration are related. Important parameters which affect the membrane performance such as crystallinity of the membrane-based polymer, porous structure, hydrophobicity/ hydrophobicity, membrane charge, and surface roughness are included. The efforts exist on to design the membrane pore structure including its surface properties and cross-section morphology by selection of excellent methods for fabrication, to create reliable membranes with anti-fouling, high-mechanical strength, and selectivity properties. This makes progress in membrane performance, further improvements are needed for fabrication techniques such as phase inversion and interfacial polymerization. In contrast, the potential of fabrication techniques such as electrospinning and track etching needs to be assessed. A comprehensive understanding between structure surface properties and performance is a key for further development and progress in membrane technology for water pollutant in the environment. Basically, the membrane is a thin layer that can separate materials depending on their physical and chemical properties when a driving force, either a gradient of chemical potential (concentration or pressure gradient) or electrical potential, is applied across the membrane.[2]

8.2 TYPES OF MEMBRANES

In general, the membranes have a discrete, thin interface that allows the permeation of chemicals pieces in contact with it. This interface may be completely uniform in composition and structure or it may be chemically or physically heterogenous, for example, they contain holes or pores of finite dimensions or is a layered structure. A normal filter meets this requirement of a membrane, but, by convention, the term filter is usually limited to structures that separate particulate suspensions larger than 1–10 μm. Generally, membranes are classified into two main types , isotropic and anisotropic as shown in Figure 8.1.

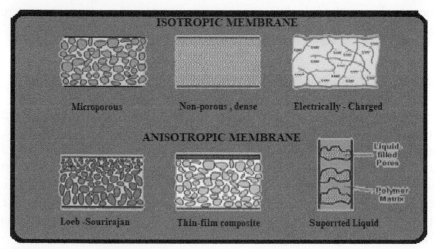

FIGURE 8.1 Types of membrane.

Technologies related to polymeric membrane are of primary interest that works on pressure-driven processes where a pressurized feed is supplied to the membrane units to produce purified product. For concentrations containing high levels of total dissolved solids (TDS), it can be used depending on the processes of cross-flow geometry.

8.2.1 ISOTROPIC MEMBRANES

The composition of isotropic membranes are chemically homogenous which includes microporous membranes, nonporous dense films, and electrically charged membranes

8.2.1.1   MICROPOROUS MEMBRANES

Microporous membrane has a rigid, highly voided structure with randomly distributed interconnected pores. The separation of solutes by using microporous membranes is mainly due to molecular size and pore size distribution. In general, only molecules that differ considerably in size can be separated effectively by microporous membranes viz. ultrafiltration (UF) and microfiltration (MF).[3] Zhao et al. prepared silicone rubber membrane with ordered micropores in the surface by means of the solvent evaporation-induced phase separation. Results indicated that the micropores were generated by removing liquid paraffin phase in the cured silicone rubber film as explained in Figure 8.2, the average pore size increases with increasing liquid paraffin concentration or the initial casting solution thickness as shown in Figure 8.3.[11]

8.2.1.2   NONPOROUS DENSE MEMBRANES

Nonporous dense membranes consist of a dense film through which permeants are transported by diffusion due to the development of pressure, concentration, or electrical potential gradient. Therefore, the separation of various components of a mixture is related directly to their relative transport rate within the membrane determined by their diffusivity and solubility in the membrane material. Thus, nonporous dense membranes can separate permeants of similar size if the permeant concentrations in the membrane material differs significantly.

8.2.1.3   ELECTRICALLY CHARGED MEMBRANES

Electrically charged membranes can be dense or microporous but are most commonly microporous having pore walls carrying fixed positively or negatively charged ions. An anion exchange membrane is considered as a membrane with fixed positively charged due to its high ability to attract anion in the surface and surrounding fluid. Membrane with fixed positively charged ions is referred to as an anion exchange membrane because it binds anions in the surrounding fluid. Similarly, a membrane containing fixed negatively charged ions is considered as cation exchange membrane. Separation of solutes with charged membranes is achieved mainly by exclusion of ions of the same charge as the fixed ions of the membrane structure, and to a much lesser extent by the pore size. The separation is affected by the charge and concentration of the ions present in the solution. For example, monovalent ions are excluded less effectively than divalent.

8.2.2 ANISOTROPIC MEMBRANES

Membranes include two main types:

1. Loeb–Sourirajan membranes
2. Composite membranes

Loeb–Sourirajan membranes are homogenous in chemical composition similar to isotropic microporous membranes, but pore sizes and porosity varies across the membrane thickness. Development of such anisotropic membranes in the early 1960s is a major breakthrough in the field of membrane technology. Composite membranes are chemically and structurally heterogenous. For example, a thin surface layer is supported by a much thicker porous structure (as mechanical support), and these structures are traditionally made of different polymeric materials.[4]

8.2.3 CERAMIC, METAL, AND LIQUID (INORGANIC) MEMBRANES

Due to their relative chemical, thermal, reusability, as well as mechanical robustness, photocatalytic activity inorganic membranes have recently received considerable attention. Ceramic membranes exhibit greater chemical stability and fouling resistance than current polymeric membranes in many wastewater treatment applications. Materials developed recently are typically in a nanocrystalline form and these membranes include porous ceramics such as ZnO, SiO_2 ,Al_2O_3 , TiO_2, and ZrO_2.[5] Composites contain two or more materials such as TiO_2–SiO_2, TiO_2–ZrO_2, and Al_2O_3–SiC, and various nanoparticle (NP) composites (Ag–TiO_2, Zn–CeO_2, and zeolites).[6,7] In the case of ceramic membranes, photocatalytic materials such as TiO_2 and composites containing TiO_2 have been studied due to their multifunctionality and find wide applications in the field of remediation of ground and wastewater.[8–10] Along with a separation function, TiO_2 offers photocatalytic ability for decomposition of organic species/microorganisms/pollutants, photolysis, and superhydrophilicity, which reduces unwanted adsorption of organic/biological species to the membrane surface. The process of TiO_2 photocatalysis is based on photo-induced charge separation on the surface of the oxide. If an incoming photon energy (hν) is greater than or equal to the band-gap energy (Eg) of TiO_2 (3.0 eV and 3.2 eV for rutile and anatase crystal structures, respectively), an electron (e⁻) will be photo excited to the conduction band (CB) of TiO_2 leaving an empty unfilled valence band (VB) resulting in the formation of an electron–hole pair (e⁻–h⁺).

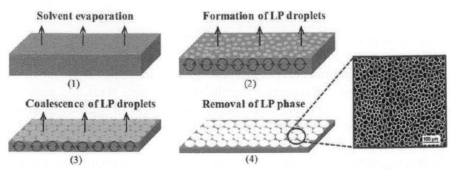

FIGURE 8.2 Explanation of the process of pore formation process in the surface of silicone rubber membrane. (Reprinted with permission from Ref. 11. © 2013 American Chemical Society.)

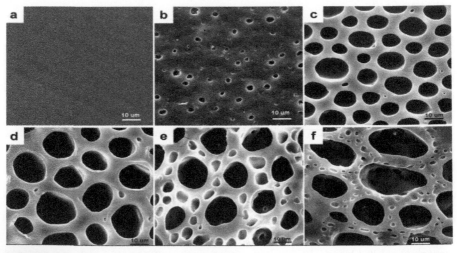

FIGURE 8.3 SEM micrographs of the surface morphologies of the porous membranes prepared at different liquid paraffin concentration: (a) 10, (b) 15, (c) 20, (d) 25, (e) 30, (f) 40 wt%. (Reprinted with permission from Ref. 11. © 2013 American Chemical Society.)

8.3   PREPARATION OF POLYMERIC MEMBRANES

Polymeric membranes contain concentrated solution of the polymer in a solvent with subsequent immersion into a liquid bath, typically water or a mixture with the solvent, in which the solvent is miscible but the polymer is not. Water vapor adsorption from humid atmosphere, solvent evaporation, or some combination of techniques may be used in place of immersion in the liquid bath. Some methods have been summarized under proper conditions in which a film is formed comprising of a continuous phase of solid polymer and an interconnecting phase of voids, chambers, or pores through which liquids can flow. The distribution of phases during solvent exchange dictates the physical structure of the solid membrane.[12]

8.3.1   ANISOTROPIC MEMBRANES

Anisotropic membranes are prepared by contacting the top surface of the cast film with the nonsolvent first creating a finely porous selective skin layer. The precipitated skin layer slows the penetration of nonsolvent into the film causing polymer below the skin layer to precipitate more slowly. This process has been called "phase inversion" as shown in Figure 8.4. Structures of this type have being studied more than a century ago. An analogous procedure is used to make fibers by wet spinning.[13,14] In the process of solidification, pore structure, pore size, shape, and volume is affected by many factors.[15–18] Early membranes made in this way consisted of a similar pore structure through the entire membrane, and because of their thickness such membranes had low fluxes. Loeb–Sourirajan introduced a solvent evaporation step prior to precipitating the polymer; the polymer concentration gradient in the nascent film leads to a gradation of pore size upon phase inversion. This gives a "thin layer" with fine pores, that is, separating layer, overlaying a substrate consisting of much larger pores that provide mechanical support but relatively little resistance to water flow. With a wet annealing step, Loeb–Sourirajan were able to make the first practical reverse osmosis (RO) membrane. The polymer solution can be casted in batch mode to make laboratory membranes or in a continuous fashion to be used as commercial membranes. The solution can be casted on a fabric or other porous substrate for additional support. An analogous process known as dry-jet wet spinning is used to prepare hollow fiber membranes. One of the most important membrane preparation methods is the Loeb–Sourirajan process described in 1963.[19] The Loeb–Sourirajan process uses water as the phase inversion nonsolvent and was originally used to produce cellulose acetate RO membranes. Nowadays, RO and nanofiltration (NF) membranes are polyamide thin-film composite nature . However, the Loeb–Sourirajan process still predominates from the rest of the method for the preparation of UF and MF membranes. Commonly, UF membrane materials include many polymers via cellulose acetate, polyacrylonitrile, poly(ether imides), aromatic polyamides, polysulfone, poly(ether sulfone.), poly(vinylidene fluoride), and poly(vinyl pyrrolidone). Early MF membranes were nitrocellulose, and cellulose acetate materials used more recently are poly(vinylidenefluoride), polysulfone, polyamide, poly(tetrafluoroethylene), and polyethylene.[20,21] Thermally induced phase separation (TIPS) bears some similarity to the phase inversion process but uses lower temperature rather than a nonsolvent to coagulate the polymer. A polymer solution is spread on a support while one face of the film is cooled, initiating phase separation. The rest of the film is gradually cooled and phase inversion gradually propagates to form an isotropic or anisotropic porous membrane as shown in Figure 8.5. In order to create selective surface layer with regard to anisotropic membranes, solvent evaporation at the selective surface is sometimes used to enhance the phase inversion process rather than only a simple thermal gradient in the solvent. TIPS also makes a number of polymers accessible for membrane formation that cannot be used in the traditional phase inversion technique. TIPS has been carried out on a number of different polymers including homopolymers such as polypropylene and diphenylether and copolymers such as poly(ethylene-co-acrylic acid).[22] Connected pore structures are formed at low concentrations of polymer; as polymer concentration and cooling rate increases, pore size tend to decrease. When evaporation is used to create anisotropic membranes, the polymer molecular weight does not significantly affect the cell size of the selective layer and therefore does not greatly influence the membrane performance.[23] The process begins with a precursor film which shows row-nucleated lamellar morphology. The precursor film is annealed to eliminate any inconsistencies in the crystal structure. Stretching is then carried out at low temperature to introduce voids and subsequently high temperature is used to enlarge those voids.[24,25]

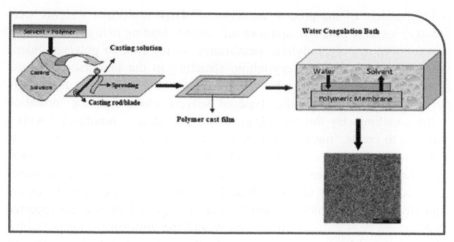

FIGURE 8.4 Preparation steps of polymeric membrane by nonsolvent phase inversion method. (Reprinted from Ref. 25. Open access.)

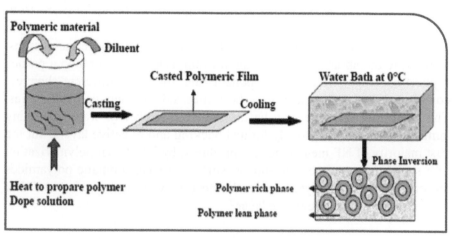

FIGURE 8.5 Preparation steps of polymeric membrane by thermal-induced phase separation method. (Reprinted from Ref. 25. Open access.)

The morphology of the precursor film is most important for the success of the stretching technique.

The crystals formed as a result of stress and elongation induced during the extrusion process and their formation is a strong function of processing conditions and most importantly the polymer molecular weight. A critical molecular weight for crystal formation is known to exist that is dependent on shear rate and temperature up to a particular shear rate after which it is independent of the process conditions.[26] High-molecular weights were found to increase pore size and pore uniformity, leading to high-water vapor transmission in polypropylene membranes. In the case of poly(vinylidene fluoride) membranes, the crystalline structure in the precursor film was found to form most readily when a blend of low- and high-molecular weight polymer was used.[27] Another type of solvent-less membrane formation is track etching. By this technique, a polymer film is bombarded with a particles to create "tracks" through the film. The film is then immersed in a chemical etchant to create straight through circular pores. Polycarbonate membranes have been formed by this technique. Unlike membranes prepared by the other methods described here, track-etch membranes are typically of uniform thickness and have defined pore diameters. As reported by Alpatova et al.,[28] the unity tortuosity and the uniform thickness (which allows the membrane to be exceedingly thin everywhere) of a track-etch membrane may be significantly lower than that of a solvent cast membrane but both types of membranes may show similar permeability. Semiporous NF membranes bear a strong compositional similarity to RO membranes. Both RO and NF membranes, though formerly produced by the Loeb–Sourirajan process from cellulose acetate, are today thin-film composite membranes. Composite membranes consist of an ultrathin selective layer atop a porous support backing. These two components are always of different chemical compositions and may, therefore, be optimized for their particular roles. The composite structure may be formed in a number of ways, including laminating together separately formed backing and selective layers, but the vast majority of NF membranes are produced by interfacial polymerization of a set of monomers on the support surface. Linear aromatic polyamides are one of the few polymers with the necessary solute rejection and flux characteristics for the selective layer.[2]

8.4 MEMBRANES FABRICATION

In addition to membrane structure, porosity, and thickness, membrane surface properties such as hydrophilicity, pore size, charge density, and roughness have a major impact on the membrane performance in terms of separation and antifouling characteristics. Modification of surface properties, therefore, could significantly improve the efficiency of membrane water treatment for surface-located nanocomposite membranes. The process of preparing this type of membranes has effects on the membrane's intrinsic structures, so there is a good potential of implementing such an approach on commercially available membranes. Surface-located nanocomposite membranes could be prepared based on methods such as self-assembly, coating/deposition, and chemical grafting. Those fabrication methods can be carried out individually or involved simultaneously, for example, during the layer-by-layer assembly process, the electrostatic attraction is a common assembly force. Those methods are listed here based on their most unique properties such as bonding force and bonding process.

8.4.1 SELF-ASSEMBLY

Self-assembly is the main process used to attach NPs onto specific membrane surfaces containing functional groups viz. COOH, -SO_3-H^+, and sulfone groups through interactions such as coordination and hydrogen bonding (Fig. 8.6).[29]

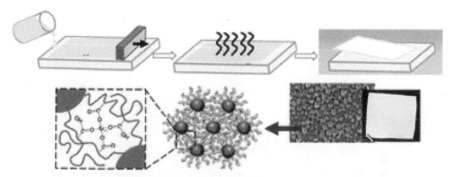

FIGURE 8.6 Self-assembly process, NPs attached onto specific membrane surfaces. (Reprinted with permission from Ref. 29. © 2003 Elsevier.)

Membrane surfaces without these functional groups could be pretreated to introduce such groups prior to the self-assembly process. For example, sulfonated PES,[30,31] polyimide-blended PES, poly(styrene-alt-maleic anhydride)-blended PVDF, and PAA-modified polypropylene membranes were all successfully used to carried out the NPs self-assembly process.

8.4.2 COATING/DEPOSITION

Coating/deposition is another widely used process to prepare surface-located nanocomposites in which straight forward dip coating or filtration deposition are applied to place nanomaterials onto membrane surface. For enhancing the features of membranes TiO_2 NPs are widely used to deposit onto different polymeric membrane (PSU, PVDF, and PAN) surfaces and compared their performance with TiO_2 entrapped membrane. Ngang et al. found TiO_2-deposited membranes showed a greater fouling mitigation effect because of a large amount of TiO_2 located on membrane surface.[32] Han et al.[33] deposited an ultrathin (22–53 nm thick) GO barrier onto a MF membrane surface and demonstrated that the prepared membrane had a much higher water flux and good rejection for organic dyes. Similarly, Ghanbari et al. fabricated a TiO_2/HNTs composite nanomaterial via a one-step solvothermal method as an additive for FO membranes. Regarding filtration performance, the nanocomposite membrane showed excellent water flux and solute rejection Figure 8.7.[34] However, one disadvantage of coating/deposition method is the potential loss of deposited nanomaterials with time due to the weak interactions between nanomaterials and membrane surface. This may seriously hinder its applications, particularly in cross-flow systems.

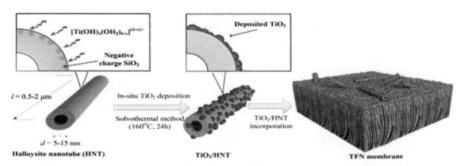

FIGURE 8.7   Schematic of TO_2–HNT deposition into the surface of membrane. (Reprinted with permission from Ref. 34. © 2015 Elsevier.)

8.4.3   ELECTROSTATIC ATTRACTION

Electrostatic attraction has been explored to attach positively charged polymer such as PEI, PSU, and PVDF-encapsulated NPs onto negatively charged membrane surfaces to enhance membrane's antimicrobial and antifouling properties.[35] The membranes covered with NPs via Ag^+ and Cu^{2+} showed a clear antimicrobial activity, while superhydrophilic silica covered membrane showed enhanced antifouling capability due to the lower adhesion force between organic foulant and membrane surface. However, the challenging task is quick release of silver or copper ions leading to a fast depletion of antimicrobial capability. Using CuNPs covered membrane, over 30% of the loaded copper will detach from the membrane surface into water either as Cu^{2+} or as NPs during the first 2 days. The detachment rate is expected to increase when the membrane is utilized under the cross-flow configuration, especially with a feed solution of low pH, high ionic strength, and high concentrations of chelating agents or organic ligands.

8.4.4 ANTIMICROBIAL ACTIVITY OF MEMBRANES

To maintain a sufficient antimicrobial activity, frequent recharge of those NPs is necessary that gradually gets depleted in order to have the desired functionality of the nanocomposite membranes and may also release nano-materials to water stream leading to potential risks to humans. As a result, stronger attachments through processes such as chemical bonding have been actively explored to extend the desirable functionality of nanocomposite membranes. Ben-Sasson et al.[35] used polyethylene glycol-grafted multi walled nano tubes (MWNTs) as a bridging structure between AgNPs and hollow fiber membrane surface. In their study, AgNPs were firstly coated onto MWNTs to prepare Ag/MWNTs and then Ag/MWNTs were covalently bonded to the external surface of a chemically modified PAN membrane. Yin et al.[36] effectively attached AgNPs onto the surface of PA TFC membrane through chemical bonding (Ag-S) by using cysteamine (H_2N-$(CH_2)_2$-SH) as a bridging agent, as shown in Figure 8.8. The prepared membrane showed good stability of the immobilized AgNPs and excellent antibacterial proper-ties, while the water flux and salt rejection were maintained. Lately, Park et al.[37] attached the AgNPs onto thiolated PVDF UF membrane surface, resulting in a stable antimicrobial membrane.

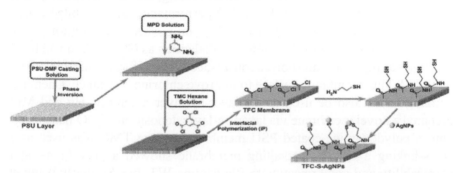

FIGURE 8.8 Silver NPs immobilized onto the surface of TFC membrane. (Reprinted with permission from Ref. 36. © 2013 Elseiver.)

8.4.5 ADSORPTION–REDUCTION

To enhance the surface of membranes, NPs could be attached onto membrane surfaces through an adsorption–reduction mechanism, where NPs can be adsorbed by the membrane surface, and then reduced by chemical agents such as formaldehyde, vitamin C, and ascorbic acid or under light irradiation. Using this approach, Zhu et al.[38] prepared chitosan-based membranes immobilized with ionic or metallic silver and then compared their anti-biofouling performances by using two typical bacteria (*E. coli* and *Pseudomonas sp.*).

8.4.6 LAYER-BY-LAYER ASSEMBLY

During the layer-by-layer assembly method, electrostatic attraction, hydrogen bonding, and chemical bonding could be involved in the attachment of multiple layers of nanomaterials onto the membrane surface. However, the electrostatic attraction is the most common force exploited, notably between polycation and polyanion species. For instance, Park et al.[39] used poly(allylamine hydrochloride) (PAH) as polycation and PAA containing carboxylated MWNTs as polyanion to do the assembly on a negatively charged PSU membrane surface. After thermal cross-linking, the membrane showed an enhanced thermal stability and chlorine resistance. Wang et al.[40] used PEI-modified GO as polycation and PAA as polyanion to prepare PAN-based NF membranes, which exhibited improved mechanical and thermal properties. Liu et al. [41] used PAH as polycation and poly(sodium 4-styrene-sulfonate) (PSS) as polyanion for the surface assembly and during the process, AgNPs were introduced by being suspended either in the PAH solution or in the PSS solution. After cross-linking, the silver nanocomposite membranes showed a good NF and FO performance and exhibited excellent antibacterial properties against both gram-positive and gram-negative bacteria. In another studies,[42,43] new materials such as GO and aminated-GO were used as polyanion and polycation, respectively, to form a chlorine barrier on the PA TFC membrane surface to improve chlorine resistance. Chemical bonding could also be used in the layer-by-layer assembly. Hu and Mi[44] prepared a novel membrane via layer-by-layer deposition of GO nanosheets onto a polydopamine-coated PSU membrane, where TMC was used as a cross-linking agent. The resulting membrane showed a very high-water permeability and good rejection to Rhodamine-WT dye. Similarly, Wang et al.[45] prepared a novel membrane by using layer-by-layer assembly of GO onto polycation membrane for dye removal as illustrated in Figure 8.9.

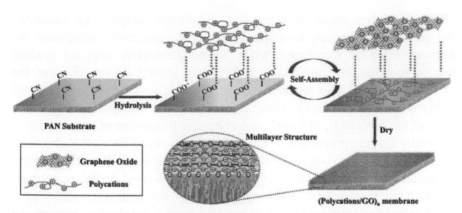

FIGURE 8.9 Schematic representation of layer-by-layer assembly method to prepare membrane of GO onto polycation. (Reprinted with permission from Ref. 45. © 2016 Elsevier.)

8.4.7 CHEMICAL GRAFTING

One concern of the surface-located nanocomposite membrane is the depletion of the deposited nanomaterials during the filtration process, especially for those attached onto the membrane surface through weak van der Waals and electrostatic forces.

8.4.8 OTHER FABRICATION METHODS

Other fabrication methods such as cross-linking, colloidal precipitation, and hydrothermal process have been successfully used to prepare surface-located nanocomposite membranes. In the cross-linking process, an additional polymer was needed to form the cross-linked matrix to wrap up nanomaterials for their incorporation onto the membrane surface. For example, Wang et al.[46] attached cellulose nanofiber onto nanofibrous UF membrane surface by using cross-linked PEG, resulting in a superior membrane with an improved water permeability and fouling resistance.

8.5 MEMBRANE CHARACTERIZATION

Membrane characterization techniques include two types: static and dynamic techniques. The dynamic techniques are of fundamental importance while investigating membrane performance. The static techniques can give information about structure and membrane morphology, chemical and physical properties. In some cases, characterization techniques are destructive for the membrane, while the nondestructive ones are applied also to monitor the membrane performance during its use. Except for bubble pressure all other reported techniques have not yet been standardized or harmonized (Table 8.1). This fact often causes confusion and can be misleading. Structural, physical, and chemical properties of membranes— especially surface properties of the membranes—must be well understood in order to develop successful membrane technologies. Understanding the surface properties of the membranes, which depend on the membrane materials, membrane type, and interactions between membrane and solute, is not only of scientific importance but also have technological importance in the water treatment industry and is critical to membrane performance metrics including permeation, flux, rejection, lifetime, and fouling. There are several number of qualitative and quantitative analytical methods for characterizing membranes. These include scanning electron microscopy (SEM), transmission electron microscopy (TEM), scanning tunneling microscopy (STM), secondary ion mass spectrometry (SIMS), contact angle measurements, zeta potential, X-ray photoelectron spectroscopy (XPS), laser scanning confocal microscopy (LSCS), electron spin resonance (ESR), neutron reflectivity (NR), thermogravimetric analysis (TGA), Fourier-transform infrared (FTIR) spectroscopy with attenuated total reflection (FTIR-ATR), Raman spectroscopy, atomic force microscopy (AFM), and X-ray diffraction (XRD). These analytical tools provide us with structural information, elemental composition, surface morphology, and fouling phenomena. The most widely used technique for structural and chemical composition characterization of membranes is SEM. A SEM image is formed by scanning a focused electron beam across the sample and recording the intensity of scattered or secondary electrons. In addition to electrons, X-rays are ejected from the sample, and can be detected using energy dispersive X-ray spectroscopy (EDX), or wavelength dispersive X-ray spectroscopy. The SEM electron beam can also generate photons in the UV-Vis-IR range and these can be recorded using cathode luminescence. Where high resolution images of membrane structures are necessary, TEM can be used. Most of these microscopes also offer qualitative and quantitative evaluation of different types of foulants and chemical composition of membranes. Various signals obtained from a collection of detectors contain information about the surface topology, pore size, pore size distribution, pore shape, chemical composition, and thickness of the membrane. If the sample is cross-sectioned (or is viewed tilted), the cross-sectional images can provide thickness and structural information. Sample preparation and imaging parameters greatly affect the resultant images, the last studies pointed out the importance of noting and reporting imaging parameters and membrane sample preparation for SEM characterization.[47] Another imaging technique often used to elucidate the surface roughness, pore size and its distribution, and aggregate size at the surface of the membrane is AFM.[48] AFM uses mechanical interactions between the sample and a probe tip mounted at the end of a cantilever scan across the sample surface. As the tip nears the sample surface, deflection of the cantilever is measured by a laser beam reflected from the cantilever onto a photodiode. AFM studies have been applied to characterize various membrane materials.[49-51] Determination of the surface roughness by AFM has become a routine analytical method as it relates to membrane fouling. In case of surface charge properties of the membranes, zeta potential measurements are commonly used. Zeta (surface) potentials are determined from electrokinetic measurements; changes in membrane surface potential can be used to study cake deposition and fouling behavior during filtration, and the measurements are also useful for NF/RO applications in which electrolyte solutions containing different pH values are used. Membrane charge electrostatically interacts with ions and results in a change in charge density near the surface of the membrane. The separation efficiency of ions is governed by the relative sign of charge on the membrane's surface, ions, colloids, or molecules. IR and Raman spectroscopy measure the vibrations of molecules and are used to identify or study structural/chemical composition of samples (Raman-active transitions require a change in the polarizability of the molecule, whereas IR-active transitions require a change in the dipole moment). These techniques can be used to characterize polymeric membranes, for example, to monitor surface modification and biofouling of the membrane.[52] Recently, surface-enhanced Raman spectroscopy has also been utilized to examine fouling of organic species on membrane surfaces.[53,54] As described earlier, it is generally considered that fouling increases for membranes with more hydrophobic, less negatively charged and rough surfaces. Hydrophilic surfaces hydrate the membrane surface by water molecules, which makes it less susceptible to initial organic fouling than hydrophobic surfaces. Analytical tools used for membrane protein-fouling characterization include SEM, TEM, radiolabeling, XPS, microspectrophotometry, EPRS (quantitative analysis of protein fouling/ denaturation by chemical attachment of spin labels to protein), SANS (in situ measurement for quantification and location of protein fouling, thickness, and structural characterization), NMR, and SIMS (differentiating adsorbed proteins and determining orientation/conformation of adsorbed proteins.[55]

TABLE 8.1 Summary of Techniques Used for Characterization of Membranes.

Techniques	Characterization	Type
Bubble pressure	Max. pore size	Dynamic nondestructive
Gas and liquid displacement	Pore size distribution	Dynamic nondestructive
Mercury porosimetry	Pore size distribution	Statistic destructive
Transmission electron microscopic (TEM) and Scanning electron microscopy	Top layer thickness surface porosity Qualitative structure Analysis Pore size distribution	Statistic destructive
Atomic force microscopic (AFM)	surface porosity	Statistic non destructive
Flux and retention measurement	Permeability Selectivity	Dynamic non destructive
Permporometry	Pore size distribution	Dynamic non destructive
Infrared spectroscopy FT-IR	Function group Analysis Surface studies	Statistic destructive
Contact angle	Surface studies	Non destructive
Stress strain measurement	Surface studies	destructive
X-Ray photoelectron spectroscopy	Chemical Analysis Surface studies	Statistic destructive
SEM + X-Ray microanalysis (EDS)	Chemical Analysis Surface studies	Statistic destructive
Thermogravimetric Analysis (TGA)	Stability studies	destructive

8.6 MEMBRANES FOR WATER TREATMENT

Water treatment processes employ several types of membranes based on their pore sizes viz. reverse osmosis (RO), nanofiltration (NF), ultrafiltration (UF), microfiltration (MF), and membrane distillation (MD).[56] Figure 8.10 summarizes various membrane filtration processes relative to common materials that would be filtered out through each process. MD has the potential to desalinate highly saline water. UF and MF follow a similar process, mode of separation is particle sieving through the pores present in the membrane. MF membranes have greater pore sizes (approx. 0.1–5 μm) and reject particles, asbestos, and various cellular materials such as red blood cells and bacteria from 0.1 to 10 μm in diameter. UF membranes have smaller pores (approx. 0.01–0.1 μm) than MF membranes such that in addition to filtering out large particles and microorganisms. They can filter dissolved biomacromolecules, such as pyrogens, proteins, and viruses (sizes ranging from 0.01 to 0.2 μm). UF has various applications in wastewater treatment, water remediation, recovery of surfactants in industrial cleaning, food processing, protein separation, gene engineering, etc. Commonly, particle sizes are characterized by their molecular weight cutoff (MWCO). MWCO value is defined by rejection of organic solutes (90% rejection by the membrane), and the particle retention is evaluated by converting MWCO to the membrane pore size.[57] A general guideline for designing UF membranes is that the MWCO must be about half of the lowest molecular weight species to be retained. As a direct quantitative means, monodisperse NPs were also used to determine pore size distributions of a variety of UF membranes. UF membranes are fashioned in an anisotropic Loeb–Sourirajan structure. NF membranes exhibit performance between RO and UF membranes. NF membranes are porous and can filter species ranging from 0.001 to 0.01 μm in size. This includes mostly organic molecules, viruses, and a range of salts. Furthermore, NF membranes can reject divalent ions, so NF is often used to soften hard water. In contrast to NF, UF, and MF membranes, RO membranes are so dense that the "pores" are considered as nonporous (approx. 3–5 Å), and they are within the range of thermal motion of the polymer chains that form membrane. Therefore, RO membranes can even filter low-molecular weight species such as aqueous inorganic solids including salt ions, minerals, and metal ions and organic molecules. The accepted mechanism of transport by RO is via diffusion through statistically distributed free volume areas. Solutes pass through the membrane by dissolving in the membrane material and diffusing down a concentration gradient when the applied pressure exerted exceeds the osmotic pressure. Separation occurs because of the difference in solubilities and mobilities of different solutes within the membrane. The most common applications of RO is the desalination of brackish groundwater or seawater and the production of potable water.

TABLE 8.2 Some Membrane Filtration Relative to Fabrication Technique.

Water treatment process	Polymers used for membrane fabrication	Fabrication techniques	Average pore size of the membrane
RO	Cellulose acetate/triacetate	Phase inversion	3–5 Å
	Aromatic polyamide	Solution casting	
	Polypiperzine		
	Polybenziimidazoline		
NF	Polyamides	Interfacial polymerization	0.001–0.01 μm
	Polysulfones		
	Polyols	Layer-by-layer deposition	
	Polyphenols	Phase inversion	
UF	Polyacrylonitrile (PAN)	Phase inversion	0.001 – 0.1 μm
	Polyethersulfone (PES)	Solution wet spinning	
	Polysulfone (PS)		
	Polyethersulfone (PES)		
	Poly(phthazine ether sulfone ketone) (PPESK)		
	Poly(vinyl butyral)		
	Polyvinylidenefluoride (PVDF)		
MF	PVDF	Phase inversion	0.1–10 μm
	Poly(tetrafluorethylene) (PTFE)	Stretching	
	Polypropylene (PP)	Track etching	
	Polyethylene (PE)		
	PES		
	Polyetheretherketone (PEEK)		
MD	PTFE	Phase inversion	0.1–1 μm
	PVDF	Stretching	
		Electrospinning	

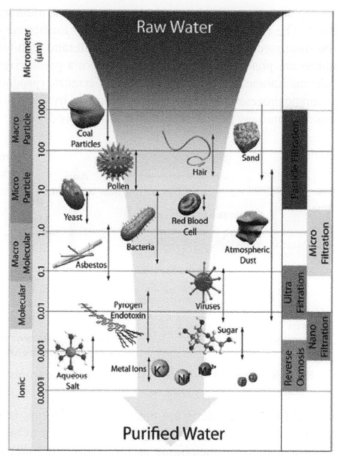

FIGURE 8.10 Schematic illustration of membrane filtration spectrum. Reverse osmosis, nanofiltration, ultrafiltration, microfiltration, and conventional particle filtration differ principally in the average pore diameter of the membranes. Reverse osmosis membranes are so dense that the pores are considered as nonporous. (Reprinted with permission from Ref. 56. © 2016 Royal Society of Chemistry.)

8.7 MEMBRANE OPERATION PROCESSES

Membrane filtration-based water purification utilizes pressure-driven, dialysis, distillation, and electropotential processes. A simplified flow diagram for a membrane-based water purification process is shown in Figure 8.11. The membrane separation steps includes a membrane pretreatment unit for removal of particulates and other macromolecules followed by a RO unit for salt removal. The flow diagram indicates several other steps related to microbial control (chlorine addition), pH control, particle flocculation, dechlorination (to protect the reverse osmosis membrane), and scaling control. The relevance of the process steps to membrane development will be discussed in detail later on. The membrane technologies of primary interest are pressure-driven processes where a pressurized feed is supplied to the membrane unit to produce purified permeate (product). Some of these membrane processes have cross-flow geometry by which a retentate (or concentrate) containing high levels of TDS is also produced.

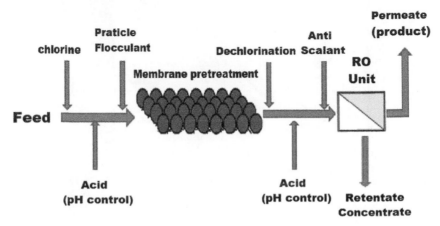

FIGURE 8.11 Simplified process flow diagram of a water purification process.

The dead-end filtration mode is the most common process in the treatment of water in the research lab. In this type, flow of water to be filtered is directed perpendicular to the membrane surface such that water is pushed through the membrane under pressure. This technique is useful if the concentration of particles or targeted pollutant is low. It is typically used in home water filtrations and also to concentrate compounds in contrast to industrial applications where amount of materials to be filtered can be as high as 30%. If the concentration of targeted species is high, the filtered materials can accumulate as a layer on the surface of the membrane. This layer formation results in a pressure drop across the membrane leading to increased resistance and reduced permeate flux. In the case of a cross-flow (or tangential flow) operation process, the feed stream is parallel to the membrane surface such that the feed water flow is perpendicular to the filtration flow as shown in Figure 8.9b. Continuous turbulent flow along the membrane surface (cross-flow velocity is typically 0.5 to 1 m/s, four to five orders of magnitude greater than the superficial water velocity toward the membrane) creates a shear force that reduces the accumulation of species. As such, the cross-flow operation mode is particularly useful for filtering high concentrations of materials or macromolecules such as cells and proteins. In case of municipal water treatment applications, surface water has dilute contamination (concentration of solids is about 0.01%). Therefore, the advantage of cross-flow operation is of less importance. The cost of the membrane system and operation cost associated with the cross-flow system is higher than those for dead-end operation.

8.8   FACTORS AFFECTING MEMBRANE

The performance of a membrane is mainly governed by the structure of its pores and the physical/chemical properties of the material. The following factors should be considered in order to design effective membranes:

1. Choice of membrane materials.
2. High-water flux, high solute rejection.
3. Module configuration.
4. Mechanical/chemical/thermal/temporal stability.
5. System design including processibility at large scale.
6. Operating conditions for cost-effectiveness.

8.9   DEVELOPMENT OF MEMBRANES

Recently, many types of membranes have been developed that are highly efficient in the process of purification and separation of particulates from wastewater.

8.9.1   NANOCOMPOSITE MEMBRANES

The new functionalities introduced by nanocomposite membranes are not only capable of filtering solute and particles from water, but also adsorbing, degrading, and/or deactivating them. Functional membranes are normally designed and optimized with specific applications in mind. Inclusion of inorganic moieties into a polymeric matrix system can offer multifunctionality beyond separation alone and can enhance hydrophilicity, mechanical strength, water permeability, rejection rate, and antifouling properties. The additives also can modify the kinetics and thermodynamics of the formation process of the polymeric membrane such as membrane surface and pore structure can be altered. In the adsorption method, NPs incorporated inside the polymer matrix to make functional membranes with specific capability to adsorb heavy metals from water. The incorporation of nanocomposite into PES matrix enhances the removal of metal ions (II) from aqueous solution. The efficient dispersion of NPs is to provide more accessible active sites for sorption of toxic metal ions. Regeneration of membrane by chelates agent via EDTA allows it to be reused. Tetala and Stamatialis[58] developed a copper removal membrane by incorporating chitosan beads (20–40 μm) inside ethylene vinyl alcohol matrix. While the particle size is significantly larger than the nanoscale, almost the same strategy was used for the membrane fabrication, resulting in a membrane with fast adsorption kinetics and high adsorption capacity (225.7 mg/g) for Cu^{2+} ions. In a similar way, hydrous manganese dioxide (HMO)[19] and Fe_3O_4 NPs (60 nm) [59] were embedded into the PES and PVC matrices, respectively, to remove lead from aqueous solution. In another study[60] Fe–Mn binary oxide was used to prepare PES nanocomposite membrane for As (III) removal. These studies confirmed the possibility of preparing nanocomposite membranes containing certain adsorbents for adsorptive removal of contaminants from water. However, relatively short contact time (high water permeability) implies that sorption could be limited by kinetics in addition to adsorption capacity that may hinder their practical applications. For photocatalysis studies, TiO_2 has been widely used for water splitting, water treatment, air purification, and self-cleaning of surfaces because of its unique photocatalytic properties (such as photodegradation and photoinducedsuperhydrophilicity), stability, commercial availability, and simplicity for its preparation.[61] In the nanocomposite membrane research area, TiO_2 has also been incorporated into various membrane matrices to provide membrane with photocatalytic activities. Rahimpour et al. [61] studied the effects of UV irradiation on the performance of TiO_2/PES nanocomposite membranes and found UV-irradiated TiO_2/PES membranes have higher flux and enhanced fouling resistance when compared to corresponding nanocomposites. They attributed this enhancement to the photocatalysis and superhydrophilicity of TiO_2 under the UV irradiation. Similarly, after applying TiO_2 into PVDF membrane matrix and providing UV irradiation, Damodar et al.[62] obtained a superior membrane with enhanced permeability and fouling resistance, as well as high antibacterial capability. In another study, UV-cleaning process was carried out by Ngang et al.[14] on their TiO_2/PVDF nanocomposite membranes, allowing the recovery of water flux after filtration process. For water biotreatment membrane, biofouling is very important due to the microbial growth and biofilm formation is one of the most challenging issues in membrane separation for water and wastewater treatment.[39] It decreases membrane permeability, reduces permeate quality, and increases energy costs of the separation process. Developing antimicrobial membranes will likely increase membrane efficiency and applications significantly. In addition, use of antimicrobial membrane helps to provide pathogen-free clean water. Silver (Ag) is the most widely explored antimicrobial agent in nanocomposite membrane due to its excellent biocidal properties and successful applications in many areas such as antimicrobial plastics, coatings, and wound and burn dressing.[63] Chou et al.[64] introduced AgNPs into cellulose acetate matrix for antibacterial applications.

8.9.2 THIN FILM WITH NANOCOMPOSITE SUBSTRATE

This type of membranes was first created to explore the impacts of nanofiller on membrane compaction conduct. In the investigation performed by Pendergast et al., silica or zeolite NPs were doped into the PSU substrate, which was then utilized to synthesis TFC membrane for RO. Membranes prepared by this method showed a stable permeability and experienced less flux decline during the compaction when compared with the original TFC membrane. The existence of nanomaterials was believed to have provided necessary mechanical support to mitigate the collapse of porous structure and thickness reduction upon compaction. Further study found that membranes with nanocomposite substrate undergone far less physical compaction and played an important role in maintaining high-water permeability.[65,66] Recently, this concept was mainly implemented to mitigate internal concentration polarization (ICP occurring inside the porous support layer), which may negatively impact on forward osmosis (FO) and RO processes because it can significantly reduce the available osmotic driving force and hence lower water flux .[67] The nanocomposite substrate may have an enhanced hydrophilicity and reduced structural parameter (S) which is controlled by thickness (l), tortuosity (τ), and porosity (ε). Generally, a large S value inevitably leads to severe ICP.

The relationship between **S** and those membrane parameters (l, τ, ε) is as follows:

$$S = S_{skin} + S_{sublayer} = \frac{\tau_{skin} l_{skin}}{\varepsilon_{skin}} + \frac{\tau_{sublayer} l_{sublayer}}{\varepsilon_{sublayer}}$$

Ma et al.[68] incorporated zeolite into PSU substrate resulting in a significant reduction of S from 0.96 mm to 0.34 mm and a higher water flux under either AL-DS (active layer facing draw solution) or AL-FS (active layer facing feed solution) condition. They attributed this phenomenon for improved porosity, better hydrophilicity, and additional water pathways through porous NPs. This is the first study that demonstrated the possibility to use porous NPs and nanocomposite substrate to control ICP in FO operation. Subsequent studies used MWNTs and TiO_2[69] to mitigate ICP problem in the nanocomposite substrate for FO process. The resulting membranes showed an enhanced FO performance with reduced S. For example, the S value of the membrane containing MWNTs decreased from 3.939 mm to 2.042 mm. Furthermore, the same substrate also showed an enhanced tensile strength.

8.10 MEMBRANE FOR REMOVAL OF BIOLOGICAL CONTENT

Membranes are useful for biological applications including separation of proteins and cell debris, purification of proteins, and removal or separations involving viruses and plasmid DNA. The key opportunities include reducing fouling in MF and UF membrane systems, increasing separation resolution so that proteins can be size selected continuously using a membrane as opposed to size exclusion chromatography; membranes for these applications need to be highly defect free, robust, and able to resist harsh chemicals and cleaning environments.[65] As some bioprocess technologies move toward smaller scale operation, single-use membranes is of high quality, inexpensive, easily produced in large quantities, and offer high product throughput. Wastewater effluents from chemical/pharmaceutical manufacturing plants often contain organic/biological contaminants that must be removed before wastewater discharge. These contaminants including pesticides mixes with groundwater requiring their removal in public drinking water treatment systems as well. Membrane treatment for removal of pharmaceuticals (including endocrine disrupters) and other low-molecular weight organic contaminants has been studied and reported extensively in the literature primarily using NF membranes, though UF and RO membranes have been studied.[70–72]

8.10.1 FICK'S LAW

The diffusive flux of a species in a mixture must be expressed relative to some reference. A fixed frame of reference or stationary coordinates is particularly appropriate for membrane systems. The following are two fully equivalent forms of Fick's law for binary diffusion in the direction :

$$n_1 = w_1\left(n_1 + n_2\right) - \rho D_{12}\frac{dw_1}{dz} \tag{8.1}$$

$$N_1 = x_1\left(N_1 + N_2\right) - CD_{12}\frac{dx_1}{dz} \tag{8.2}$$

Where n1 is the mass flux of species i relative to stationary coordinates, N_i is the molar flux, w_1 is the mass fraction of "i," X_1 is the mole fraction, q is the mass density of the mixture, C is molar density, and D_{12} is the binary diffusion coefficient. These relations can be adapted for diffusion of penetrants in a membrane. First, it is useful to let the membrane component be identified with the subscript m and the penetrant with a subscript, for example, 1; this is particularly useful for the cases when there are other penetrants that can be identified as 2, 3, etc. In some cases, the binary form of Fick's law can be used when there are more than two components but not always. For all practical cases in steady state, the membrane itself is stationary, so the flux of this component is zero. Finally, it is important to recognize that for polymeric membranes molar concentrations and terms like x_1 or C in Eq . 8.2 are at best ill defined and at worst not meaningful since the molecular weight of the polymer may not be unique or may even be infinite. The earliest thermodynamic treatments of polymer mixtures revealed that mole fractions were not an appropriate concentration scale in such systems. Thus, Eq. 8.5 provides a more useful form of Fick's first law for membrane systems. With the simplifications noted above, it becomes

$$n_1 = -\frac{\rho D_{1m}}{1 - w_1}\frac{dw_1}{dz} = -\frac{\rho D_{1m}}{w_m}\frac{dw_1}{dz} \tag{8.3}$$

$$n_1 = -D_{1m}\frac{dC_1}{dz} \tag{8.4}$$

$$N_1 = -D_{1m}\frac{dC_1}{dz} \tag{8.5}$$

KEYWORDS

- fabrication strategies
- membrane technology
- water purification
- polymeric membrane
- pollutants removal

REFERENCES

1. Ng, L. Y.; Mohammad, A. W.; Leo, C. P.; Hilal, N. Polymeric Membranes Incorporated with Metal/Metal Oxide Nanoparticles: A Comprehensive Review. *Desalination* **2013,** *308,* 15–33. doi:10.1016/j.desal.2010.11.033.
2. Geise, G. M.; Lee, H. -S.; Miller, D. J.; Freeman, B. D.; McGrath, J. E.; Paul, D. R. Water Purification by Membranes: The Role of Polymer Science. *J. Polym. Sci. Part B Polym. Phys.* **2010,** *48,* 1685–1718. doi:10.1002/polb.22037.
3. Baker, R. W. Updated by Staff. *Membrane Technology. Kirk-Othmer Encycl. Chem. Technol.,* 2005. doi:10.1002/0471238961.1305130202011105.a01.pub2.
4. Lee, A.; Elam, J. W.; Darling, S. B. Membrane Materials for Water Purification: Design, Development, and Application. *Environ. Sci. Water Res. Technol.* **2016,** *2,* 17–42. doi:10.1039/C5EW00159E.
5. DeFriend, K. A.; Wiesner, M. R.; Barron, A. R. Alumina and Aluminate Ultrafiltration Membranes Derived from Alumina Nanoparticles. *J. Memb. Sci.* **2003,** *224,* 11–28. doi:10.1016/S0376-7388(03)00344-2.
6. Mohmood, I.; Lopes, C. B.; Lopes, I.; Ahmad, I.; Duarte, A. C.; Pereira, E. Nanoscale Materials and their Use in Water Contaminants Removal—A Review. *Environ. Sci. Pollut. Res.* **2013,** *20* 1239–1260. doi:10.1007/s11356-012-1415-x.
7. Kumar, S.; Ahlawat, W.; Bhanjana, G.; Heydarifard, S.; Nazhad, M. M.; Dilbaghi, N. Nanotechnology-Based Water Treatment Strategies. *J. Nanosci. Nanotechnol.* **2014,** *14,* 1838–1858. doi:10.1166/jnn.2014.9050.
8. Zhang, X.; Wang, D. K.; Diniz da Costa, J. C. Recent Progresses on Fabrication of Photocatalytic Membranes for Water Treatment. *Catal. Today* **2014,** *230,* 47–54. doi:10.1016/J.CATTOD.2013.11.019.
9. Liu, Z.; He, Y.; Li, F.; Liu, Y. Photocatalytic Treatment of RDX Wastewater with Nano-Sized Titanium Dioxide. *Environ. Sci. Pollut. Res.* **2006,** *13,* 328–332. doi:10.1065/espr2006.08.328.
10. Pan, J. H.; Zhang, X.; Du, A. J.; Sun, D. D.; Leckie, J. O. Self-Etching Reconstruction of Hierarchically Mesoporous F-TiO$_2$ Hollow Microspherical Photocatalyst for Concurrent Membrane Water Purifications. *J. Am. Chem. Soc.* **2008,** *130,* 11256–11257. doi:10.1021/ja803582m.
11. Zhao, J.; Luo, G.; Wu, J.; Xia, H. Preparation of Microporous Silicone Rubber Membrane with Tunable Pore Size via Solvent Evaporation-Induced Phase Separation. *ACS Appl. Mater. Interfaces* **2013,** *5,* 2040–2046. doi:10.1021/am302929c.

12. Pinnau, I. Recent Advances in the Formation of Ultrathin Polymeric Membranes for Gas Separations. *Polym. Adv. Technol.* **1994,** 733–744. doi:10.1002/pat.1994.220051106.

13. Jomekian, A.; Mansoori, S. A. A.; Monirimanesh, N. Synthesis and Characterization of Novel PEO–MCM-41/PVDC Nanocomposite Membrane. *Desalination* **2011,** 276, 239–245. doi:10.1016/J.DESAL.2011.03.058.

14. Ngang, H. P.; Ooi, B. S.; Ahmad, A. L.; Lai, S. O. Preparation of PVDF–TiO2 Mixed-Matrix Membrane and its Evaluation on Dye Adsorption and UV-Cleaning Properties. *Chem. Eng. J.* **2012,** *197*, 359–367. doi:10.1016/J.CEJ.2012.05.050.

15. Huang, J.; Zhang, K.; Wang, K.; Xie, Z.; Ladewig, B.; Wang, H. Fabrication of Polyethersulfone-Mesoporous Silica Nanocomposite Ultrafiltration Membranes with Antifouling Properties. *J. Memb. Sci.* **2012,** *423–424*, 362–370. doi:10.1016/J. MEMSCI.2012.08.029.

16. Pakizeh, M.; Moghadam, A. N.; Omidkhah, M. R.; Namvar-Mahboub, M. Preparation and Characterization of Dimethyldichlorosilane Modified SiO2/PSf Nanocomposite Membrane. *Korean J. Chem. Eng.* **2013,** *30*, 751–760. doi:10.1007/s11814-012-0186-x.

17. Wu, H.; Tang, B.; Wu, P. Development of Novel SiO2–GO Nanohybrid/Polysulfone Membrane with Enhanced Performance. *J. Memb. Sci.* **2014,** *451*, 94–102. doi:10.1016/J. MEMSCI.2013.09.018.

18. Csetneki, I.; Filipcsei, g.; Zrínyi, M. Smart Nanocomposite Polymer Membranes with On/Off Switching Control. *Macromolecules* **2006,** 39 (5), 1939–1942. doi:10.1021/ MA052189A.

19. Gohari, R. J.; Lau, W. J.; Matsuura, T.; Halakoo, E.; Ismail, A. F. Adsorptive Removal of Pb(II) from Aqueous Solution by Novel PES/HMO Ultrafiltration Mixed Matrix Membrane. *Sep. Purif. Technol.* **2013,** *120*, 59–68. doi:10.1016/J.SEPPUR.2013.09.024.

20. Gohari, R. J.; Halakoo, E.; Nazri, N. A. M.; Lau, W. J.; Matsuura, T.; Ismail, A. F. Improving Performance and Antifouling Capability of PES UF Membranes via Blending with Highly Hydrophilic Hydrous Manganese Dioxide Nanoparticles. *Desalination* **2014,** *335*, 87–95. doi:10.1016/J.DESAL.2013.12.011.

21. Alhoshan, M.; Alam, J.; Dass, L. A.; Al-Homaidi, N. Fabrication of Polysulfone/ZnO Membrane: Influence of ZnO Nanoparticles on Membrane Characteristics. *Adv. Polym. Technol.* **2013,** *32*. doi:10.1002/adv.21369.

22. Zodrow, K.; Brunet, L.; Mahendra, S.; Li, D.; Zhang, A.; Li, Q.; Alvarez, P. J. J. Polysulfone Ultrafiltration Membranes Impregnated with Silver Nanoparticles Show Improved Biofouling Resistance and Virus Removal. *Water Res.* **2009,** *43*, 715–723. doi:10.1016/J.WATRES.2008.11.014.

23. Liao, C.; Yu, P.; Zhao, J.; Wang, L.; Luo, Y. Preparation and Characterization of NaY/ PVDF Hybrid Ultrafiltration Membranes Containing Silver Ions as Antibacterial Materials. *Desalination* **2011,** *272*, 59–65. doi:10.1016/J.DESAL.2010.12.048.

24. Jewrajka, S. K.; Haldar, S. Amphiphilic Poly(acrylonitrile-co-acrylic acid)/Silver Nanocomposite Additives for the Preparation of Antibiofouling Membranes with Improved Properties. *Polym. Compos.* **2011,** *32*, 1851–1861. doi:10.1002/pc.21218.

25. Zahid, M.; Rashid, A.; Akram, S.; Rehan, Z. A.; Razzaq, W. A Comprehensive Review on Polymeric Nano-Composite Membranes for Water Treatment. *J. Membr. Sci. Technol.* **2018,** *8*, 1–20. doi:10.4172/2155-9589.1000179.

26. Zhang, M.; Zhang, K.; De Gusseme, B.; Verstraete, W. Biogenic Silver Nanoparticles (bio-Ag0) Decrease Biofouling of Bio-Ag0/PES Nanocomposite Membranes. *Water Res.* **2012,** *46*, 2077–2087. doi:10.1016/J.WATRES.2012.01.015.

27. Liu, Y.; Rosenfield, E.; Hu, M.; Mi, B. Direct Observation of Bacterial Deposition on and Detachment from Nanocomposite Membranes Embedded with Silver Nanoparticles. *Water Res.* **2013**, *47*, 2949–2958. doi:10.1016/J.WATRES.2013.03.005.

28. Alpatova, A.; Kim, E. -S.; Sun, X.; Hwang, G.; Liu, Y.; Gamal El-Din, M. Fabrication of Porous Polymeric Nanocomposite Membranes with Enhanced Anti-Fouling Properties: Effect of Casting Composition. *J. Memb. Sci.* **2013**, *444*, 449–460. doi:10.1016/J. MEMSCI.2013.05.034.

29. Kim, S. H.; Kwak, S.- Y. ; Sohn, B.-H.; Park, T. H. Design of TiO2 Nanoparticle Self-Assembled Aromatic Polyamide Thin-Film-Composite (TFC) Membrane as an Approach to Solve Biofouling Problem. *J. Memb. Sci.* **2003**, *211*, 157–165. doi:10.1016/ S0376-7388(02)00418-0.

30. Bae, T. –H. Tak, T. -M. Preparation of TiO2 Self-Assembled Polymeric Nanocomposite Membranes and Examination of Their Fouling Mitigation Effects in a Membrane Bioreactor System. *J. Memb. Sci.* **2005**, *266*, 1–5. doi:10.1016/J.MEMSCI.2005.08.014.

31. Luo, M. -L.; Zhao, J. -Q.; Tang, W.; Pu, C. -S. Hydrophilic Modification of Poly(ether sulfone) Ultrafiltration Membrane Surface by Self-Assembly of TiO2 Nanoparticles. *Appl. Surf. Sci.* **2005**, *249*, 76–84. doi:10.1016/J.APSUSC.2004.11.054.

32. Ngang, H. P.; Ooi, B. S.; Ahmad, A. L.; Lai, S. O. Preparation of PVDF–TiO2 Mixed-Matrix Membrane and its Evaluation on Dye Adsorption and UV-Cleaning Properties. *Chem. Eng. J.* **2012**, *197*, 359–367. doi:10.1016/J.CEJ.2012.05.050.

33. Han, Y.; Xu, Z.; Gao, C. Ultrathin Graphene Nanofiltration Membrane for Water Purification. *Adv. Funct. Mater.* **2013**, *23*, 3693–3700. doi:10.1002/adfm.201202601.

34. Ghanbari, M.; Emadzadeh, D.; Lau, W. J.; Matsuura, T.; Davoody, M.; Ismail, A. F. Super Hydrophilic TiO2/HNT Nanocomposites as a New Approach for Fabrication of High Performance Thin Film Nanocomposite Membranes for FO Application. *Desalination* **2015**, *371*, 104–114. doi:10.1016/J.DESAL.2015.06.007.

35. Ben-Sasson, M.; Zodrow, K. R.; Genggeng, Q.; Kang, Y.; Giannelis, E. P.; Elimelech, M. Surface Functionalization of Thin-Film Composite Membranes with Copper Nanoparticles for Antimicrobial Surface Properties. *Environ. Sci. Technol.* **2014**, *48*, 384–393. doi:10.1021/es404232s.

36. Yin, J.; Yang, Y.; Hu, Z.; Deng, B. Attachment of Silver Nanoparticles (AgNPs) onto Thin-Film Composite (TFC) Membranes Through Covalent Bonding to Reduce Membrane Biofouling. *J. Memb. Sci.* **2013**, *441*, 73–82. doi:10.1016/J.MEMSCI.2013.03.060.

37. Park, S. Y.; Chung, J. W.; Chae, Y. K.; Kwak, S. -Y. Amphiphilic Thiol Functional Linker Mediated Sustainable Anti-Biofouling Ultrafiltration Nanocomposite Comprising a Silver Nanoparticles and Poly(vinylidene fluoride) Membrane. *ACS Appl. Mater. Interfaces.* **2013**, *5*, 10705–10714. doi:10.1021/am402855v.

38. Zhu, X.; Bai, R.; Wee, K. -H.; Liu, C.; Tang, S. -L. Membrane Surfaces Immobilized with Ionic or Reduced Silver snd Their Anti-Biofouling Performances. *J. Memb. Sci.* **2010**, *363*, 278–286. doi:10.1016/J.MEMSCI.2010.07.041.

39. Park, H. J.; Kim, J.; Chang, J. Y.; Theato, P. Preparation of Transparent Conductive Multilayered Films Using Active Pentafluorophenyl Ester Modified Multiwalled Carbon Nanotubes. *Langmuir* **2008**, *24*, 10467–10473. doi:10.1021/la801341t.

40. Wang, N.; Zhang, G.; Ji, S.; Qin, Z.; Liu, Z. The Salt-, pH- and Oxidant-Responsive Pervaporation Behaviors of Weak Polyelectrolyte Multilayer Membranes. *J. Memb. Sci.* **2010**, *354*, 14–22. doi:10.1016/J.MEMSCI.2010.03.002.

41. Liu, X.; Qi, S.; Li, Y.; Yang, L.; Cao, B.; Tang, C. Y. Synthesis and Characterization of Novel Antibacterial Silver Nanocomposite Nanofiltration and Forward Osmosis Membranes Based on Layer-By-Layer Assembly. *Water Res.* **2013,** *47,* 3081–3092. doi:10.1016/J.WATRES.2013.03.018.

42. Liu, Q.; Xu, G. -R. Graphene Oxide (GO) as Functional Material in Tailoring Polyamide Thin Film Composite (PA-TFC) Reverse Osmosis (RO) Membranes. *Desalination* **2016,** *394,* 162–175. doi:10.1016/J.DESAL.2016.05.017.

43. Freger, V.; Gilron, J.; Belfer, S. TFC Polyamide Membranes Modified by Grafting of Hydrophilic Polymers: An FT-IR/AFM/TEM Study. *J. Memb. Sci.* **2002,** *209,* 283–292. doi:10.1016/S0376-7388(02)00356-3.

44. Hu, M.; Mi, B. Enabling Graphene Oxide Nanosheets as Water Separation Membranes. *Environ. Sci. Technol.* **2013,** *47,* 3715–3723. doi:10.1021/es400571g.

45. Wang, L.; Wang, N.; Li, J.; Bian, W.; Ji, S. Layer-By-Layer Self-Assembly of Polycation/ GO Nanofiltration Membrane with Enhanced Stability and Fouling Resistance. *Sep. Purif. Technol.* **2016,** *160,* 123–131. doi:10.1016/J.SEPPUR.2016.01.024.

46. Wang, Z.; Ma, H.; Hsiao, B. S.; Chu, B. Nanofibrous Ultrafiltration Membranes Containing Cross-Linked Poly(ethylene glycol) and Cellulose Nanofiber Composite Barrier Layer. *Polymer (Guildf)* **2014,** *55,* 366–372. doi:10.1016/J.POLYMER.2013.10.049.

47. Tung, K. -L.; Chang, K. -S.; Wu, T. -T.; Lin, N. -J.; Lee, K. -R.; Lai, J. -Y. Recent Advances in the Characterization of Membrane Morphology. *Curr. Opin. Chem. Eng.* **2014,** *4,* 121–127. doi:10.1016/J.COCHE.2014.03.002.

48. Johnson, D. Characterisation and Quantification of Membrane Surface Properties Using Atomic Force Microscopy: A Comprehensive Review. *Desalination* **2015,** *356,* 149–164. doi:10.1016/J.DESAL.2014.08.019.

49. Wyart, Y.; Tamime, R.; Siozade, L.; Baudin, I.; Glucina, K.; Deumié, C.; Moulin, P. Morphological Analysis of Flat and Hollow Fiber Membranes by Optical and Microscopic Methods as a Function of the Fouling. *J. Memb. Sci.* **2014,** *472,* 241–250. doi:10.1016/J.MEMSCI.2014.08.012.

50. Khanukaeva, D. Y.; Filippov, A. N.; Bildyukevich, A. V. An AFM Study of Ultrafiltration Membranes: Peculiarities of Pore Size Distribution. *Pet. Chem.* **2014,** *54,* 498–506. doi:10.1134/S0965544114070068.

51. Singh, K.; Devi, S.; Bajaj, H. C.; Ingole, P.; Choudhari, J.; Bhrambhatt, H. Optical Resolution of Racemic Mixtures of Amino Acids through Nanofiltration Membrane Process. *Sep. Sci. Technol.* **2014,** *49,* 2630–2641. doi:10.1080/01496395.2014.911023.

52. Khulbe, K. C.; Matsuura, T.; Kim, H. J. Raman Scattering of PPO Membranes. *J. Appl. Polym. Sci.* **2000,** *77,* 2558–2560. doi:10.1002/1097-4628(20000912)77:11<2558::AID-APP25>3.0.CO;2-Y.

53. Lamsal, R.; Harroun, S. G.; Brosseau, C. L.; Gagnon, G. A. Use of Surface Enhanced Raman Spectroscopy for Studying Fouling on Nanofiltration Membrane. *Sep. Purif. Technol.* **2012,** *96,* 7–11. doi:10.1016/J.SEPPUR.2012.05.019.

54. Chen, P.; Cui, L.; Zhang, K. Surface-Enhanced Raman Spectroscopy Monitoring the Development of Dual-Species Biofouling on Membrane Surfaces. *J. Memb. Sci.* **2013,** *473,* 36–44. doi:10.1016/J.MEMSCI.2014.09.007.

55. Chan, R.; Chen, V. Characterization of Protein Fouling on Membranes: Opportunities and Challenges. *J. Memb. Sci.* **2004,** *242,* 169–188. doi:10.1016/J.MEMSCI.2004.01.029.

56. Lee, A.; Elam, J. W.; Darling, S. B. Membrane Materials for Water Purification: Design, Development, and Application. *Environ. Sci. Water Res. Technol.* **2016**, *2*, 17–42. doi:10.1039/C5EW00159E.

57. Jonsson, G. Molecular Weight Cut-Off Curves for Ultrafiltration Membranes of Varying Pore Sizes. *Desalination* **1985**, *53*, 3–10. doi:10.1016/0011-9164(85)85048-7.

58. Tetala, K. K. R.; Stamatialis, D. F. Mixed Matrix Membranes for Efficient Adsorption of Copper Ions From Aqueous Solutions. *Sep. Purif. Technol.* **2013**, *104*, 214–220. doi:10.1016/J.SEPPUR.2012.11.022.

59. Gholami, A.; Moghadassi, A. R.; Hosseini, S. M.; Shabani, S.; Gholami, F. Preparation and Characterization of Polyvinyl Chloride Based Nanocomposite Nanofiltration-Membrane Modified by Iron Oxide Nanoparticles for Lead Removal from Water. *J. Ind. Eng. Chem.* **2014**, *20*, 1517–1522. doi:10.1016/J.JIEC.2013.07.041.

60. Gohari, R. J.; Lau, W. J.; Matsuura, T.; Ismail, A. F. Fabrication and Characterization of Novel PES/Fe–Mn Binary Oxide UF Mixed Matrix Membrane for Adsorptive Removal of As(III) from Contaminated Water Solution. *Sep. Purif. Technol.* **2013**, *118*, 64–72. doi:10.1016/J.SEPPUR.2013.06.043.

61. Rahimpour, A.; Madaeni, S. S.; Taheri, A. H.; Mansourpanah, Y. Coupling TiO2 Nanoparticles with UV Irradiation for Modification of Polyethersulfone Ultrafiltration Membranes. *J. Memb. Sci.* **2008**, *313*, 158–169. doi:10.1016/J.MEMSCI.2007.12.075.

62. Damodar, R. A.; You, S. -J.; Chou, H. -H. Study the Self Cleaning, Antibacterial and Photocatalytic Properties of TiO2 Entrapped PVDF Membranes. *J. Hazard. Mater.* **2009**, *172*, 1321–1328. doi:10.1016/J.JHAZMAT.2009.07.139.

63. Liu, Y.; Wang, X.; Yang, F.; Yang, X. Excellent Antimicrobial Properties of Mesoporous Anatase TiO2 and Ag/TiO2 Composite Films. *Microporous Mesoporous Mater.* **2008**, *114*, 431–439. doi:10.1016/J.MICROMESO.2008.01.032.

64. Chou, W. -L.; Yu, D. -G.; Yang, M. -C. The Preparation and Characterization of Silver-Loading Cellulose Acetate Hollow Fiber Membrane for Water Treatment. *Polym. Adv. Technol.* **2005**, *16*, 600–607. doi:10.1002/pat.630.

65. van Reis, R.; Zydney, A. Bioprocess Membrane Technology. *J. Memb. Sci.* **2007**, *297*, 16–50. doi:10.1016/J.MEMSCI.2007.02.045.

66. Pendergast, M. M.; Ghosh, A. K.; Hoek, E. M. V. Separation Performance and Interfacial Properties of Nanocomposite Reverse Osmosis Membranes. *Desalination* **2013**, *308*, 180–185. doi:10.1016/J.DESAL.2011.05.005.

67. McCutcheon, J. R.; Elimelech, M. Influence of Concentrative and Dilutive Internal Concentration Polarization on Flux Behavior in Forward Osmosis. *J. Memb. Sci.* **2006**, *284*, 237–247. doi:10.1016/J.MEMSCI.2006.07.049.

68. Ma, N.; Wei, J.; Qi, S.; Zhao, Y.; Gao, Y.; Tang, C. Y. Nanocomposite Substrates for Controlling Internal Concentration Polarization in Forward Osmosis Membranes. *J. Memb. Sci.* **2013**, *441*, 54–62. doi:10.1016/j.memsci.2013.04.004.

69. Emadzadeh, D.; Lau, W. J.; Ismail, A. F. Synthesis of Thin Film Nanocomposite Forward Osmosis Membrane with Enhancement in Water Flux Without Sacrificing Salt Rejection. *Desalination* **2013**, *330*, 90–99. doi:10.1016/J.DESAL.2013.10.003.

70. Nghiem, L. D.; Schäfer, A. I.; Elimelech, M. Removal of Natural Hormones by Nanofiltration Membranes: Measurement, Modeling, and Mechanisms. *Environ. Sci. Technol.* **2004**, *36* (6), 1888–1896. doi:10.1021/ES034952R.

71. Kimura, K.; Amy, G.; Drewes, J.; Watanabe, Y. Adsorption of Hydrophobic Compounds Onto NF/RO Membranes: An Artifact Leading to Overestimation of Rejection. *J. Memb. Sci.* **2003**, *221*, 89–101. doi:10.1016/S0376-7388(03)00248-5.
72. Snyder, S. A.; Adham, S.; Redding, A. M.; Cannon, F. S.; DeCarolis, J.; Oppenheimer, J.; Wert, E. C.; Yoon, Y. Role of Membranes and Activated Carbon in the Removal of Endocrine Disruptors and Pharmaceuticals. *Desalination* **2007**, *202*, 156–181. doi:10.1016/J.DESAL.2005.12.052.

CHAPTER 9

MICROFILTRATION, ULTRAFILTRATION, AND OTHER MEMBRANE SEPARATION PROCESSES: A CRITICAL OVERVIEW AND A VISION FOR THE FUTURE

SUKANCHAN PALIT*

43, Judges Bagan, Post-Office - Haridevpur, Kolkata-700082, India

Corresponding author. E-mail: sukanchan68@gmail.com, sukanchan92@gmail.com

ABSTRACT

The domain of environmental engineering science and chemical engineering is moving at a drastic pace and is in the midst of deep scientific rejuvenation. Stringent environmental engineering regulations, the loss of ecological biodiversity, and the ever-growing concerns for global climate change has urged the scientific domain to gear forward toward newer scientific innovations and environmental engineering techniques. Thus the need of novel separation processes such as membrane science and traditional and non-traditional techniques such as advanced oxidation processes. Scientific vision, vast scientific ingenuity, and scientific provenance will all today lead a long and visionary way in the true emancipation of environmental engineering and environmental sustainability. In this well researched treatise, the author deeply elucidates on membrane separation processes particularly microfiltration and ultrafiltration. Other membrane science techniques are also discussed and described in minute details. The other salient features of this treatise are the areas of heavy metal and arsenic groundwater remediation. A critical overview of environmental sustainability and the visionary area of sustainable development stands as major pillars of this treatise. The author of this treatise depicts profoundly the scientific success and the scientific inquiry in the field of various membrane separation processes.

9.1   INTRODUCTION

Human civilization and human scientific progress are today in the path of newer rejuvenation and successful regeneration. Environmental protection science and environmental remediation technologies are today challenging the vast scientific firmament of deep might and vision. Global climate change, global water scarcity, and the vast issue of loss of ecological biodiversity are the veritable forerunners towards a newer world of scientific innovation and deep scientific instinct in the field of environmental engineering science. Arsenic and heavy metal groundwater contamination are deeply challenging the global scientific community. There are practically no answers to this monstrous scientific issue. Thus the imminent need of traditional and non-traditional environmental engineering techniques. In this chapter, the author elucidates the salient features, the techniques, the applications, and the futuristic vision of membrane science applications in environmental engineering and chemical process engineering. Science, technology, and engineering have no answers to the monstrous and marauding issues of arsenic and heavy metal drinking water and groundwater contamination in South Asia. Thus comes the importance of membrane science and advanced oxidation processes.

9.2   THE AIM AND OBJECTIVE OF THIS STUDY

Technology and engineering science of environmental protection and membrane science are in the vistas of deep scientific introspection. Today, membrane science and other novel separation processes are veritably challenging the vast scientific fabric of might and vision. The aim and objective of this study is to delineate the needs of conventional and non-conventional environmental engineering techniques in mitigating global water scarcity and industrial water pollution. Arsenic and heavy metal groundwater contamination are the monstrous scientific and engineering issues in many developing and developed nations around the world. This chapter will target these issues. The authors in this treatise rigorously points toward the scientific success, the scientific vision, and the vast scientific profundity in environmental engineering applications in water purification and wastewater treatment.

Ultrafiltration (UF), microfiltration (MF), and other membrane separation processes are the other pivots of this research endeavor. The vision and the scientific challenges has no bounds as regards environmental protection. The salient features of this treatise is to target novel separation processes and other non-conventional environmental engineering tools. Human civilization's immense scientific prowess, scientific imagination, and vast scientific sagacity will thus open new doors of innovation in environmental remediation.

9.3 WHAT DO YOU MEAN BY MICROFILTRATION?

Microfiltration is a type of separation process where a contaminated fluid is passed through a pore sized membrane to separate microorganisms and suspended particles from the liquid stream and wastewater. It is prominently used in conjunction with reverse osmosis and UF. The array of applications are water treatment, sterilization, petroleum refining, and dairy processing. The subtleties of science, engineering, and technological validation and the futuristic vision of membrane science will all lead a long and visionary way in the true realization of environmental remediation. For these reasons, MF stands as a pre-eminent technique. In the food and beverage industry, MF has vital applications. Process design and process engineering of membrane separation processes are today in the vistas of new scientific regeneration. This chapter will be a veritable eye opener to the vast scientific intricacies of fouling phenomenon of membrane science.

9.4 WHAT DO YOU MEAN BY ULTRAFILTRATION?

Ultrafiltration is a variety of membrane filtration in which forces like pressure and concentration gradients lead to separation through a semipermeable membrane. Suspended solutes of high-molecular weight are retained on the so-called retentate, while water and low-molecular solutes pass through the membrane in the permeate. The range of applications are drinking water purification, protein concentration, filtration of effluent from paper pulp mill, enzyme recovery, and waste water treatment. Scientific and engineering validation and prowess will all be the forerunners toward a new era in chemical process engineering. The vision, the subtleties, and the ingenuity of chemical process technology will surely go a long and effective way in the true realization of science and engineering.

9.5   THE VISION AND THE SCIENTIFIC DOCTRINE OF MEMBRANE SEPARATION PROCESSES

Membrane science and water purification are the challenges and the vision of environmental protection globally today. The scientific doctrine and the scientific sagacity of environmental protection science needs to be reorganized and revamped with the passage of time. Membrane separation processes are the marvels of science and are directly linked with greater emancipation of environmental sustainability. Technological candor, vast scientific profundity, the vast world of scientific validation and prowess will lead a long and visionary way in the true realization of environmental sustainability globally. A membrane is a selective barrier, it allows some things to pass through but stops others. Such substances can be molecules, ions, or other small particles. Biological membranes include cell membranes, nuclear membranes, and tissue membranes. Today, membrane science is linked with environmental engineering science and environmental protection. The challenges, the scientific success, and the vast scientific ingenuity in membrane science applications will surely open up new doors of innovation and scientific intuition in decades to come. Membrane technology encompasses all engineering approaches for the transport of substances between two fractions with the help of permeable membranes. In general, mechanical separation processes for separating gaseous and liquid streams use membrane technology. The challenges of membrane science and membrane engineering are immense, versatile, and ground-breaking today in the global scientific scenario. Membrane separation processes operate without heating and less energy than conventional thermal separation processes such as distillation, sublimation, or crystallization. This is the target and the vision of research endeavor in membrane science today. The basic models of mass transfer in membrane science are the (1) the solution–diffusion model and (2) the hydrodynamic models. Diffusion phenomenon in membranes is still unclear but far reaching. Here comes the importance of membrane science applications in environmental protection and chemical process engineering. Human civilization's vast scientific prowess and knowledge, the futuristic vision of chemical process engineering, and the needs of human society will go a long and effective way in the true emancipation of key global issues such as environmental sustainability. Environmental remediation and sustainability are two opposite sides of the visionary coin. This chapter will surely open a newer vista in the field of environmental sustainability, sustainable development, and environmental remediation.

9.6   THE WORLD OF CHALLENGES IN THE FIELD OF WATER PURIFICATION

Water purification and drinking water treatment are the imminent needs of the hour. The challenges and the vision of environmental and energy sustainability are immense and far reaching. Water purification science and environmental remediation are the veritable needs of human civilization and human scientific progress today. Teeming millions around the developing and developed world have no access to pure drinking water. In Bangladesh and the state of West Bengal, India, the environmental situation is grave and ever concerning as arsenic heavy metal groundwater poisoning destroys and devastates the vast scientific firmament and the scientific might. The success of technology and engineering science, the candor and scientific subtleties of environmental engineering, and the futuristic vision of membrane technology will all lead a long and visionary way in the true emancipation of energy and environmental sustainability. Renewable energy scenario in the global scientific platform is immensely grave and thought provoking. Technology and engineering has practically no answers to the intricate questions of drinking water treatment, groundwater remediation, and renewable energy. Here comes the importance of energy and environmental sustainability. The visionary words of Dr Gro Harlem Brundtland, former Prime Minister of Norway on the science of "sustainability" needs to be envisioned and reorganized as human civilization surges forward.

9.7   SCIENTIFIC VISION AND DEEP SCIENTIFIC INGENUITY IN THE FIELD OF GROUNDWATER REMEDIATION

Groundwater remediation and drinking water treatment are challenging the face of human civilization today. Scientific provenance, scientific subtleties, and vast scientific cognizance stands as veritable pillars of research and development initiatives in the global scenario today. Technology and engineering science needs to be redrafted and re-envisioned if water scarcity and global warming are to be ameliorated. Human civilization and human scientific progress today stands in the midst of deep scientific and engineering vision and scientific forbearance. Environmental engineering and chemical process engineering in the similar vein are in the midst of ingenuity, engineering vision, and vast validation of science. Water treatment mainly groundwater remediation today needs to be rejuvenated as regards application of environmental sustainability to human society. The challenge of science in the field of biotechnology applications in water treatment also needs to be re-envisaged and reorganized with the passage of scientific history and scientific vision. Technological advancement in the poor and developing countries around the world are backward and requires lot of re-envisioning. Here comes the importance of newer innovations which are cost-effective and groundbreaking. Membrane science and other novel separation processes will surely lead a long and visionary way toward the larger emancipation of environmental engineering science and environmental sustainability. Membrane science and groundwater remediation should be the opposite sides of the visionary coin.

9.8   RECENT SCIENTIFIC RESEARCH PURSUIT IN THE FIELD OF MEMBRANE SEPARATION PROCESSES

Membrane separation processes and other novel separation processes are the necessities of human scientific progress globally today. Research and development initiatives in the field of environmental remediation are highly challenging and in the similar vein groundbreaking. The author deeply elucidates the marvels of environmental engineering science, the futuristic vision, and the vast world of novel separation processes.

Huehmer)[1] discussed with vast scientific vision and far-sightedness MF/UF pretreatment trends in seawater desalination. The use of MF/UF has been studied by researchers and the scientific domain since the mid-1990s.[1] In the mid-2000, cost reduction in these technologies resulted in the installation of MF/UF pretreatment in seawater desalination plants.[1] Since 2007, the author has been maintaining a database documenting installations of the MF/UF pretreatment in desalination technique. The pursuit of science in membrane science is today groundbreaking and surpassing vast and versatile scientific frontiers. The database includes design flux, chemical additives, backwash and cleaning frequencies, chemical usage, and recovery.[1] The true vision and forbearance of science and engineering, the success of research and development initiatives, and the world of scientific challenges will surely open up new doors of innovation in environmental remediation.[1] Pretreatment has historically been the most important technical issue of desalination plants utilizing open or direct intakes. A deep comprehension and an autopsy shows that 51% of membrane failures are associated with deficient pretreatment, an additional 12% related to the chemical dose of coagulants and flocculants and 30% is due to oxidative processes.[1] The futuristic vision of environmental engineering science, the scientific prowess of environmental remediation, and the scientific perseverance will all lead a long and visionary way in the true emancipation of science and engineering today.[1]

Saboyainsta et al.[2] discussed with deep scientific insight current developments of MF technology in the dairy industry. This chapter vastly deals with the most recent developments of cross-flow MF, some of them are patented in the dairy industry. Integration of the use of uniform transmembrane hydraulic pressure concept with its different ways of carrying out microfiltrate recirculation and longitudinal porosity gradient.[2] Pretreatment by cross-flow MF of incoming milk is used for the production of low heated fluid milks having a distinct and similar flavor similar to that of raw milk and a veritable shelf life 3–5 times longer than that of classical products.[2] Engineering science of MF are in the midst of vision, provenance, and deep scientific revelation. Numerous other applications of cross-flow MF are in the process of immense research and development such as the removal of residual fat from whey or the clarification and the removal of bacteria from cheeses brine.[2] With the vast array of products obtained on both sides of the MF membrane, dairy technology will have the immense vision to improve yield and quality of many dairy products.[2] The authors in this treatise reviewed with vast scientific vision membranes, MF equipment considerations, cleaning and sanitation of the MF equipment, and the vast applications of MF in the dairy industry.[2] A brief idea of selective separation of micellar casein and selective fractionation of globular fat are dealt with minute comprehension in this paper. The other pillars of this treatise are the area of cheese brine purification. With the use of membrane MF technology, the dairy industry has today a powerful and economical environmental and chemical process engineering tool.[2] The challenges and the vision of chemical process engineering, the futuristic vision of membrane science and MF are immense and groundbreaking. The authors with immense scientific far sightedness pronounce the success of cross-flow MF in dairy industry. The other areas of this well researched treatise are the fouling of MF membranes.[2]

Cheryan[3] discussed and deliberated with vast scientific vision the concepts of UF and MF. The development of the Sourirajan-Loeb membrane in 1960 provided a valuable and visionary environmental and chemical process engineering tool. In the beginning it faced immense hurdles.[3] The situation today is immensely different: membranes are more robust, modules and equipment are better designed, and the scientific community has a better understanding of the fouling phenomenon. The authors described with vision and scientific ingenuity membrane chemistry, structure, function, membrane properties, performance and engineering models, fouling and cleaning, process design, and the vast and varied applications of UF and MF.[3] Filtration is defined as the separation of two or more components from a fluid stream totally based on size and shape of the particles. The primary role of a membrane is to act as a veritable selective barrier. Membranes can be further classified by (1) nature of the membrane, (2) structure of the membrane, (3) application of the membrane, and (4) mechanism of the membrane action.[3]] The other function of the membrane is that it physically or chemically modifies the permeating species, conducts electric current, prevents permeation, or regulates the rate of permeation. In this entire treatise, the author deeply comprehends with vision and insight the necessity of membrane science in dairy industry, food and beverage industry, and the areas of water purification. Science of chemical process engineering and environmental remediation are in the path of newer scientific regeneration. This book will surely be a scientific eye opener toward the needs of membrane science in the true emancipation of environmental sustainability.[3]

9.9   RECENT SCIENTIFIC RESEARCH ENDEAVOR IN THE FIELD OF GROUNDWATER REMEDIATION

Groundwater remediation and environmental protection are today in the vistas of deep scientific introspection and vast scientific vision. In this well researched treatise, the author with immense lucidity and scientific farsightedness deals with the wide array of environmental engineering and chemical process engineering tools. Human scientific progress and the academic rigor behind groundwater remediation need to be re-envisioned and reorganized as civilization, mankind, and modern science surges forward. In this section, the author describes in minute details some of the environmental engineering problems faced by the scientific community in the field of groundwater remediation, global water shortage, and the monstrous issue of global climate change.

Mukherjee et al.[4] discussed with cogent insight assessment and remedies of contemporary environmental issues of landfill leachate. Landfills are the predominant option for waste disposal all around the world.[4] Most of the landfill sites around the world are old and not engineered to prevent contamination of the underlying soil and groundwater by the polluted leachate. The polluted leachates have accumulative and detrimental effect on the ecology and food chains leading to serious health effects and carcinogenic effects to human beings. Management of these environmental engineering problems is challenging to the regulatory authorities who have set specific regulations regarding the maximum limits of contaminants and hazardous materials. This review pointedly focuses on: (1) leachate composition, (2) Plume migration, (3) contaminant fate, (4) leachate plume monitoring techniques, (5) risk assessment techniques, (6) mathematical modeling techniques, and (7) recent innovations and recent scientific intuition in leachate treatment technologies.[4] The vast scientific prowess of environmental remediation, the futuristic vision of chemical process engineering, and integrated waste management techniques and the needs of human civilization will lead a long and visionary way in the true emancipation of environmental sustainability.[4] Leachate management and industrial pollution control today stands in the midst of deep scientific comprehension and vast scientific vision. Due to seasonal fluctuations in leachate composition, flow rate, and leachate volume, the management options also need to show variations.[4] Here comes the best management option and a sound environmental engineering technique. The world of challenges and the definite vision of environmental remediation will surely open new doors of innovation and scientific instinct in the field of waste management techniques globally today.[4] The authors in this review discussed with lucid insight landfill leachate, characteristics and regulatory limits, leachate plume migration, monitoring of plume generation and migration, environmental impact of landfill leachate and its assessment, hazard assessment of landfill leachate, and recent technological advancements in landfill leachate treatment and remediation.[4] The areas of further discussion include advanced oxidation treatments, adsorption, coagulation–flocculation, electrochemical treatment and filtration, and membrane bioreactors. Landfill leachate is defined as any liquid effluent containing undesirable materials percolating through waste and emitted within a landfill and dump site. Often, its route of exposure and toxicity remains unknown and a deep matter of prediction is the utmost need of the hour. This is due to complicated geochemical processes in the landfill and the soil layers.[4] Leachate presents high values of biochemical oxygen demand, chemical oxygen demand, total organic carbon, total suspended solids, total dissolved solids, recalcitrant organic and inorganic compounds, sulfur compounds, and dissolved organic matter.[4] The vision of science, the challenges of industrial waste management, and the success of environmental sustainability will surely be the forerunners toward a new era in environmental engineering.[4] Around 200 hazardous compounds have already been identified in landfill leachate such as aromatic compounds, halogenated compounds, phenols, pesticides, heavy metals, and ammonium. This well researched treatise opens up newer future thoughts, futuristic vision, and futuristic recommendations in the landfill leachate management.[4]

Hashim et al.[5] discussed and described with cogent insight remediation technologies for heavy metal contaminated groundwater. The widespread contamination of groundwater by heavy metal originating either from natural soil resources or from anthropogenic sources is a matter of immense concern to public health engineering.[5] Technological and scientific validation, the success of remediation science, and the futuristic vision of environmental protection will surely be the forerunners toward a newer era in the field of science and technology.[5] The authors in this deeply pronounce the scientific success and the scientific needs of groundwater remediation in human society in many developing countries around the world. In this well researched treatise, the authors deeply discussed sources, chemical property, and speciation of heavy metals in groundwater and drinking water. The other focal points of this research endeavor are chemical treatment technologies, in situ treatment by using reductants, reduction by dithionite, reduction by gaseous hydrogen sulphide, reduction by using iron-based technologies, removal of chromium by ferrous salts, soil washing, in situ soil washing, in situ chelate flushing, biological, biochemical, and biosorptive treatment technologies.[5] The other areas of research pursuit are enhanced biorestoration and physicochemical treatment technologies. Adsorption, filtration, and absorption mechanisms are the other veritable pillars of this treatise.[5] Heavy metals are extremely toxic for living beings and they highly pollute drinking water and groundwater. Complex speciation chemistry is the reason behind the concept of recalcitrance of the organic and inorganic compounds. Once they get into the soil subsurface or in groundwater, it becomes extremely difficult to handle them due to the complex organic chemistry of the recalcitrant organic and inorganic compounds. In this chapter, the authors discussed various treatment techniques such as chemical, biological/biochemical/biosorptive, and physicochemical treatment techniques.[5] There was a critical discussion on the various chemical treatment technologies, biological, biochemical, and biosorptive technologies.[5] Groundwater treatment technologies have come a long way since the beginning of research pursuit in this field. Much research and scientific deliberations have been done on numerous technologies ranging from simple ex situ physical separation techniques to complex in situ microbiological and adsorption techniques.[5]

Shannon et al.[6] discussed with vast scientific foresight science and technology for water purification in years to come. One of the ever-growing problems afflicting people around the world is inadequate access to clean drinking water and proper sanitation. Problems with water will be worse in future with water scarcity going to be alarming in both developing as well as developed nations around the world.[6] Thus the need of scientific vision, vast scientific grit, and determination. In order to address these problems, tremendous amount of research endeavor needs to be done as human civilization and scientific acumen moves forward toward a newer visionary era. Robust new methods of purifying water at lower cost and with less energy while at the same time minimizing the use of chemicals will be the major objective of research endeavor in the global scenario. Thus the need of the application of environmental sustainability and environmental engineering in human civilization and human society.[6] The problems associated with the lack of provision of clean drinking water are well known: 1.2 billion people lack safe access to drinking water, 2.6 billion people have little or no sanitation, millions of people die annually, 3900 children die annually from disease and contamination. Science, technology, and engineering have practically no answers to the ever-growing concerns of arsenic or heavy metal groundwater and drinking water contamination.[6] Thus human scientific progress and academic rigor in the field of water purification stands in the midst of deep engineering vision and immense scientific validation. The authors in this chapter discussed with vision disinfection, decontamination, and reuse and reclamation. Desalination is the other pillar of this entire treatise.[6] The vast scientific research pursuit highlighted here plus the tremendous amount of additional research endeavor being conducted in every continent that could not be mentioned is sowing the seeds of a new revolution in the field of environmental engineering science and chemical process engineering. The vast enormity of the problem facing the world from the lack of adequate water supply and sanitation means lot of research pursuit needs to be done considering the socioeconomical–political–traditional constraints and also require a broader approach toward the application of energy and environmental sustainability in human society. Water science and technology today stands in the crucial juncture of vision, scientific forbearance, and immense scientific comprehension. The crucial success of science and engineering of water purification are deeply discussed in this chapter.[6]

Hassan[7] deeply discussed with lucid and cogent insight poisoning and risk assessment in the field of arsenic contamination of groundwater and drinking water. The author deeply discussed the areas of the global scenario and groundwater arsenic catastrophe, groundwater arsenic discontinuity, spatial mapping, spatial planning, and public participation. Environmental health concerns due to chronic arsenic exposure to drinking water are also discussed in this book.[7] Arsenic induced health and social hazard and survival strategies as well as experiences from arsenicosis patients are deeply deliberated in this well researched treatise.[7] Arsenic poisoning in Bangladesh and legal issues of responsibility also are discussed with vast scientific vision and deep scientific provenance. Groundwater is the main source of safe drinking water in many countries of the world but much of that groundwater is contaminated with arsenic and heavy metal. It is highly ironic that so many tubewells have been installed for drinking water that are safe from water-borne diseases but highly contaminated with high levels of arsenic. Here comes the irony of science and technology in the global scenario. It is estimated that more than 300 million people in 70 countries worldwide are at a risk of arsenic groundwater poisoning. Thus the need of scientific innovations and deep scientific intuition. The contribution of civil society and the scientific community is thus the need of the hour. This treatise is an eye opener toward the scientific needs of groundwater remediation techniques with a sole vision toward furtherance of science and engineering globally.[7]

Human mankind and human scientific progress today are in the path of newer scientific regeneration. Water science and technology are in the similar vein in the new vistas of regeneration and vision. Groundwater heavy metal contamination is a bane of human civilization today. The author in this book deeply deals with the necessities of water purification tools such as membrane science and advanced oxidation processes which will surely widen future thoughts and future recommendations.

9.10 THE VAST SCIENTIFIC DOCTRINE IN DRINKING WATER TREATMENT

The vast scientific doctrine and the scientific sagacity of drinking water treatment and groundwater treatment stands in the midst of deep scientific introspection. Environmental and energy sustainability are the imminent needs of human civilization today. Immense academic and scientific rigor in the field of sustainability science, the scientific adjudication of science and technology, and futuristic vision of environmental engineering and systems engineering will surely open up newer avenues of environmental remediation in decades to come. The provision of basic human needs such as water and energy are today facing vast scientific challenges and engineering hurdles. In South Asia particularly India and Bangladesh, arsenic drinking water contamination is challenging the vast scientific firmament. Technology, engineering, and science have few answers to the ravaging and ever-growing concerns of drinking water heavy metal contamination. Thus the need of effective environmental engineering techniques such as membrane science and advanced oxidation processes. In the similar vein and similar vision, industrial wastewater treatment and industrial pollution control assumes immense importance as human civilization and human scientific progress moves forward. The application of the science of sustainability also is the scientific imperative of research and development initiatives in water science and technology. The vast scientific prowess of human mankind, the sagacity of engineering science and the futuristic vision of environmental engineering tools will surely open new doors of innovation and intuition in the field of environmental sustainability. Scientific sagacity and vast scientific ingenuity are the veritable pillars of research and development initiatives in water science and technology today. Health effects of arsenic groundwater contamination are disastrous and thought provoking to the global scientific community. In this well researched treatise, the author pointedly focuses on the scientific success, the technological and engineering inquiry, and the futuristic vision of both chemical process engineering and environmental engineering. The applications of chemical process engineering in environmental protection are groundbreaking and surpassing vast scientific frontiers. This treatise will surely go a long and visionary way in the true emancipation of membrane science, its various branches, the true realization of energy and environmental sustainability.

9.11 ENVIRONMENTAL SUSTAINABILITY, BIOREMEDIATION, AND GROUNDWATER REMEDIATION

Environmental sustainability, bioremediation, and groundwater remediation are the visions and the challenges of environmental engineering science today. Chemical process engineering, traditional and non-traditional environmental engineering tools, and novel separation processes are the marvels of engineering science and technology today. There are today few answers to energy and environmental sustainability. Application of biotechnology and biological sciences in environmental science will surely open new doors of innovation and scientific instinct in days to come. Today biological sciences and nanotechnology needs to be integrated for the greater emancipation and furtherance of science and engineering of environmental protection. Human factor engineering, technology management, and integrated water resource management are the fountainheads of research pursuit today. The vast scientific challenges associated with groundwater remediation needs to be overcome and re-envisioned with the passage of scientific vision and visionary timeframe. Human civilization and human scientific progress are today in the path of newer scientific regeneration and vast scientific rejuvenation. Provision of basic human needs such as water, electricity, energy, food, shelter, and education are highly as human civilization moves in the path of scientific might and vision. Bioremediation and biological wastewater treatment are the scientifically relevant areas of science and engineering of environmental protection in today's modern science. In the similar vision, modern science should target the areas of provision of basic human needs. In this treatise, the author deeply comprehends the success of science and technology of biological wastewater treatment, bioremediation, and biotechnology applications in environmental remediation and the success of engineering science. The world of science and engineering thus stands envisioned and mesmerized with the ever-growing concerns for drinking water contamination, global water scarcity, and global warming. Thus the need of a detailed description of the areas of novel separation processes such as membrane science in confronting global water issues.[8–11]

9.12   HUMAN FACTOR ENGINEERING, RELIABILITY ENGINEERING, AND WASTE MANAGEMENT

Technology and engineering science are today in the avenues of deep scientific introspection and vision. The scientific acumen and the vast scientific acuity in the field of human factor engineering reliability engineering needs to be re-envisioned and revamped with the passage of human history and time. Waste management, global water hiatus, and global climate change are today destroying the vast scientific fabric of immense might and vision. Today waste management and integrated water resource management should be merged with reliability engineering. Integrated waste management techniques and water resource management are replete with immense scientific intricacies. The vast challenges and the scientific vision of water purification and industrial wastewater treatment has no bounds and veritably crossing visionary frontiers. In solving the crucial issues of water resource management and waste management scientific imagination and scientific ingenuity of technology management and reliability engineering are of vital importance. In this chapter, the author deeply comprehends the success of human factor engineering and reliability engineering in tackling heavy metal drinking water contamination, waste management, and water purification.[8-11]

9.13 FUTURE SCIENTIFIC RECOMMENDATIONS AND FUTURE FLOW OF THOUGHTS

Future scientific recommendations need to target newer scientific intuition and newer scientific innovations. Technology and engineering science of environmental remediation are today in the path of immense scientific catastrophe. The challenges and the targets of environmental protection are groundbreaking and surpassing vast and versatile scientific boundaries. Today scientific barriers and scientific hurdles need to be surpassed as environmental engineering science and chemical process engineering surges forward. Future scientific recommendations should be targeted toward more application of membrane science in environmental protection. Human factor engineering, reliability engineering, and systems science needs to be integrated with water pollution control and water purification science. Integrated water resource management and its scientific vision should be the fountainheads of engineering science and technology today. Man's immense scientific grit and prowess, mankind's vast scientific determination, and the futuristic vision of environmental remediation will surely open new doors of innovation in environmental sustainability. Environmental sustainability and its applications are the needs of science and engineering today. Thus a deep discussion and a deep deliberation in the field of science of sustainability are the utmost need of the hour. The other challenges of human civilization and human scientific progress are application of human factor engineering and reliability engineering to the vast areas of environmental protection and water purification. Application of renewable energy in human society and the success of energy sustainability should be the other cornerstones of research pursuit globally today. Human mankind's immense scientific prowess, scientific acuity, and deep scientific alacrity will today surely lead a long and visionary way in the true realization of energy and environmental sustainability. This chapter will be a veritable march toward a newer generation and a newer rejuvenation in the field of environmental remediation. Developing countries such as Bangladesh and the state of West Bengal, India are in the throes of world's largest environmental catastrophe—arsenic groundwater contamination. A deep rendezvous of science and engineering and interdisciplinary areas of engineering science will surely open the veritable doors of novel separation processes and chemical process engineering.

Humanity today is in the threshold of a devastating scientific crisis—the arsenic and heavy metal groundwater and drinking water contamination in many developing and developed nations around the world. The author rigorously and rigidly focuses on the scientific needs, the scientific inquiry, and vast scientific forbearance of application of human factor engineering and systems science in environmental remediation. Thus industrial engineering merged with water resource management and water purification will be forerunners toward a newer emancipation in science and technology.[8–11]

9.14 CONCLUSION, SUMMARY, AND ENVIRONMENTAL PERSPECTIVES

Scientific grit, scientific determination, and vast scientific vision are the hallmarks of research and development initiatives in water science and technology today. Environmental engineering perspectives today stands in the midst of deep scientific introspection and ingenuity. Technology and engineering science of water purification, drinking water treatment, and industrial wastewater treatment today needs to be re-envisioned and reorganized with the passage of scientific history, the visionary timeframe, and the future vision. The challenges and the vision of water purification and groundwater remediation are vast, versatile, and far reaching. In this chapter, the author deeply urges the scientific community to gear forward toward newer innovation and newer visionary tools at the earliest. The developing world is in the midst of immense scientific hurdles and introspection. Developed and developing nations around the world in the similar vein are in the midst of a massive environmental engineering crisis—the arsenic drinking water contamination. Here comes the importance of scientific innovation and deep scientific truth. Scientific alacrity, vast scientific acuity, and the vision of engineering science will all be the cornerstones toward a newer redefining era in the field of membrane science, chemical process engineering, and environmental sciences. Health effects due to arsenic drinking water poisoning are challenging the human society and the human scientific firmament. This well researched treatise opens up newer thoughts and newer vision in the field of drinking water treatment and industrial wastewater treatment in decades to come. Human scientific trials and tribulation and engineering barriers are devastating the scientific knowhow and scientific inquiry. Environmental perspectives will thus widen the world of membrane science and novel separation processes as human civilization surges forward.

IMPORTANT WEBSITES FOR REFERENCE

https://en.wikipedia.org/wiki/Microfiltration
https://emis.vito.be/en/techniekfiche/microfiltration
www.waterprofessionals.com › Learning Center
https://www.youtube.com/watch?v=WWIzoyAwccA
https://www.academia.edu/Documents/in/Microfiltration
https://hal.archives-ouvertes.fr/hal-00895452/document
https://www.sciencedirect.com/science/article/pii/S0263876297715648
https://www.researchgate.net › ... › Blotting Techniques › Western Blot › Transfer
https://www.omicsonline.org/microfiltration-nano-filtration-and-reverse-osmosis-for-t...
www.scielo.br/pdf/bjce/v33n4/0104-6632-bjce-33-04-0783.pdf
https://researchbank.rmit.edu.au/eserv/rmit:160308/Nguyen.pdf
www.academia.edu/Documents/in/Ultrafiltration
https://www.academia.edu/Documents/in/Ultrafiltration_membrane
www.ijesd.org/papers/376-A023.pdf
www.jbc.org/content/94/2/373.full.pdf
www.insituarsenic.org/
www.insituarsenic.org/details.html

KEYWORDS

- **membrane**
- **separation**
- **microfiltration**
- **ultrafiltration**
- **water**
- **groundwater**
- **remediation**

REFERENCES

1. Huehmer, R. Microfiltration/Ultrafiltration Pretreatment Trends in Seawater Desalination. World Congress/Perth Convention and Exhibition Centre (PCEC), Perth, Western Australia, September 4–9, 2011.
2. Saboyainsta, L.; Maubois. J. –L. Current Developments of Microfiltration Technology in the Dairy Industry. *Le Lait* **2000,** *80*(6), 541–553.
3. Cheryan, M. *Ultrafiltration and Microfiltration Handbook.* Technomic Publishing Company. Inc., Lancaster, Pennsylvania, USA, 1998.

4. Mukherjee, S.; Mukhopadhyay, S.; Ali Hashim, .M.; Sen Gupta, B. Contemporary Environmental Issues of Landfill Leachate: Assessment and Remedies. *Crit. Rev. Environ. Sci. Technol.* **2015,** *45*(5), 472–590.
5. Ali Hashim, M; Mukhopadhyay, S.; Sahu, J. N.; Sengupta, B. Remediation Technologies for Heavy Metal Contaminated Groundwater. *J. Environ. Management* **2011,** *92*, 2355–2388.
6. Shannon, M. A.; Bohn, P. W.; Elimelech, M.; Georgiadis, J. G.; Marinas, B. J.; Mayes, A. M. *Science and Technology for Water Purification in the Coming Decades.* Nature Publishing Group, London, United Kingdom, pp. 301–310.
7. Hassan, M. M. *Arsenic in Groundwater: Poisoning and Risk Assessment.* CRC Press, Taylor and Francis Group, Boca Raton, USA, 2018.
8. Kunduru, K. R.; Nazarkovsky, M.; Farah, S.; Pawar, R. P.; Basu, A.; Domb, A. J. *Nanotechnology for Water Purification: Applications of Nanotechnology Methods in Wastewater Treatment, Chapter 2,* Water Purification, Elsevier/Academic Press, USA, pp. 33–74.
9. Pandey, P.; Dahiya, M. Carbon Nanotubes: Types, Methods of Preparation and Applications. *Int. J. Pharm. Sci. Res.* **2016,** *1* (4), 15–21.
10. The Danish Environmental Protection Agency Report. Carbon nanotubes: types, products, market and provisional assessment of the associated risks to man and the environment, Environmental Project No. 1805, 2015.
11. Das, R.; Hamid, S. B. A.; Ali, M. E.; Ismail, A. F.; Annuar, M. S. M.; Ramakrishna, S.Multifunctional Carbon Nanotubes in Water Treatment: The Present, Past and Future. *Desalination* **2014,** *354*, 160–179.

CHAPTER 10

BIOREMEDIATION, WASTE MANAGEMENT, AND NANOTECHNOLOGY APPLICATIONS: A CRITICAL OVERVIEW AND THE VISION FOR THE FUTURE

SUKANCHAN PALIT

43, Judges Bagan, Post-Office - Haridevpur, Kolkata 700082, India

*Corresponding author. E-mail: sukanchan68@gmail.com,
sukanchan92@gmail.com*

ABSTRACT

The world of environmental engineering science and waste management are witnessing drastic challenges. Scientific vision, deep scientific sagacity, and the futuristic vision of engineering and technology will all lead a long and visionary way in the true realization of environmental sustainability and environmental protection. The status of waste management globally today is extremely dismal and thought provoking. Human civilization and human scientific progress today stands in the midst of deep scientific vision and vast scientific introspection. Science and technology has advanced so much in this century yet environmental engineering science is today in the path of scientific difficulties and barriers. In this chapter, the author deeply elucidates the need of bioremediation in the furtherance of science and engineering of environmental remediation. Drinking water crisis and heavy metal groundwater contamination are devastating the vast scientific fabric of human civilization. Arsenic groundwater contamination is veritably a monstrous issue in Bangladesh, India, and other developing and developed nations around the world. The author in this paper elucidates on the various bioremediation techniques in tackling global waste management and drinking water issues. Various bioremediation techniques are described in minute details in this chapter. Today is the world of energy and environmental sustainability. The linkage between environmental sustainability and environmental engineering science is elucidated in minute details in this chapter. This chapter will be a veritable eye-opener toward the scientific success and the scientific vision of waste management, bioremediation and the world of environmental engineering. Nanotechnology and nano-engineering are the other visionary areas of this scientific research pursuit. The challenge and the vision of nanotechnology applications in bioremediation and industrial wastewater treatment are the other areas of scientific endeavor of this treatise.

10.1   INTRODUCTION

Science and technology in the present century are moving at a rapid pace. Human civilization and human scientific endeavor today stands in the midst of deep scientific vision and vast scientific ingenuity. Provision of basic human needs such as water and energy are still today neglected areas of scientific progress globally. Energy and environmental sustainability are the needs of human civilization and human scientific progress. Biological sciences and biotechnology needs to be integrated with water resource engineering and water resource management in global scientific scenario today. Thus, the imminent need of bioremediation and wastewater management. Today renewable energy and energy sustainability are the coin words of human civilization. In this chapter, the author deeply comprehends and envisions the scientific success, the scientific vision, and the profundity in the application areas of bioremediation and wastewater management to human society. The techniques of novel separation processes, traditional and nontraditional environmental engineering techniques are dealt in minute details in this chapter. Nontraditional environmental engineering technique such as advanced oxidation processes is today revolutionizing the vast scientific firmament globally. Zero-discharge norms in industrial wastewater treatment, stringent environmental restrictions, and loss of ecological biodiversity are the issues facing human civilization today. The necessity of a deep discussion in the field of biological treatments and bioremediation is immense and far reaching. The author in this chapter elucidates in minute details the need of environmental sustainability in the unraveling of the scientific truth of environmental engineering, bioremediation and advanced oxidation processes. The scientific success and the vast scientific provenance of traditional and nontraditional environmental engineering techniques are dealt in detail in this chapter. Nanotechnology and nano-engineering are the visionary areas of scientific endeavor today. In this chapter, the author deeply elucidates the application and the vision of nanotechnology applications in groundwater and industrial wastewater treatment.

10.2 THE AIM AND OBJECTIVE OF THE TREATISE

Human scientific progress and environmental engineering tools applications in human society are today in the midst of deep scientific vision and revelation. The primary aim and objective of this chapter is to delineate the vast application areas of bioremediation, biological sciences, and biotechnology in the field of environmental protection. The status of human environment globally today is extremely dismal. In this chapter, the author rigorously points toward the success of science and the vast scientific vision in the field of bioremediation and waste management. Drinking water treatment and industrial wastewater treatment are necessities of human scientific progress today. Scientific and academic rigor in the field of industrial wastewater treatment and wastewater management are the other pillars of this well researched treatise. Technology and engineering science of environmental remediation today stands in the midst of deep scientific introspection and contemplation. The concerted efforts of environmental scientists, environmental engineers, and the civil society are the utmost needs of humanity today. Thus, also the need of integrated waste management techniques, human factor engineering, reliability engineering, and technology management. This chapter will surely open up new doors of scientific innovation and scientific instinct in the field of biological treatment of industrial wastewater and drinking water treatment. Thus human civilization's immense scientific prowess, the futuristic vision of environmental engineering science, and the vast scientific ingenuity of environmental sustainability will surely lead a long and visionary way in the true emancipation of biological treatments and bioremediation of industrial wastewater. This chapter is a veritable eye-opener toward the newer age and a newer genre in the field of environmental engineering techniques particularly bioremediation and other biological treatments. The other aims and objective of this chapter are the nanotechnology applications in environmental protection. The world today is moving immensely fast and human mankind is in the path of newer scientific regeneration. The success of science and engineering and its futuristic vision will all lead a long and visionary way in the true emancipation and true realization of nanotechnology.

10.3   WHAT DO YOU MEAN BY BIOREMEDIATION?

Bioremediation is a process to treat contaminated media, including water, soil, and subsurface substances by altering environmental conditions to stimulate and enhance the growth of microorganisms and degrade the recalcitrant pollutants. In many scientific and technological cases, bioremediation is less expensive and more sustainable than other remediation technologies. Biological treatments are similar approach used to treat wastes such as industrial wastes, industrial wastewater, and solid wastes. In today's scientific world, groundwater treatment and drinking water treatment can be done with the help of bioremediation and biotechnology. In this chapter, the author pointedly focuses on the merger of nanotechnology with biological sciences in the mitigation of global environmental engineering challenges. Most bioremediation processes involves oxidation-reduction reactions to reduce oxidized pollutants such as nitrate, perchlorate, oxidized metals, chlorinated solvents, explosives, and propellants. Technological and engineering profundity and scientific sagacity are the utmost needs of the hour. This well researched treatise elucidates on the success of biological sciences and biotechnology in environmental protection. Some examples of bioremediation techniques are phytoremediation, mycoremediation, bioventing, bioleaching, landfarming, bioreactor, composting, bioaugmentation, rhizofiltration, and biostimulation. Today, globally biological sciences are moving at a drastic pace. Engineering science and technology are the veritable answers to the scientific intricacies and scientific barriers of waste management. This chapter opens up new vision and new thoughts in the field of biotechnology and biological sciences.

10.3.1   TYPES OF BIOREMEDIATION

Bioremediation is the present generation and next generation environmental engineering technique. Scientific vision and vast academic rigor are the pillars of scientific research pursuit today. Bioremediation can be classified into various types. Genetic engineering and biotechnology are the wonders of science in present day human civilization. All the branches of bioremediation surpass vast and versatile scientific frontiers. On the basis of place where wastes are removed and cleaned, there are principally two methods of remediation:

1.   In Situ Bioremediation:

Mostly, in-situ bioremediation is applied to remove pollutants in contaminated soils, groundwater, and drinking water. It is highly superior and important for the cleaning of contaminated environment and surroundings because it highly saves transportation costs and uses harmless microorganisms to degrade the recalcitrant contaminants. These microorganisms are better for positive chemotactic affinity toward contaminants and hazardous pollutants. This method is preferred as it causes the least disruption of the contaminated zone. Another advantage of in-situ bioremediation is the feasibility of synchronous treatment of soil, groundwater, and drinking water. However in-situ bioremediation poses severe negative challenges—the method is time consuming compared with other remedial methods. Two types of in-situ bioremediation are distinguished based on the origin of the microorganisms applied and targeted as bioremediants:

i)   Intrinsic bioremediation: This type of in situ bioremediation is carried out without direct applications and through intermediation in ecological conditions of the contaminated region.
ii)   Engineered in situ bioremediation: This type of bioremediation is conducted through the introduction of certain microorganisms to a contamination site.

2.   Ex Situ Bioremediation:

The process of bioremediation here takes place somewhere out from contamination site, and therefore requires transportation of contaminated and polluted soil or pumping of groundwater to the site of bioremediation and the areas of environmental remediation.

Technology, engineering, and science today stand in the midst of deep scientific inquiry and vast scientific fortitude. The rendezvous of science and engineering globally today is environmental remediation and mitigation of climate change and water scarcity. The author with immense scientific conscience and foresight targets the environmental engineering profundity and vision. The world of science today stands mesmerized and inspired with the new innovations and newer directions. In this chapter, the author deeply elucidates the success of biological sciences and environmental biotechnology in the scientific progress of human society.

10.3.2 TECHNIQUES OF BIOREMEDIATION

Techniques of bioremediation are today surpassing vast and versatile scientific frontiers. Technological and scientific validation, the success of research endeavor and the futuristic vision of environmental engineering science will all lead a long and visionary way in the true emancipation of global climate change mitigation. There are several bioremediation techniques:

1. Bioaugmentation
2. Biofilters
3. Bioreactors
4. Biostimulation
5. Bioventing
6. Composting
7. Landfarming
8. Biopiling

Scientific vision and deep scientific acuity are the challenges of human scientific progress today. Bioremediation is the veritable wonder of environmental engineering science today. Developing nations around the world are confronting the testing time of this century—the global water scarcity and industrial pollution. Water pollution and groundwater heavy metal contamination is troubling and destroying the vast scientific firmament. This chapter widely opens the success of science and engineering in tackling global climate hiatus and drinking water issues.

10.3.3 ROLE OF BIOTECHNOLOGY AND ENVIRONMENTAL BIOTECHNOLOGY IN ENVIRONMENTAL PROTECTION

Biotechnology and biological sciences are the branches of science which has diverse applications. Scientific and academic rigor in the field of biotechnology and environmental biotechnology are at its zenith in the present century. Environmental protection, water purification, and groundwater remediation stands as major scientific imperatives in the pursuit of science and technology globally. Today bioremediation and other biological treatments are the challenges and the vision of human civilization and human scientific progress. The futuristic vision of biological sciences and the immense scientific prowess of bioremediation will veritably lead a long and effective way in the true realization and the true emancipation of environmental sustainability. Environmental as well as energy sustainability are the marvels of human scientific progress today. These are the answers to water purification and groundwater remediation. Nanotechnology is on the other side of the visionary coin.

10.4 THE VAST SCIENTIFIC DOCTRINE OF WASTE MANAGEMENT

Waste management is the imminent need of the hour in global scientific pursuits. Engineering science and technology has practically no answers to the scientific intricacies of water purification and water technology. The challenges and the vision of environmental engineering and chemical engineering techniques need to be revamped and re-organized with the passage of time. Today nano-science and nanotechnology are the wonders of human civilization. This chapter opens up newer thoughts and newer directions in the field of bioremediation and other biological treatments. The success of waste management lies in the hands of environmental engineering science and biotechnology. The author rigorously pronounces the scientific success, the scientific inquiry, and the deep scientific divination in the field of nanotechnology applications in environmental protection. Provision of basic human needs such as energy, water, food, shelter, and education are in a state of immense disaster as science and engineering gears forward. Energy and environmental sustainability today are in a dismal state. Here comes the importance of new technologies and newer innovations such as membrane separation processes and advanced oxidation processes. The author rigorously points toward the success of science and technology in the mitigation of global water challenges. Thus human civilization and human scientific progress will witness a newer beginning and drastic changes.

10.5 RECENT SCIENTIFIC RESEARCH PURSUIT IN THE FIELD OF BIOREMEDIATION

Technological and scientific prowess in the field of bioremediation are at its zenith in present day human civilization. The vast challenges and the vision of environmental engineering are in a dismal state. Here comes the importance of environmental sustainability, water purification, and industrial wastewater treatment. In this section, the author deeply elucidates the recent research pursuit in the bioremediation and other biological treatments of industrial wastewater and groundwater.

Sardrood et al. (2013)[1] discussed with immense scientific conscience and lucid insight the vast world of bioremediation. The explosive rise of global population has led to the enhanced exploitation of natural resources around the world.[1] Natural resource management is highly neglected throughout the world. Food, energy, and water are the challenged areas of human civilization today. Industrial revolution has resulted in the production of huge number of organic and inorganic chemicals in wastewater. The duration of the contamination is due to difficult biodegradability of recalcitrant chemicals. The vast trend of environmental pollution is today of immense extent globally. Based on the estimations made by environmental protection agencies around the world, only around 10% of all the wastes are disposed off. Here comes the important of innovations and scientific vision.[1] Serious occurrence of environmental disasters such as Exxon Valdez oil spill, the Union Carbide (Dow) Bhopal disaster, large-scale contamination of the Rhine river, the progressive deterioration of the aquatic habitats in Northeastern USA and Canada, or the release of radioactive material in the Chernobyl accident and most recently the crisis resulting from crude oil pollution of Mexico gulf waters and the leakage of the radioactive material from Fukushima reactor in Japan are devastating the global scientific firmament.[1] The authors in this chapter discussed in minute details the role of environmental biotechnology in industrial pollution management, bioremediation, types of bioremediation, bioremediation techniques, landfarming and land treatment, organisms involved in bioremediation process, mycoremediation, and the vast environmental perspective of bioremediation.[1] Today, technology and engineering science are highly advanced globally. In the similar vein, environmental remediation is in the path of newer scientific regeneration.[1] Bioremediation is an interdisciplinary branch of science and engineering. Based on the scientific principles of bioremediation and biotechnology, the science of mycoremediation will surely open up newer knowledge dimensions in years to come.[1] The author deeply pronounces these facts of biological sciences.

Azubuike et al. (2016)[2] deeply discussed with cogent insight bioremediation techniques and classification based on site of applications, principles, advantages, limitations, and prospects. Environmental pollution has been on the rise in the past few years owing to the increase of human activities and enhanced industrialization.[2] Amongst the pollutants are heavy metals, nuclear wastes, pesticides, green house gases, and the vast array of hydrocarbons.[2] Remediation of polluted sites using microbial processes has been proven to be effective and groundbreaking. Bioremediation can be carried out ex-situ or in-situ, depending on several factors which vastly include but not limited to cost, site characteristics, type, and concentration of the hazardous pollutants and wastewater.[2] The authors discussed in deep details ex-situ bioremediation techniques, biopile, windrows, bioreactor, landfarming, enhanced in-situ bioremediation, bioslurping, biosparging, phytoremediation, and intrinsic bioremediation.[2] The foremost step of a successful and enhanced bioremediation is site characterization which helps establish the most suitable and feasible bioremediation technique (ex-situ and in-situ). Geological characteristics of polluted sites including soil type, pollutant depth and type, site location and performance characteristics of each bioremediation techniques should be widely incorporated in order to result in effective bioremediation.[2] This well researched treatise deeply comprehends the scientific success, the scientific ingenuity, and the vast scientific profundity in bioremediation technique applications in wastewater treatment.

Mary Kensa (2011)[3] discussed with immense scientific conscience and foresight bioremediation. Bioremediation is an ecologically sound and state-of-the-art technique in industrial wastewater treatment and groundwater remediation. The author in this treatise discussed in minute details various bioremediation technologies such as bioventing, landfarming, bioreactor, composting, bioaugmentation, rhizofiltration, and biostimulation.[3] Technological verve, scientific validation, and scientific sagacity are today the pillars of environmental engineering science. This research paper discusses in minute details principles of bioremediation, factors of bioremediation, strategies, types, genetic engineering and biotechnology approaches, monitoring bioremediation and advantages and disadvantages of bioremediation.[3] The highlight of this paper is the genetic engineering and biotechnology approaches in bioremediation. Bioremediation technique is far less expensive than other technologies that enhances industrial wastewater treatment and environmental remediation. This paper is a veritable eye-opener toward the success of bioremediation science and the larger emancipation of biotechnology and genetic engineering.[3]

Genetic engineering and biotechnology are the challenges and the vision of science and engineering today. The author in this chapter deeply comprehends the imminent need of biological treatments of wastewater and groundwater with clear view toward the furtherance of science and engineering.

10.6 RECENT SIGNIFICANT RESEARCH ENDEAVOR IN THE FIELD OF WASTE MANAGEMENT

Waste management and integrated water resource management are the visionary coin words of scientific research pursuit today. Technological and engineering profundity and ingenuity, the sagacity of science, and the futuristic vision of science of environmental sustainability will all lead an effective way in the true emancipation of environmental remediation globally.

Australian Government Productivity Commission Inquiry Report (2006)[4] discussed in deep details the status of waste management in Australia and other OECD countries. Technologies and processes to avoid, reduce and recover waste are not used as extensively in Australia and other OECD countries. Here comes the importance of scientific inquiry, scientific vision, and profundity.[4] The objective of this report is to identify policies that will direct Australia to address market failures associated with generation and disposal of industrial wastes including resource use efficiency and natural resource management. As a developed country, Australia is in the verge of newer scientific direction in natural resource management and industrial pollution control.[4] This inquiry will involve resources including solid waste, municipal waste, commercial and industrial waste. This report deeply discusses waste management in Australia, government policy responses, the costs and benefits of waste management, a waste policy framework, and the stringent regulations and environmental restrictions.[4] The world of science and engineering are today in the path of newer scientific rejuvenation. This entire report deeply opens up newer directions and newer knowledge dimensions in waste management techniques in developed nations.

The Energy and Resources Institute, India Report (2014)[5] discussed with lucid and cogent insight waste to resources in a waste management handbook. The vision, the targets and the challenges of global waste management scenario are immense and far reaching. This treatise titled "Waste to resources" provides a detailed overview of urban liquid and solid waste management, construction and demolition debris management, industrial solid and waste management biomedical waste management, and e-waste management.[5] Science and engineering of waste management are today surpassing visionary frontiers. This report is a veritable eye-opener toward the needs of waste management toward furtherance of science and technology.[5] Basel convention of United Nations Environmental Programme defines wastes "as substances or objects, which are disposed of or intended to be disposed of or are required to be disposed of by the present national law."[5] Scientific vision and deep scientific ingenuity are the pillars of environmental engineering research pursuit. This report details the vast domain of industrial wastes, the areas of e-waste, and sustainable biomedical waste management. Industrial waste management and water resource management are in the threshold of a new visionary era. This report rigorously points toward the success of science and engineering in confronting environmental sustainability and environmental remediation.[5]

Sustainable development whether it is social, economic, energy, and environmental are today challenging the scientific fabric of might and vision. Developing and developed nations around the world are in the verge of an unmitigated crisis. The status of environmental protection in developing nations around the world is extremely dismal. Thus, the need of a concerted effort from engineers, scientists, civil society, and governments around the globe.

10.7   RECENT RESEARCH ENDEAVOR IN THE FIELD OF NANOTECHNOLOGY AND ENVIRONMENTAL PROTECTION

The world of science and technology are moving at a drastic pace. In the similar vein nanotechnology is gearing forward toward a newer visionary era. In this section, the author touches upon some research endeavor in the field of nanotechnology and environmental remediation. Today, nanomaterials and engineered nanomaterials are the marvels of engineering science. A brief description on nanomaterials applications in environmental protection science is given here.

Palit (2018)[6] with deep scientific conscience and farsightedness elucidates on environmental engineering applications of carbon nanotubes. Technological and scientific validation and inquiry, the futuristic vision of nano-science and nanotechnology and the vast world of scientific divination will all be the forerunners toward a newer visionary dimension of science and engineering globally. Environmental sustainability is the utmost need of the hour in global scientific scenario today.[6] The author depicts profoundly the success of applications of carbon nanotubes in environmental engineering science. Engineered nanomaterials are the other side of the visionary coin of material science. This chapter is a veritable eye-opener toward the success and vision of nano-science, nanotechnology, and nano-engineering.[6]

Khan et al. (2014)[7] described with vast scientific vision nanotechnology for environmental remediation. Environmental protection is of major concern in the global scientific scenario today. This review investigates the different types of nanomaterials, including various types of metal nanoparticles for remediation purposes. The success of application of nanomaterials and engineered nanomaterials are effectively dealt with in this paper. Iron nanoparticles are of immense use in the chemical, electronics, and other industries. Recently their applications have been extended to the treatment of toxic and hazardous pollutants in industrial wastewater.[7] These technologies and innovations are vastly dealt with in this paper. A sound deliberation on the future prospects of material science and nanotechnology is dealt effectively in this paper.[7]

Human civilization's immense scientific stance and prowess, the vast world of scientific validation, and the futuristic vision of engineering science are all the forerunners toward a newer world of sustainability science. Today, environmental and energy sustainability, the vast world of nanotechnology, and reliability engineering are integrated toward a greater emancipation of environmental protection. In this chapter, the author pointedly focuses on the vision, the targets, and the scientific revelation in the field of nanotechnology.

10.8 THE SCIENTIFIC DOCTRINE AND THE VISION BEHIND WASTEWATER MANAGEMENT

Wastewater management today is in the path of vast scientific rejuvenation. In the global scientific scenario, the world of challenges and the vision lies in the hands of environmental engineers and energy technologists. The developing and the third world are in the midst of deep scientific division and vast environmental catastrophe. Energy and environmental sustainability stands in the midst of scientific chaos and vast scientific abyss. Here comes the need of concerted effort of civil society and the environmental engineering domain. Nanotechnology thus needs to be re-envisioned as human civilization confronts the crisis of global climate change and water scarcity. Integrated water resource management and wastewater management are the new technologies of the future. The vision of wastewater management will be widened if nanotechnology is merged with environmental engineering, biological sciences and chemical process engineering. This is the main vision of this chapter. Bioremediation will emerge as an effective process in treatment of industrial wastewater and mitigation of groundwater contamination. The author repeatedly urges the need of biological science and biotechnology in the advancement of global science and engineering. This chapter will surely open new doors of scientific innovation and engineering instinct in the field of biotechnology, nanotechnology, and environmental engineering.

The world will thus witness a new beginning and a newer vision in the field of environmental remediation.

10.9 HEAVY METAL AND ARSENIC GROUNDWATER CONTAMINATION AND THE VISION FOR THE FUTURE

Heavy metal and arsenic groundwater contamination is human civilization's largest environmental engineering disaster. Scientific vision and vast scientific cognizance are the utmost needs of the hour for global scientific progress. Industrial wastewater treatment, drinking water treatment, and water purification are today replete with immense scientific and academic rigor. The vision for the future in the field of water purification science and nanotechnology are vast and far reaching. Energy and environmental sustainability are in the similar vein needs to be revamped if the global scientific domain really confronts global environmental catastrophes. The scientific onus and the vast responsibility lie in the hands of environmental engineers and chemical process engineers. Developing and developed nations around the world are in the critical juncture of deep scientific introspection and scientific contemplation. In this chapter, the author deeply elucidates the scientific success, the vast scientific inquiry, and the needs of environmental engineering science globally today. Bangladesh and India are veritably in the threshold of an unbelievable environmental engineering crisis that is arsenic groundwater contamination. The success and the vision of environmental engineering science are ground breaking and surpassing vast scientific frontiers. In this chapter, the author also delineates the scientific success of application of nanotechnology in environmental remediation. This chapter will surely be an eye-opener toward the world of environmental sustainability and unravel scientific truths of environmental engineering techniques.

10.10 TECHNOLOGICAL VALIDATION AND THE VISION BEHIND ENVIRONMENTAL SUSTAINABILITY

Technological and scientific validation in the field of environmental engineering and chemical process engineering today stands in the crucial juncture of strong scientific vision and scientific profundity. The areas of sustainable development in the global scenario need to be re-envisioned and revamped with the passage of scientific history and the global timeframe. Today nanotechnology has applications in every branch of scientific endeavor.

The author deeply elucidates the immense importance of nanotechnology in global scientific understanding and scientific discernment. Globally, renewable energy technology and energy sustainability are in the path of scientific rejuvenation. The author deeply stresses on the scientific success of nanotechnology applications in water pollution control, wastewater treatment, and water resources management. Technological profundity, scientific inquiry, and vast vision of engineering science will surely open the doors of environmental sciences and environmental sustainability in years to come. This well researched treatise targets the areas of nanotechnology applications in environmental protection and the new area of bioremediation. The vision behind applications of environmental sustainability in human society is vast, versatile, and far reaching. Sustainable development whether it is social, economic, environmental, or energy is the imminent need of scientific progress of humanity today. Nanofiltration, reverse osmosis, ultrafiltration, and microfiltration have tremendous applications in environmental engineering and chemical process engineering. Thus, the immense scientific need of novel separation processes such as membrane science.

10.11   HUMAN FACTOR ENGINEERING, RELIABILITY ENGINEERING AND WASTE MANAGEMENT

Human factor engineering, reliability engineering, and systems engineering are the tomorrow's vision of science and engineering. These branches of engineering should be integrated with nanotechnology applications and water science and technology. The world of environmental engineering in the similar vein should be merged with human factor engineering. Integrated waste management principles should also be merged with human factor engineering and reliability engineering for a greater emancipation of science and engineering of environmental protection. Global water shortage and global climate change are veritably destroying the scientific firmament of might and vision. Every nation around the world is taking immense positive steps and sound research and development initiatives toward the mitigation of global water scarcity and global climate change. The scientific onus lies in the hands of environmental engineers and environmental scientists. The need of industrial engineers is of immense importance as global science and engineering surges forward. Human factor engineering will surely open newer thoughts and newer scientific understanding in the field of groundwater remediation and environmental protection as a whole. This is a practical innovation of science and engineering today. Industrial systems and production engineering will surely open up newer vistas in environmental remediation and the greater and larger field of environmental sustainability.

10.12 SCIENTIFIC RESEARCH PURSUIT IN THE FIELD OF GROUNDWATER REMEDIATION

Groundwater remediation and environmental protection are today replete with scientific farsightedness and scientific divination. Technology and engineering science today has few answers to the burning question of pure drinking water provision. There are tremendous health effects of arsenic and heavy metal contamination to human society. Here comes the importance of scientific innovations and vast scientific profundity. In this section, the author deeply comprehends the success of science and engineering in tackling groundwater contamination.

Hashim et al. (2011)[8] discussed with lucid and cogent insight remediation technologies for heavy metal groundwater. Scientific prudence, scientific vision, and vast scientific profundity stand as major pillars in research pursuit in groundwater remediation. The contamination of groundwater by heavy metal originating from natural soil resources or from anthropogenic sources is a matter of immense concern to human health and public health engineering. The sagacity of science and technology and its futuristic vision will all lead a long and effective way in the true emancipation of environmental engineering science and public health engineering.[8] Remediation of contaminated wastewater is of highest priority as science and engineering globally surges forward. In this paper, 35 approaches for groundwater treatment are reviewed and classified under three large categories mainly, chemical, biochemical, and physico-chemical treatment processes.[8] Selection of suitable technology is of utmost importance in the groundwater remediation process. In this paper, the author immensely points out toward the future of environmental remediation today.[8] This well researched treatise elucidates heavy metals in groundwater, technologies of heavy metal contaminated groundwater, chemical treatment technologies, iron-based technologies, soil washing, in-situ chelate flushing, in-situ chemical fixation, biosorptive treatment technologies, and physico-chemical treatment technologies.[8] Today, there is an immense need of groundwater remediation as human civilization surges forward. Technological and scientific validation and the imminent need of water and wastewater treatment will sure lead a long and visionary way in the true realization of environmental protection science.[8]

Shannon et al.[9] discussed with vast scientific farsightedness science and technology for water purification in the coming decades of this century. Global climate change and water shortage are challenging the vast scientific firmament. Tremendous amount of research and development initiatives around the world in water science and technology are changing the face of human scientific endeavor and the scientific stance of human civilization.[9] This well researched treatise targets the areas of disinfection, decontamination, re-use and reclamation and the vast world of desalination.[9] This is a watershed text in the field of environmental engineering. The author deeply reviews the necessities and the vision of water purification, industrial wastewater treatment and drinking water treatment.[9]

The challenges and the targets of science and engineering globally are in the path of newer scientific regeneration. Public health engineering and reliability engineering are the needs of science today. The author pointedly focuses on the scientific success, the scientific ingenuity and the scientific provenance in water treatment areas.[10,11,12,13]

10.13 GROUNDWATER AND SUBSURFACE REMEDIATION AND THE FUTURISTIC VISION

Groundwater and subsurface remediation are the immediate needs of human scientific progress today. Arsenic and heavy metal groundwater and drinking water contamination are the monstrous and thought-provoking issues facing global scientific progress and human civilization today. Thus, deep scientific revelation and vast scientific vision are the necessities of the hour. Developing and developed countries around the world are severely affected by the ever-growing concern of heavy metal groundwater contamination. The scientific success, the scientific inquiry, and the scientific sagacity will all lead a long and visionary way in the true realization of environmental sustainability and environmental protection today. Human civilization today stands in the midst of deep scientific introspection and deep scientific ingenuity. The author in this chapter deeply discusses the contribution of nanotechnology in groundwater and industrial wastewater treatment. Nano-science and nano-engineering are the forerunners toward a greater vision of environmental engineering and chemical process engineering.[10,11,12,13]

10.14   BIOREMEDIATION AND ENVIRONMENTAL SUSTAINABILITY

Bioremediation and environmental sustainability are today integrated with each other in the global scientific scenario. The true vision of Dr Gro Harlem Brundtland, the former Prime Minister of Norway as regards the definition of sustainability and sustainable development needs to be re-organized and revamped as science and engineering surges forward. Energy and environmental sustainability are the true vision of human civilization today. In the similar vein, bioremediation and biological treatments of drinking water and industrial wastewater are the fountainheads of science and engineering today. Science and technology of energy and environmental sustainability needs to be in the forefront of human civilization today. Renewable energy technology and energy sustainability are also the needs of human civilization today. The global energy hiatus and global water crisis are the forerunners toward a newer era in science and engineering today. The social status of global human civilization, the futuristic vision of science and engineering, and the provision of basic human needs will surely be the torchbearers toward a larger emancipation of energy and environmental sustainability globally today. This chapter is a veritable eye-opener toward a newer visionary era in waste management, industrial wastewater treatment, and nanotechnology.

10.15   THE STATE OF SCIENCE AND ENGINEERING OF BIOREMEDIATION GLOBALLY

Globally science and technology are moving at a rapid pace. In the similar manner environmental protection is an absolute need of the hour. The status of global environment is extremely dismal and thought provoking. Bioremediation and biological treatment of groundwater, drinking water, and industrial wastewater are the utmost need of the human civilization and human scientific progress. The vision of environmental sustainability needs to be re-organized with the passage of scientific history and time. Biotechnology and the vast domain of biological sciences will lead a true and effective way in the visionary emancipation of science and technology today. Arsenic and heavy metal groundwater poisoning in many developing and developed nations around the world are challenging the vast scientific fabric globally. In this chapter, the author rigorously points toward the vision and success of environmental engineering and biological sciences.[10-13]

10.16   FUTURE RESEARCH RECOMMENDATIONS AND THE DOCTRINE OF ENVIRONMENTAL REMEDIATION

Future research recommendations in the field of environmental remediation should be targeted toward more zero-discharge norms and a newer order in the area of global climate change. The doctrine of environmental remediation today is far reaching and surpassing vast and versatile scientific frontiers. Energy and environmental sustainability are the utmost needs of the hour. The provision of basic human needs such as energy, electricity, water, food, shelter, and education are in the midst of deep scientific crisis. The world of chemical process engineering and environmental engineering stands mesmerized and puzzled with the ever-growing concerns of loss of ecological biodiversity. In this chapter, the author deeply comprehends the success of technology and engineering science in tackling environmental protection problems and the vast area of chemical process engineering. Future of science and engineering in the global scenario is advancing at a rapid pace. Thus, future research recommendations should be targeted toward more scientific emancipation in the field of environmental remediation globally. In the similar vein, more scientific regeneration in the field of environmental sustainability is the utmost need of the hour. Other future research recommendations and future research thoughts are the greater research and development initiatives in the field of traditional and nontraditional environmental engineering techniques. Membrane separation processes, advanced oxidation processes, and bioremediation thus are the present generation and next generation environmental engineering tools. The unequivocal definition of sustainability as defined by visionaries around the world thus needs to be re-envisioned and re-organized with the passage of scientific history and time. Biological sciences, biotechnology, and bioremediation thus are the areas of scientific regeneration in the present global framework. In this well researched treatise, the technologies of bioremediation are deeply elucidated with vast scientific conscience. Nanotechnology is the present and futuristic vision of science and engineering in the global scenario. The author also deeply elucidates the success of science and engineering of nanotechnology in the true emancipation of science and technology.[10,11,12,13]

10.17   CONCLUSION AND FUTURE ENVIRONMENTAL PERSPECTIVES

Human scientific progress and human scientific rigor today stands in the midst of deep scientific forbearance and vast scientific revelation. Future environmental perspectives in the similar vein are in the midst of vision and ingenuity. Modern science and environmental protection needs to be re-envisioned and reorganized with the progress of human civilization. Technology and engineering science has today practically no answers to arsenic and heavy metal groundwater and drinking water contamination. Here comes the importance of traditional and nontraditional environmental engineering techniques. Advanced oxidation processes and novel separation processes such as membrane science are the fountainheads of environmental engineering science today. In this chapter, the author rigorously points toward the scientific intricacies and the scientific barriers in the field of applications of bioremediation. The world today is advancing rapidly surpassing one scientific barrier over another. This chapter opens up newer thoughts and newer vision in the field of bioremediation, biological sciences, biotechnology, and chemical process engineering. Human factor engineering, reliability engineering, and systems engineering also needs to be integrated with environmental engineering science and environmental remediation. This chapter also targets these areas of scientific research pursuit. Future scientific perspectives and future research recommendations in the field of bioremediation and biological treatments should target zero-discharge norms and stringent environmental paradigm. This well researched chapter widely projects the success of science, the visionary avenues, and the futuristic vision of environmental remediation particularly bioremediation and biological treatments. Nano-science and nanotechnology are changing the face of human civilization and the vast global scenario of human scientific progress. This chapter opens up the scientific needs and the scientific vision in the field of nanotechnology applications in environmental applications and the greater emancipation of environmental and energy sustainability.

IMPORTANT WEBSITES FOR REFERENCE

https://en.wikipedia.org/wiki/Bioremediation
https://www.conserve-energy-future.com/what-is-bioremediation.php
https://www.sciencedirect.com/topics/earth-and-planetary-sciences/bioremediation
https://www.nature.com › subjects
https://www.nap.edu/read/2131/chapter/4
www.imedpub.com/scholarly/bioremediation-journals-articles-ppts-list.php
www.hawaii.edu/abrp/biordef.html
https://www.thebalance.com › Investing › Biotech Industry › Biotech Glossary
https://study.com/academy/lesson/soil-bioremediation.html
ei.cornell.edu/biodeg/bioremed/
waterquality.montana.edu/energy/cbm/lit-reviews/bioremed-soil.html

https://en.wikipedia.org/wiki/Waste_management
https://www.sciencedirect.com/journal/waste-management
https://www.wm.com/us
www.wm.com/location/california/north-county/oceanside/residential/tips.jsp
https://www.conserve-energy-future.com/waste-management-and-waste-disposal-meth...
www.delhi.gov.in › Home › Environmental Issues
https://journals.sagepub.com/home/wmr
www.insituarsenic.org/
www.insituarsenic.org/origin.html
https://www.crcpress.com › Engineering - Environmental › General
www.understandingnano.com/environmental-nanotechnology.html
https://www.nanowerk.com/nanotechnology-and-the-environment.php
https://www.nano.gov/node/110

KEYWORDS

- **waste**
- **water**
- **bioremediation**
- **industry**
- **vision**
- **human**
- **factor**
- **reliability**

REFERENCES

1. Sardrood, B. P.; Goltapeh, E. M.; Varma, A. An Introduction to Bioremediation, Soil Biology 32. In *Fungi as Bioremediators*; Goltapeh, E. M., et al. Eds.; Springer-Verlag Berlin Heidelberg: Germany, 2013; pp 3–27.
2. Azubuike, C. C.; Chikere, C. B.; Okpokwasili, G. C. Bioremediation Techniques-Classification Based on Site of Application: Principles, Advantages, Limitations and Prospects. *World J. Microbiol. Biotechnol.* **2016,** *32* (180), 1–18.
3. Mary Kensa, V. Bioremediation—an Overview. *J. Ind. Pollut. Control* **2011,** *27* (2), 161–168.
4. Australian Government Productivity Commission Inquiry Report (2006). Waste Management, No.38, 20 October, 2006.
5. The Energy and Resources Institute, India, Report (2014). Waste to Resources: A Waste Management Handbook; pp 1–75.
6. Palit, S. Environmental Engineering Applications of Carbon Nanotubes: A Critical Overview and a Vision for the Future. In *Engineered Carbon Nanotubes and Nanofibrous*

Materials: Integrating Theory and Technique, Apple Academic Press Research Notes on Nanoscience and Nanotechnology; Haghi, A. K., Praveen, K. M., Sabu Thomas, Eds.; Apple Academic Press, USA (Taylor and Francis Group): Waretown, New Jersey, USA, 2018; pp 101–126.

7. Khan, I.; Farhan, M.; Singh, P.; Thiagarajan, P. Nanotechnology for Environmental Remediation. *Res. J. Pharmaceutical, Biol. Chem. Sci.* **2014,** *5* (3), 1916–1927.

8. Hashim, M. A.; Mukhopadhyay, S.; Sahu, J. N.; Sengupta, B. Remediation Technologies for Heavy Metal Contaminated Groundwater. *J. Environ. Manag.* **2011,** *92*, 2355–2388.

9. Shannon, M. A.; Bohn, P. W.; Elimelech, M.; Georgiadis, J. G.; Marinas, B. J.; Mayes, A. M. *Science and Technology for Water Purification in the Coming Decades*; Nature Publishing Group: London, United Kingdom, 2008; pp 301–310.

10. Palit, S.; Hussain, C. M. Engineered Nanomaterial for Industrial Use. In *Handbook of Nanomaterials for Industrial Applications*; C. M., Hussain, Ed.; Elsevier: Netherlands, 2018; pp 3–12.

11. Palit, S. Recent Advances in the Application of Engineered Nanomaterials in the Environment Industry—A Critical Overview and a Vision for the Future. In *Handbook of Nanomaterials for Industrial Applications*; Hussain, C. M., Ed.; Elsevier: Netherlands, 2018; 883–893.

12. Elimelech, M.; Phillip, W. A. The Future of Seawater Desalination: Energy, Technology and the Environment. *Science* **2011,** *333*, 712–717.

13. Palit, S.; Hussain, C. M. Frontiers of Application of Nanocomposites and the Wide Vision of Membrane Science: A Critical Overview and a Vision for the Future. In *Nanocomposites for Pollution Control*; Hussain, C. M., Mishra, A. K., Eds.; Pan Stanford Publishing Pte. Ltd.: Singapore, 2018; pp 441–476.

PART 2

Informatics: Modern Tool in Applied Chemistry

CHAPTER 11

PROGRESS IN CHEMICAL INFORMATION TECHNOLOGY

HERU SUSANTO, LEU FANG-YIE, and
ANDRIANOPSYAH MAS JAYA PUTRA

[1]*The Indonesian Institute of Sciences, Indonesia*

[2]*Tunghai University, Taichung, Taiwan*

[3]*University of Brunei, Bandar Seri Begawan, Brunei*

Corresponding author. E-mail: heru.susanto@lipi.go.id

ABSTRACT

This study consist of an important feature of information system (IS) and technology, as it explores the many different technologies inherent in the field of information technology and their impact on ISs design, functionality, operations, and management to its people, organization as well as science today. Sciences tend to make full usage of IS and technology to seek a description and understanding of the natural world and its physical properties. Hence this report examines the concept of IS, technology, and how it supports science namely in terms of cheminformatic.

11.1 INTRODUCTION

Information system (IS) is similar to information and communication technology. It consists of raw data collection (input); that can be processed into an additional value for an organization (output). After being analyzed and processed, it would turn into information data or facts that could be used to answer questions, solve problems, or to conduct a project. Moreover, IS also known as a software which is being used in organizations to help organize and analyze databases into a useful information that can be used in the decision-making.

Information technology has been around for quite a long time. Essentially the length of individuals have been around, information technology has been around on the grounds that there were always ways of communicating through innovation accessible at that point in time. There are four main ages that divide up the historical background information of information technology. Just the latest age (electronic) and a percentage of the electromechanical age absolutely influence us today. Information technology can be referred to as comprising of three fundamental parts: computational information processing, decision support, and business programming. Information technology or IT is broadly utilized as a part of business and the field of computing. Individuals use the terms generically when alluding to different sorts of PC-related work.

11.2   LITERATURE REVIEW

The usage of technology is very crucial in this modern world. The newly improved advanced technology has changed our life dramatically and made it much easier to anticipate the demand of our needs and wants. IS is practically being used everywhere and it has been evolving ever since in the form of device such as smartphone. Now it has been moving to home appliances such as television, which is currently called as a "smart television."

Information system or IS is integrated components that collect, manipulate, store, and disseminate data, information, and provide a feedback mechanism to meet objectives for individual, group, and organization. It is a vital processor that every computer needs because without it, information technology in short IT and information communication technology or ICT will not be able to function properly or follow the instruction given by us. This is because IS is build with software that comes from manufacturer itself where it has its own function and follows the task given by us.

In today's world, technology has taken a huge part in improving our life. Now, information system in short IS could take part in helping science to improvise their medical methods into a new whole dimension of science in order to improve the development and assimilation of human life and capabilities of technology. Not only those, but it could also help scientists to conduct a practical research to find a new discovery in science to find solutions to the problems of what human has been facing as such, new remedies are yet to be created to fight diseases. Everyday, scientists all around the world are working hard to uncover mysterious solutions to what we, the human, are actually facing not only in the terms of medicine but also new methods in the field of surgeon and technology.

How IS could help science would depend on the characteristic and how it is being used. Development in advanced technology and human capabilities; by using IS software it could help to gather all the raw data and molecule from the past research by synthesis with the new finding. Hence, it would uncover the mysterious of what it might be a new discovery in medicine or drugs to cure unsolved diseases. Especially by emerging technology combine with high instrumentation will help in advance the process without any complication. For instance, how research is being conducted for the past couple of decade was immensely complicated because they have to go through every practical research and books unlikely how it is now because we have advance machinery, technology, information, and Internet to help us.[26, 14–17]

11.3 DISCUSSION

11.3.1 INFORMATION SYSTEM

Range of definition given for this concept; however information does not have a specific and uniform definition. The definition is stated by N. Winer who determines the content of the information gleaned from the outside world in the process of our adjustment to it and adapt it to our senses (McGarry M., 2008).

IS is the combination of people's innovation and computer that processes or unravels information. Information and communication technology in short ICT, that organization uses, as well as a way of communication people use to interact which leads in support for business processes. There are no specific and clear distinction between ISs, computer system, and business processes. Nowadays ISs are mainly used to gain maximum benefits by processing data from inputs to generate IS, a combination of hardware, coordination, and decision-making in an organization. It is also known as decision-making support system. It is a collective combination of collection of people, procedure, software, database, and device that support problem specific decision-making.

The main usage of IS in an organization is communication, which allows people to communicate easily, for instance the method of communication which is electronic mail or E-mail, which are delivered extremely fast, it can be sent and received from any devices all around the world which has an Internet connection. In addition, the availability of cloud function which allows people to make changes, add information, and share with one another in a community cloud. Thus, it makes communication better and faster, as most information is now accessible, which people could gain the right information at the right time. In order to meet one's need and want as it is important to obtain accurate and complete information.

Business process may be difficult, as entrepreneurial culture and the degree to which the existing ISs tends to represent compatibility and application functionality significantly affect a firm's propensity to adopt cloud computing technologies. The discovery supports our abstract development and suggests complementarities between innovation diffusion theory and the information processing view. Industry professionals to aid in making more informed adoption decisions in regard to cloud computing technologies in order to support of the supply chain.

Due to the fact that IS play a major role mainly in business by creating new products and services, make it possible for managers to use real time data when making a decision; therefore, information must be relevant and reliable in order to help people in their organization to achieve their goals and perform tasks more effectively that then lead to competitive advantage and these system make their jobs easier as possible. Z. Messner stated that information as data on monetary phenomena and processes used in decision-making processes (Messner Z., 1991) namely for human resource management, marketing, and administration.

Moreover, IS could stores documents, histories, communication records as well as operational data, that could be use in the future for better references, as it acts as a useful historical information. It improves efficiency of certain organization's operation in order to achieve aims, objectives, and goals, namely to achieve higher profitability.[18,20] IS needs to be flexible, which is able to accommodate certain amount of variation, regarding the requirement in terms of supporting business process. The impact of IS being flexible, enables cost efficiency for the business.[21,22]

Information technology tends to be applied within business operations that can save a great deal of time during the fulfillment of daily tasks. Business tends to include knowledge management, artificial intelligence, expert system, multimedia, and virtual reality system. Paperwork is processed immediately, and financial transactions are automatically calculated by IS. Although businesses may view this expediency as a boon, there are untoward effects to such levels of automation. As technology amends, tasks that were formerly performed by human employees are now carried out by computer and ISs. This leads to the elimination of jobs and, in some cases, alienation of clients, thus lack of face to face interaction. Unemployed specialists and once loyal employees may have difficulty securing future employment.

Therefore, as much as IS could help make tasks easier for an organization, it has certain issues and obstacles, which for an organization to have to face. First, are culture challenges, to state the fact that each country and regional area has their own intellectual awareness or culture as well as their own customs that could significantly affect individuals and organization involved in global trade, by means that it is a contagious issues with an influence affecting our culture. An organization should also be aware of the best way to approach global demographics, which had profound on the global landscape as well as on profession of globalization.

Second, limitation of language usage, by means that language differences create an issue that can make it difficult to translate the actual meaning of a conversation for instance. Due to the fact that meanings from another language may be different which leads to misunderstandings, irritations, feelings of exclusion, and a sense of inferiority, are daily challenges for not having spoken the language of English speakers trying to communicate in the language of global business. As English is mainly used all across the world for conversing. Moreover, on the Internet there are no facial expressions, body language, or other nonverbal cues, which makes communication even more complex.

Third, time and distance challenges where by this issues can be difficult to overcome for an individual and organizations involved with global trade in remote locations. Moreover large time differences make it difficult to communicate directly to people on the other side of the world. With long distance, it can take days to get the products or part from one location to another. Although IS as well and technology may be able to make communication faster and easier to use for individuals and organization, it may lead to difficulty in terms of different time zone, and the distance, where schedule for instance a meeting will be hardly organized to get group of individuals from all over the country to have a meeting through video conversation.[12,9–10,13]

Fourth, technology transfer issues, where most government does not allow certain military-related equipment and systems to be sold in some countries. As access to capital is limited, the capital costs of ESTs are generally higher than those of standard technologies. Also, the fact that risks of identifying existence for new technologies, financing costs will tend to be higher. Moreover, the availability of FDI is restricted and unevenly distributed around the world. Although many countries review their trade policies in order to loosen restrictions in terms of the markets, substantial tariff barriers as an obstacle remain in many cases for imports of external technologies including energy supply equipment. This limits exposure to energy in terms of productivity, resulting improvement pressures from foreign competition on national suppliers, and avert early introduction of sustainable energy that are able to maintain on certain level by alternation from abroad, where foreign exchange of restrictions and public revenue deliberation make across-the-board tariff removal difficult.

Last, regarding trade agreement—an international agreement on condition of trade in goods and services. Countries often enter into trade agreements with each other. Although it creates a dynamic business climate where business are protected by the agreement, lower government spending where numbers of government may put the fund for better use, increase in number of expertise that could develop local resources that then helps local entrepreneurs, as a consequence it results in an increase in job outsourcing, reduce tax revenue as without tariff and fees some countries may have to find a way to replace the revenue.

Moreover, one main problem is that ISs may not function properly which affects the running of the business, system may conflict with business strategies, the system analysis, or design may perform incorrectly, software development that inherits properties such as complexity, conformity, changeability, and invisibility. It can result in system breakdown, which interrupts smooth operations and which leads to consumer dissatisfaction. As has been noted, defective ISs can deliver wrong information to other systems which could create problems for the business and its customers. In other words, ISs are also vulnerable to hackers and frauds.

11.3.2 INFORMATION AND TECHNOLOGY

Information technology is considered as a subset of data frameworks. It manages innovation part of any data framework that is equipment, servers, working frameworks, and programming. Personal computer-based device that individuals use to work with data and backing the information and data processing is the need of an association. Concentrates on innovation and how it can help in spreading data. A business can utilize information technology to create organization database applications which can permit employees access information at any given moment. They can also use information technology instruments to set up networks that permit departments share information without any problem or wastage of time. With information technology, most associations have made a decentralized enlisting structure which unites the entire scope of the business' information in a methodical manner with the objective that it can be gotten to and used by any person who needs it. This structure of information is frequently a database, which is planned to clearly support the idea of shared information.

To state that the advantages of information technology include, that information technology has united the world, as well as it has permitted the world's economy to end up a solitary reliant framework. This implies we can share data rapidly and proficiently, as well as it cuts down hindrances of phonetic and geographic limits. The world has formed into a worldwide town because of the assistance of data innovation permitting nations like Chile and Japan who are isolated by separation as well as by dialect to impart thoughts and data to one another. Also, with the assistance of data innovation, correspondence has additionally gotten to be less expensive, faster, and more proficient. We can now correspond with anybody around the world by basically message informing them or sending them an e-mail for a practically prompt reaction. The web has additionally opened up face to face direct correspondence from various parts of the world because of the aides of video conferencing.

In addition, data innovation has electronic the business handle subsequently streamlining organizations to make them amazingly practical cash making machines. This thusly expands profitability which at last offers ascend to benefits that implies better pay and less strenuous working conditions. Data innovation has spanned the social crevice by peopling from various societies to correspond with each other, and take into account the trading of perspectives and thoughts, along these lines expanding mindfulness and decreasing partiality. Information technology also has made it feasible for organizations to be open 24×7 everywhere throughout the globe. This implies a business can be open at whatever time any place, making buys from various nations less demanding and more helpful. It additionally implies that merchandises conveyed right to the doorstep with moving a solitary muscle.

In terms of education, information technology makes it conceivable to have online education. Unlike in the past when education was tied to particular limits, now the education sector has changed. With the establishment of online education services, students can learn from anywhere utilizing Internet. This has helped in spreading of vital education materials to all individual mainly students across the globe.[7,5] Online education is also being improved by the making of portable application which empowers students' access to educational material via their mobile phones.[6,8]

In terms of agriculture, information technology plays a major part in advancing the agricultural sector. These days farmers can sell their products right from the homestead utilizing the Internet. All they need to do is create a site to advertise their products, orders will be placed directly by means of the site and the farmers will deliver fresh goods to the consumers once orders have been made. This gets rid of the middle men who tend to increase cost of farming products with the aim of making profits. In this case, information technology benefits both the farmer and the consumer. The consumer gets the product at a low price when it is still fresh, and the farmer gains additional income.

Somehow, most likely the best point of interest of data innovation is the making of new and intriguing occupations. PC software engineers, systems analyzers, hardware and software designers and web fashioners are only a portion of the numerous new business opportunities made with the assistance of information technology. In addition to that, job posting sites usually use information technology as a classification in their databases. The class incorporates an extensive variety of employments across architecture, engineering, and management functions. Individuals with occupations in these areas commonly have a progress on education mainly in software engineering and or ISs which they may also possess related industry certifications. Short courses in information technology fundamentals can also be discovered online and are particularly helpful for the individuals who want to get some introduction to the field before focusing on it as a career. A career in information technology can include working in or leading information technology departments, product advancement teams, or research groups.

Headways in data innovation have had numerous huge points of interest on society, and might be the crown gem of our time and indicate the progression of humankind, be that as it may, this has not come without its disadvantages to information technology or ISs that leave individuals thinking about whether the great exceeds the terrible. A number of weaknesses of data innovation incorporate while data innovation might have streamlined the business process it has additionally made occupation redundancies, cutting back, and outsourcing. This implies a ton of lower and center level occupations have been done away with bringing on more individuals to end up unemployed. Despite the fact that unemployment made and occupation made due to information technology is no place close in examination as information technology has unquestionably delivered inconceivable amount of employments. Change tragically is constant and in business terms you need to move with movement or be abandoned.

Despite the fact that data innovation might have made correspondence speedier, less demanding and more advantageous, it has additionally purchased along security issues.[1-3,7,25] From wireless sign block attempts to e-mail hacking, individuals are currently agonized over their once private data getting to be open information. Significant measure of individuals is uninformed of the endeavors substantial organizations go to gather information on individuals and the utilization and offering of this information. By and large thoughts, for example, online treats which promote the web client's hobbies can be seen as something to be thankful for yet one could think about whether faculty data processing is something worth being thankful for in the hands of extensive organizations whose essential premium is to inspire you to spend your well deserved cash.

On the other hand, industry specialists trust that the web has made professional stability a major issue as since alternation continues changing with every day. This suggests one must be in a steady learning mode, on the off chance that he or she wishes for their business to be secured. This of the disservices of information technology or IS has been around since the presentation of all innovation and one must not overlook that life is a ceaseless learning cycle and that you should stick to it or be abandoned. Moreover, as computing systems and capabilities keep growing worldwide, information overload has turned into an undeniably critical issue for many information technology experts. Efficiently processing immense measures of information to produce beneficial business intelligence necessitates a lot of processing power, sophisticated software, and human analytic expertise.

As data innovation might have made the world a worldwide town, it has likewise added to one society overwhelming another weaker one. For instance, it is currently contended that Unites States impacts how most youthful young people everywhere throughout the world now act, dress, and carry on. Dialects too have gotten to be dominated, with English turning into the essential method of correspondence for business and everything else. Loss of dialect and society is never something worth being thankful for, yet information technology or IS additionally impressive at holding learning of society, dialect, and one of a kind practices, so it is hence down to individuals to hold their character or social personality.

Over dependence on innovation where computers and the Internet has turned into a fundamental part of this current life, a few individuals, particularly youths who grow up with it, would not have the capacity to work without it. The Internet is conceivably making individuals sluggish, especially with regards to task or venture research as opposed to perusing books in a library, individuals can simply do a Google search. Also, the usage of technology in an organization, company, or business decreases the number of hours that a human works at that company. This may even result in some people losing their work because technology is doing it for them. However, this is advantageous for the organization as their profit will increase because they do not need to pay their workers as much because they are not needed as much.

With the ever growing variety of social networking sites such as Facebook and Twitter, it is not impossible that the traditional communication skills will be lost. Especially children who always engrossed in these websites because exchange of information and responsive skills are not important with computers. E-mails and instant messaging have replaced the old tradition of handwritten letters. Although this is advantageous considering time constraints but a personal touch and sense of feeling are lost in comparison to consuming the time to sit down and handwrite a letter.

In terms of health, studies have proved that technology can create an amount of problems with a person's health. Many scientists, doctors, and researchers are worried about potential links between technology and heart problems, eye strain, obesity, muscle problems, and deafness. Waste released from technology can contaminate the environment which not only makes people ill but also harms the environment.

11.3.3 INFORMATION SYSTEM AND TECHNOLOGY SUPPORT SCIENCE

The availability of open access and online chemical databases has made it easier for the people to know more about what is going on in the chemistry world and to keep updated with the recent findings. Integrating databases with other resources, including journal writing, it was important for advance scientific progress.[9–11] Enhancement of data and information integration specifically in scientific software system has become an issue of awareness among the chemists and the cheminformatics community for the past few years, but the issue has been solved by the development of the Semantic Web techniques.[24]

In the field of sciences, in order to develop a new research, it is always based on the previous findings. Therefore, it is crucial to keep record of the previous concepts of sciences, so that it can be used in the future whether for improvements on the finding or for the benefit of references.

The Semantic Web technique itself is the inclusion of machine-processable data in web documents and it is aimed to transform information which has not been structured or only semistructured into a fully organized web document which will be made accessible both to humans and machines. There are actually three major sections in the Semantic Web technique, they are the Dublin Core, Open Archives Initiative Object Reuse, and Exchange in short OAI-ORE and finally Simple Knowledge Organization in short SKOS .In this modern world, IS plays a very important role especially in collecting, organizing, and storing data or information. It is almost impossible for an organization to not have IS department. IS also consists of a series of advanced technology such as the latest software and hardware. In order to achieve information faster and in an organized manner, software such as Microsoft Excel might be of help in such situation. These sophisticated technologies can only be functioned to its full potential by people who have the accomplishment in handling those technologies. Basically IS is initially created for supporting operations, management, and decision-making in an organization. It also helps an organization in terms of communication networking, processing, and interpreting data.

Science is the body of knowledge of the physical and natural world which often requires the support of technology. The advancement of science is largely maintained by the frequently updated technology. Especially in the field of research, technology is probably the most significant necessity in ensuring the success of that certain research. Even tools and equipment used in science cannot be developed without the help of technology. The results of a science research which is made possible by the existence of technology are usually used for the benefits of the society such as drug discovery in the medical world.

For instance, drug discovery nowadays requires proper management system and the enhancement of the accessibility of potentially useful data. This can only be achieved by the existence of an IS and technology. If IS and technology were to be combined in the advancement of science, surely they will produce an efficient and a more effective finding in any research. This is due to the requirements needed for a research to be conducted, such as the complicated comprehensions of words in chemistry which can only be interpreted using a certain program such as the Semantic Web techniques (Borkum and Frey, 2014).

Semantic as we know it is the study of meanings of words, phrases, signs, and symbols. In chemistry alone there are plenty of words, phrases, signs, and symbols which are not familiar to the people who are not in the field of chemistry. Fortunately, with this newly founded technology, it is easier for the consumers to search for the meaning of a certain word, phrases, signs, or symbols. A Semantic Web technique is especially created for the purpose of smoothing of any chemistry research.[9–11]

The Dublin Core specifically focuses on definitions of specifications, vocabularies, and best practice for the assertion of metadata on the web. Dublin core is an initiative to create a digital "library card catalog" for the Web the elements that offer extended categorizes information and improved document indexing.

Similarly OAI-ORE Open Archives Initiative Object Reuse and Exchange in short OAI-ORE, is when some resources have meaningful relationship with other resources or becoming a part of other resources, such as a figure or a table which belongs to another resource of a scientific publication. Another example is when a resource is being associated with another resource such as when a review is made; it is related back to its original text of the scientific publication. The automated software system will then manipulate these sources as a whole instead of separating them. This technique can definitely make a certain research or reference a lot more efficient, effective, and less time consuming.

Then finally, the Simple Knowledge Organization System, in short SKOS, is a project that encouraged publication of controlled vocabularies on the Semantic Web, it is also highly dependent upon informal methods, including natural language. Another few important aspects of IS and technology in terms of supporting science are validity, accountability, and value proposition. Validity is deeply important, for instance in a laboratory environment an invalid risk assessment could have negative consequences, endangerment of human life inclusive. In the case of accountability, an organization or an individual is accountable or responsible for the validity of the information that they provided.

Value proposition depends on individual perspective and also organizational perspective. From an individual perspective, it is less time consuming in doing work and the data or information provided have been standardized so that it is easier to carry out a research. From an organizational perspective, it is actually risky to provide a source of information as an unauthorized access can easily condemn or leak out valuable information on the website if the website has a weak safety system.

Science, innovation, and advancement each speaks to a progressively bigger classification of exercises which are profoundly associated however particular. Science adds to innovation in no less than six ways.

First, new learning which serves as an immediate wellspring of thoughts for new mechanical potential outcomes. Second, wellspring of devices and methods for more proficient building outline and an information base for assessment of attainability of plans. Third, research instrumentation, lab strategies, and expository techniques utilized as a part of examination that in the end discover their way into configuration or mechanical practices, frequently through middle of the road disciplines.

Next, routine of exploration as a hotspot for improvement and absorption of new human aptitudes and abilities in the end helpful for innovation. Moreover, formation of an information base that turns out to be progressively critical in the appraisal of innovation as far as its more extensive social and natural effects. Last, information base that empowers more productive methodologies of connected examination, advancement, and refinement of new innovations.

The opposite effect of innovation on science is of in any event measure up to significance that through giving a ripe wellspring of novel exploratory inquiries and in this manner additionally defending the assignment of assets expected to address these inquiries in an effective and auspicious way, amplifying the motivation of science and as a wellspring of generally inaccessible instrumentation and methods expected to address novel and more troublesome investigative inquiries all the more effectively.

Particular illustrations of each of these two-way connections are stated. Due to numerous backhanded and also coordinate associations in the middle of science and innovation, the exploration arrangement of potential social advantage is much more extensive and more looking so as to differ than would be recommended just at the immediate associations in the middle of science and innovation.

The Institute has previously distributed contextual investigations one of advancements that have been come regarding the interest driven material science research. The studies show the long time scales over which the procedure between the first disclosures and the advancement of items that use the examination can happen. Four key advances highlighted in these distributions that empowered a number of the innovation based developments found in the administration areas are portrayed beneath.

First example is Fiber Optics where the advancement of fiber optic advances has took into consideration broadband web associations and quick overall correspondence of data, empowering online developments in the administration areas, for example, virtual interfaces, online medicinal services checking, and remote systems administration. The innovation has its roots in material science research by John Tyndall in the 1800s and in later research into photonics.

Second, the utilization of lasers has taken into consideration both quick correspondence through broadband systems, furthermore for quick information stockpiling and recovery through Compact Disc in short CD and Digital versatile Disc or Digital Video Disc, DVD advances. The guideline behind the laser was produced by Albert Einstein and it took over 40 years before the principal obvious wavelength laser was built. Analysts at the University of Surrey are as of now attempting to control quantum course lasers which could be utilized for restorative analysis, for instance, glucose observing for diabetics.

Third, liquid crystal display in short as LCD innovation empowers cell phones through light-weight, low-control utilization, and ease screens. The first exploratory examination that supports LCD innovation was led over 100 years prior, with further advancement in the second 50% of the most recent century. Basic LCD showcases are found in watches and number crunchers with more intricate shows now in cell telephones, personal computer, or PC screens and televisions.

Last is GPS where the capacity to precisely decide the position of an article or individual has empowered advancements, for example, web-based robbery following of autos and satellite route. GPS is supported by an extensive variety of material science research, from nuclear timekeepers to the hypothesis of general relativity, consolidated with space science and innovation.

11.3.4   CHEMINFORMATICS

Cheminformatics, otherwise called cheminformatics and compound informatics, is the utilization of personal computer and instructive procedures connected to a scope of issues in the field of science. These in silico strategies are utilized as a part of, for instance, pharmaceutical organizations during the time spent medication disclosure.[29,34]

The utilization of cheminformatics instruments is picking up significance in the field of translational exploration from medicinal chemistry to neuro-pharmacology. Specifically, require for it the investigation of substance data on huge datasets of bioactive mixes. These mixes frame vast multi-target complex systems; it is a drug-target interactome system that brings about an extremely difficult information examination issue. Counterfeit neural network in short for CNN, calculations may offer some support with anticipating the collaboration of medications and focuses in CNS interactome.

Cheminformatics can be simple with the right devices, as it begins with database. It keeps all your basic data and delegate information readily available. In any case, all the more imperatively, a cheminformatics framework like CDD Vault. CCD is a hosted database solution for secure analysis, management, and sharing chemical biological data, where it let intuitively organize chemical structure and biological structure data, and allows collaboration within the organization internal or external partners through web interface and it incorporates every one of the instruments important for a simple to utilize, end-to-end arrangement. CDD Vault even serves as the storehouse for more than two million open mixes and examined results. CDD Vault gives a complete medication revelation informatics framework, effortlessly information for action, compound similitude, and selectivity.

Perform Structure Activity Relationship (SAR) examination to recognize vital basic components.

Center for obtaining and screening endeavors on the ideal one of a kind arrangement of mixes survey drug suitability with Lipinski rules, Oprea Lead-like measurements, and other custom descriptors

CDD Vault makes the testing objectives of cheminformatics less troublesome where commonplace web interface makes cheminformatics more available to learners. In addition, incorporated diagrams offer some assistance with visualizing the relationship in the middle of action and properties. Information base instructional exercises also offer some assistance with navigating the cheminformatics scene such as submerge seek on exercises, structure, or naturally figured physical properties.

What is more, as a protected, facilitated cloud application, CDD Vault is a practical arrangement that is ideal for scholastic gatherings, charities, and little organizations. Current facilitated cloud design gives cost investment funds, additionally makes it conceivable to fabricate more natural interfaces, better than legacy frameworks. At the point when cheminformatics work includes others, CDD Vault makes coordinated effort simple with implicit correspondence and sharing capacities. The precise cut of the information with just the accomplices determined. For instance, cheminformatics analysis of organic substituents which identify most of common substituents, it calculates substituent properties, and as well as mechanical recognition of drug-like bioisosteric groups.

Namely, Enterprise Capability as an industrial strength database with all the benefits of the cloud which is easy to use. State the fact that there are zero footprint web interference run on all major browser, served from the certified cloud for immediate turnkey deployment. As for the cloud technology provides private, multi-tenant architecture that are able to provide affordable security and has demonstrated 99.98% history availability. Moreover it provides management tools that are easy to track in real time changes, showing full details for new compounds and data. The fact that it is customizable, where it could define protocols, data fields, preferred graphs, and even chemical registration business rules; however, customization services are available for more complex requests.

The entire combinatorial science is likewise taking into account the idea of substituents, acting for this situation under the name building squares. The present study concentrates on natural substituents from the perspective of cheminformatics and tries to reply questions about the aggregate number of substituents in known natural science space and the ramifications of this number for the span of virtual natural science space. The portrayal of substituents by ascertained properties is too examined, including a technique for computing substituent drug resemblance in view of an examination of the dissemination of substituents in a vast database of medications versus an extensive database of non drugs. At long last an illustration of the application of a vast database of medication such as substituents with computed properties presented a web-based instrument for programmed distinguishing proof of bioisosteric gatherings.

Cheminformatic is most likely to be beneficial for science field today equally to the world globally. The integration of chemistry and IS has proven to be advantageous as it assist chemist to comprehended and gather all the feature of molecules with certain pharmacological has improved.[31,2–3,29] This innovation and accessible database has helped to measure and analyze in the discovery of potential new drug.[33,35] Moreover, because of the unlimited accessible in the field of cheminformatics, chemists have the ultimate pleasure in the use enormous tools and methods that are advanced, safe, and bioavailable. Not only these but also different view of comprehension and predicting the bioactivity have their strength in partial least-square (PLS) or genetic algorithms.[30,32]

Unfortunately, every advantage has its own implication such as the discovery period of finding new drug will take quite some time. The tool and approach of cheminformatic might not be potential and accurate enough to reveal new discovery and it might be risky. Moreover, it might not utilize to its fullest potential and impacts than we realize. An increase in the cost of drug discovery has brought a rose in the efficiency industry only hence, it only encourages people to conduct a research only if it is profitable. Last, it requires a huge capital and maintenance.

11.4 CONCLUSION

The advancement of information technology achieved a turning point with the improvement of the Internet. Through the course of its improvement, specialists started finding different utilizations for the network, and utilization of the technology spread around the world. Access to the Internet today by people, organizations, and institutions alike has created a worldwide business sector for Internet service and has spurned an increase in productivity in the technological communication field. Information technology continuously developed in order to enhance today's organization system to be more manageable, productive, and systematic. Constant improvements can possibly create new applications of information technology that can affect all the areas of the society which includes the economy, households, government, and private sectors. Therefore, it is important to always be aware of the latest update of the technology in order to use the IS and technology to their maximum potential; it will definitely results in improvement, not just in an organization but also among individuals of the society.

KEYWORDS

- information and communication technology
- data frameworks
- computer-based device
- software engineering
- open archives initiative object reuse
- simple knowledge organization
- semantic web technique

REFERENCES

1. Susanto , H.; Almunawar, M. N. *Information Security Management Systems: A Novel Framework and Software as a Tool for Compliance with Information Security Standard*; CRC Press, 2018.

2. Susanto, H.; Chen, C. K. Macromolecules Visualization through Bioinformatics: An Emerging Tool of Informatics. *Appl. Phys. Chem. Multidiscip. Approaches* **2018**, *383*.

3. Susanto, H.; Chen, C. K. Informatics Approach and its Impact for Bioscience: Making Sense of Innovation. *Appl. Phys. Chem. Multidiscip. Approaches* **2018**, *407*.

4. Susanto, H. Smart Mobile Device Emerging Technologies: An Enabler to Health Monitoring System. *Kalman Filter. Tech. Radar Track.* **2018**, *241*.

5. Liu, J. C.; Leu, F. Y.; Lin, G. L.; Susanto, H. An MFCC-Based Text-Independent Speaker Identification System for Access Control. *Concurr. Comput. Pract. Exp.* **2018**, *30* (2), e4255.

6. Almunawar, M. N.; Anshari, M.; Susanto, H.; Chen, C. K. How People Choose and Use Their Smartphones. In *Management Strategies and Technology Fluidity in the Asian Business Sector*; IGI Global, 2018; pp 235–252.

7. Susanto, H.; Chen, C. K.; Almunawar, M. N. Revealing Big Data Emerging Technology as Enabler of LMS Technologies Transferability. In *Internet of Things and Big Data Analytics Toward Next-Generation Intelligence;* Springer: Cham, 2018; pp 123–145.

8. Almunawar, M. N.; Anshari, M.; Susanto, H. Adopting Open Source Software in Smartphone Manufacturers' Open Innovation Strategy. In *Encyclopedia of Information Science and Technology,* 4th ed.; IGI Global, 2018; pp 7369–7381.

9. Susanto, H. Cheminformatics—The Promising Future: Managing Change Of Approach Through Ict Emerging Technology. In *Applied Chemistry and Chemical Engineering, Principles, Methodology, and Evaluation Methods*, Vol. 2; 2017; 313.

10. Susanto, H. Biochemistry Apps as Enabler of Compound and DNA Computational: Next-Generation Computing Technology. *Applied Chemistry and Chemical Engineering, Experimental Techniques and Methodical Developments*, Vol. 4; 2017; 181.

11. Susanto, H. Electronic Health System: Sensors Emerging and Intelligent Technology Approach. In *Smart Sensors Networks;* 2017, pp 189–203.

12. Leu, F. Y.; Ko, C. Y.; Lin, Y. C.; Susanto, H.; Yu, H. C. Fall Detection and Motion Classification by Using Decision Tree on Mobile Phone. In *Smart Sensors Networks;* 2017; pp 205–237.

13. Susanto, H.; Chen, C. K. Information and Communication Emerging Technology: Making Sense of Healthcare Innovation. In *Internet of Things and Big Data Technologies for Next Generation Healthcare;* Springer: Cham; 2017; pp 229–250.

14. Susanto, H.; Almunawar, M. N.; Leu, F. Y.; Chen, C. K. Android vs iOS or Others? SMD-OS Security Issues: Generation Y Perception. *Int. J. Techno. Diffus. (IJTD)* **2016**, *7* (2), 1–18.

15. Susanto, H. Managing the Role of IT and IS for Suppoting Business Process Reengineering, 2016.

16. Susanto, H.; Kang, C.; Leu, F. Revealing the Role of ICT for Business Core Redesign, 2016.

17. Susanto, H.; Almunawar, M. N. Security and Privacy Issues in Cloud-Based E-Government. In *Cloud Computing Technologies for Connected Government;* IGI Global, 2016; pp 292–321.

18. Leu, F. Y.; Liu, C. Y.; Liu, J. C.; Jiang, F. C.; Susanto, H. S-PMIPv6: An Intra-LMA Model for IPv6 Mobility. *J. Netw. Comput. Appl.* **2015**, *58*, 180–191.

19. Susanto, H.; Almunawar, M. N. Managing Compliance with an Information Security Management Standard. In *Encyclopedia of Information Science and Technology*, 3rd ed.; IGI Global, 2015; pp 1452–1463.

20. Almunawar, M. N.; Susanto, H.; Anshari, M. The Impact of Open Source Software on Smartphones Industry. In *Encyclopedia of Information Science and Technology*, 3rd ed.; IGI Global, 2015; pp. 5767–5776.

21. Almunawar, M. N.; Anshari, M.; Susanto, H. Crafting Strategies for Sustainability: How Travel Agents Should React in Facing a Disintermediation. *Oper. Res.* **2013,** *13* (3), 317–342.

22. Nabil Almunawar, M.; Susanto, H.; Anshari, M. A Cultural Transferability on IT Business Application: iReservation System. *J. Hosp. Tour. Technol.* **2013,** *4* (2), 155–176.

23. Power, D. J.; Sharda, R.; Burstein, F. *Decision Support Systems*; John Wiley & Sons, Ltd., 2015.

24. Bajdor, P.; Grabara, I. The Role of Information System Flows in Fulfilling Customers' Individual Orders. *J. Studies Soc. Sci.* **2014,** *7* (2).

25. Grabara, J.; Kolcun, M.; Kot, S. The Role of Information Systems in Transport Logistics. *Int. J. Educ. Res.* **2014,** *2* (2), 1–8.

26. Tran, S. T.; Le Ngoc Thanh, N. Q. B.; Phuong, D. B. Introduction to Information Technology. In *Proc. of the 9th Inter. CDIO Conf. (CDIO)*, 2013.

27. Chaves, C. V.; Moro, S. Investigating the Interaction and Mutual Dependence Between Science and Technology. *Res. Policy* **2007,** *36* (8), 1204–1220.

28. Jamal, S.; Scaria, V. Cheminformatic Models Based on Machine Learning for Pyruvate Kinase Inhibitors of Leishmania Mexicana. *BMC Bioinform.* **2013,** *14* (1), 329.

29. Loging, W.; Rodriguez-Esteban, R.; Hill, J.; Freeman, T.; Miglietta, J. Cheminformatic/ Bioinformatic Analysis of Large Corporate Databases: Application to Drug Repurposing. *Drug Discov. Today Ther. Strategies* **2012,** *8* (3), 109–116.

30. Zhang, X.; Zheng, N.; Rosania, G. R. Simulation-Based Cheminformatic Analysis of Organelle-Targeted Molecules: Lysosomotropic Monobasic Amines. *J. Computer-Aided Molecular Design* **2008,** *22* (9), 629–645.

31. Emmanuel, A.; Sanya, J. O.; Olubiyi, O. O. Preliminary Identification of Lactate Dehydrogenase Inhibitors towards Anticancer Drug Development. *J. Dev. Drugs* **2015,** *2015*.

32. Romero-Durán, F. J.; Alonso, N.; Yañez, M; Caamaño, O.; García-Mera, X.; González-Díaz, H. Brain-Inspired Cheminformatics of Drug-Target Brain Interactome, Synthesis, and Assay of TVP1022 Derivatives. *Neuropharmacology* **2016,** *103*, 270–278.

33. Muchmore, S. W.; Edmunds, J. J.; Stewart, K. D.; Hajduk, P. J. Cheminformatic Tools for Medicinal Chemists. *J. Med. Chem.* **2010,** *53* (13), 4830–4841.

34. Stratton, C. F.; Newman, D. J.; Tan, D. S. Cheminformatic Comparison of Approved Drugs from Natural Product Versus Synthetic Origins. *Bioorganic Med. Chem. Lett.* **2015,** *25* (21), 4802–4807.

CHAPTER 12

CHEMICAL DATA MINING AND CHEMINFORMATICS

HERU SUSANTO*, LEU FANG-YIE, and
ANDRIANOPSYAH MAS JAYA PUTRA

¹Computational Science, The Indonesian Institute of Sciences, Indonesia

²Computer Science Department, Tunghai University, Taiwan

³School of Business and Economics, Universiti Brunei Darussalam, Brunei

*Corresponding author. E-mail: heru.susanto@lipi.go.id

ABSTRACT

The important features of information system and technology, as it explores the many different technologies inherent in the field of information technology and impact on information systems' design, functionality, operations, and management to its people, organization as well as science today. Sciences tend to make full usage of information system and technology to seek a description and understanding of the natural world and its physical properties. Hence, this study examines the concept of information system, technology, and how it affects science namely cheminformatics. The utilization of cheminformatics instruments is picking up significance in the field of translational exploration from medicinal chemistry to neuropharmacology. Specifically, require for it the investigation of substance data on huge datasets of bioactive mixes. These mixes frame vast multitarget complex systems; it is a drug-target interactome system, that bringing about an extremely difficult information examination issue. Counterfeit Neural Network, in short for CNN, calculations may offer some support with anticipating the collaboration of medications and focuses in CNS interactome. Here, this study finds out the effective and efficient solutions through cheminformatics approach.

12.1   INTRODUCTION

12.1.1   INFORMATION SYSTEM

Information system (IS) is similar to information and communication technology. It consists of raw data collection (input) that can be processed into an additional value for an organization (output). After being analyzed and processed, it would turn into information data or facts that could be used to answer questions, solve problems, or to conduct a project. Moreover, IS, also known as a software, is being used in an organization to help organize and analyze databases into a useful information that can be used in the decision making.

12.1.2   INFORMATION TECHNOLOGY

Information technology has been around for quite a long time. Essentially the length of individuals has been around, information technology has been around on the grounds that there were always ways of communicating through innovation accessible at that point in time. There are four main ages that divide up the historical background information of information technology. Just the latest age (electronic) and a percentage of the electromechanical age absolutely influence us today. Information technology can be referred to as comprising of three fundamental parts: Computational information processing, decision support, and business programming. Information technology or IT is broadly utilized as a part of business and the field of computing. Individuals use the terms generically when alluding to different sorts of PC-related work.

12.2   LITERATURE REVIEW

The usage of technology is very crucial in this modern world. The newly improved advanced technology has changed our life dramatically and made it much easier to anticipate the demand of our needs and wants. ISis practically being used everywhere and it has been evolving ever since in the form of devices, such as smartphones. Now it has been moving to home appliances, such as television, which is currently called as a "smart television."

ISs are integrated components that collect, manipulate, store, and disseminate data, information, and provide a feedback mechanism to meet objectives for individual, group, and organization. It is a vital processor that every computer needs because, without it, information technology and information and communication technology (ICT) will not be able to function properly or follow the instruction given by us. This is because IS is built with software that comes from manufacturer itself where it has its own function and follows the task given by us.

In today's world, technology has taken a huge part in improving our life. Now, IS could take part in helping science to improvise their medical methods into a new whole dimension of science in order to improve the development and assimilation of human life and capabilities of technology. Not only those, but it could also help scientists to conduct a practical research, to find a new discovery in science, to find solutions to the problems of what humans have been facing as such, new remedies are yet to be created to fight diseases. Every day, scientists all around the world are working hard to uncover mysterious solutions to what we, the human, are actually facing not only in the terms of medicine but also, new methods in the field of surgeon and technology.

How IS could help science would depend on the characteristics and how it is being used. Development in advanced technology and human capabilities; by using IS software it could help to gather all the raw data and molecules from the past research by synthesis with the new finding. Hence, it would uncover the mysteries of what it might be a new discovery in medicine or drugs to cure unsolved diseases. Especially by emerging technology combined with high instrumentation will help in advance the process without any complication. For instance, how research being conducted for the past couple of decades was immensely complicated because they have to go through every practical research and books, unlike how it is now because we have advanced machinery, technology, information, and the internet to help us.

12.3 DISCUSSION

12.3.1 INFORMATION SYSTEM

Range of definition given for this concept; however, information does not have a specific and uniform definition. The definition stated by N. Winer, who determines the content of the information gleaned from the outside world in the process of our adjustment to it and adapt it to our senses (McGarry, 2008).

IS is the combination of people's innovation and computer that processes or unravels information. ICT, that organization uses, as well as a way of communication people use to interact that leads in support for business processes. There is no specific and clear distinction between ISs, computer systems, and business processes. Nowadays, ISs mainly use to gain maximum benefits by processing data from inputs to generate IS, a combination of hardware, coordination, and decision-making in an organization. It is also known as decision-making support system. It is a collective combination of collection of people, procedure, software, database, and device that support problem specific decision making.

The main usage of IS in an organization is communication. It allows people to communicate easily, for instance, the method of communication that is electronic mail, or E-mail, which are delivered extremely fast, it can be sent and received from any device all around the world that has the internet connection. In addition, the availability of cloud function that allows people to make changes, add information, and share with one another in a community cloud. Thus, it makes communication better and faster, as most information are now accessible, and people could gain the right information at the right time. In order to meet one's need and want, as it is important to obtain accurate and complete information.

Business process may be difficult but as entrepreneurial culture and the degree to which the existing ISs tend to represent compatibility and application functionality significantly affect a firm's propensity to adopt cloud computing technologies. The discovery supports our abstract development and suggests complementarities between innovation diffusion theory and the information processing view to industry professionals to aid in making more informed adoption decisions in regard to cloud computing technologies to support of the supply chain.

Due to the fact that IS plays a major role mainly in business by creating new products and services, make it possible for managers to use real-time data when making a decision; therefore, information must be relevant and reliable in order to help people in their organization to achieve their goals and perform tasks more effectively that then leads to competitive advantage and these systems make their jobs easier as possible. Z. Messner stated that information as data on monetary phenomena and processes used in decision-making processes (Messner., 1991), namely, for human resource management, marketing, and administration.

Moreover, IS could store documents, histories, communication records as well as operational data that could be used in the future for better references, as it acts as a useful historical information. It improves efficiency of certain organization's operation in order to achieve aims, objectives, and goals. To achieve higher profitability.IS needs to be flexible, which helps to accommodate certain amount of variation, regarding the requirement in terms of supporting business process. The impact of IS being flexible enables cost efficiency for the business.

Information technology tends be to applied within business operations that can save a great deal of time during the fulfillment of daily tasks. Business tends to include knowledge management, artificial intelligence, expert system, multimedia, and virtual reality system. Paperwork is processed immediately, and financial transactions are automatically calculated by IS. Although businesses may view this expediency as a boon, there are untoward effects to such levels of automation. As technology amends, tasks that were formerly performed by human employees are now carried out by computer and ISs. This leads to the elimination of jobs and, in some cases, alienation of clients, thus lack of face to face interaction. Unemployed specialists and once-loyal employees may have difficulty securing future employment.

Therefore, as much as IS could help make tasks easier for an organization, it has certain issues and obstacles, which for an organization has to face. First is culture challenges, to state the fact that each country and regional areas have their own intellectual awareness or culture as well as their own customs that could significantly affect individuals and organizations involved in global trade, by means that it is a contagious issue with an influence affecting our culture. An organization should also be aware of the best way to approach global demographics, which had profound on the global landscape as well as on profession of globalization.

Second, limitation of language usage, means that language differences create an issue that can make it difficult to translate the actual meaning of a conversation for instance. Due to the fact that meanings from another language may be different that leads to misunderstandings; irritation, feelings of exclusion, and a sense of inferiority are daily challenges for not having spoken the language of English speakers trying to communicate in the language of global business. As English is mainly used all across the world for conversation, moreover, on the internet, there are no facial expressions, body language, or other nonverbal cues that makes communication even more complex.

Third, time and distance challenges, whereby these issues can be difficult to overcome for an individual and organizations involved with global trade in remote locations. Moreover, large time differences make it difficult to communicate directly to people on the other side of the world. With long distances, it can take days to get the products or parts from one location to another. Although IS, as well and technology, may be able to make communication faster and easier to use for individuals and organization, it may lead to difficulty in terms of different time zones, and the distance, where schedules, for instance, a meeting will be hardly organize to get group of individuals from all over the country to have a meeting through video conversation.

Fourth, technology transfer issues, where most governments do not allow certain military-related equipment and systems to be sold in some countries. As access to capital is limited, the capital costs of ESTs are generally higher than those of standard technologies. Also, the fact that risks of identifying existence for new technologies, financing costs will tend to be higher. Moreover, the availability of FDI is restricted and unevenly distributed around the world. Although many countries review their trade policies in order to loosen restrictions in terms of the markets, substantial tariff barriers as an obstacle remain in many cases for imports of external technologies including energy supply equipment. This limits exposure to energy in terms of productivity, resulting improvement pressures from foreign competition on national suppliers and avert early introduction of sustainable energy that is able to maintain on certain level by alternation from abroad, where foreign exchange of restrictions and public revenue deliberation make across-the-board tariff removal difficult.

Last, regarding trade agreement, an international agreement on condition of trade in goods and services, countries often enter into trade agreements with each other. Although, it creates a dynamic business climate where business is protected by the agreement, lower government spending, increase in number of expertise that could develop local resources that then helps local entrepreneurs, resulting an increase in job outsourcing, reduce tax revenue.

Moreover, one main problem is that ISs may not function properly, which affects the running of the business, system may conflict with business strategies, the system analysis or design may perform incorrectly, software development that inherits properties such as complexity, conformity, changeability, and invisibility. It can result to system breakdown, interrupting smooth operations and leading to consumer dissatisfaction. As has been noted, defective ISs can deliver wrong information to other systems that could create problems for the business and its customers. In other words, ISs are also vulnerable to hackers and frauds.

12.3.2 INFORMATION AND TECHNOLOGY

Information technology is considered as a subset of data frameworks. It manages innovation part of any data framework that is equipment, servers, working frameworks, and programming. Personal computer-based devices that individuals use to work with data and backing the information and data processing need of an association that concentrates on innovation and how it can help in spreading data. A business can utilize information technology to create organization database applications that can permit employees to access information at any given moment. They can also use information technology instruments to set up networks that permit departments to share information without any problem or wastage of time. With information technology, most associations have made a decentralized enlisting structure that unites the entire scope of the business information in a methodical manner with the objective that it can be gotten to and used by any person who needs it. This structure of information is frequently a database, which is planned to clearly support the idea of shared information.

The advantages of information technology include that information technology has united the world as well as it has permitted the world's economy to end up a solitary reliant framework. This implies we can share data rapidly and proficiently, as well as it cuts down hindrances of phonetic and geographic limits. The world has formed into a worldwide town because of the assistance of data innovation permitting nations like Chile and Japan, who are isolated by separation as well as by dialect, to impart thoughts and data to one another. Also, with the assistance of data innovation, correspondence has additionally gotten to be less expensive, faster, and more proficient. We can now correspond with anybody around the world by basically message informing them or sending them an email for a practically prompt reaction. The web has additionally opened up face to face direct correspondence from various parts of the world because of the aides of video conferencing.

In addition, data analytics and innovation has shifting electronically how the business handle subsequently, to make them amazingly practical cash revenue for organization improvement and sharing the market. This thusly expands profitability, which at last offers ascend to benefits that imply better pay and less strenuous working conditions. Data innovation has spanned the social crevice by peopling from various societies to correspond with each other and take into account the trading of perspectives and thoughts, along these lines expanding mindfulness and decreasing partiality. Information technology also has made it feasible for organizations to be open twenty-four seven everywhere throughout the globe. This implies a business can be open at whatever time, any place, making buys from various nations less demanding and more helpful. It additionally implies that merchandise conveyed right to the doorstep with moving a solitary muscle.

In terms of education, information technology makes it conceivable to have online education. Unlike in the past, when education was tied to particular limits, now the education sector has changed. With the establishment of online education services, students can learn from anywhere utilizing the internet. This has helped in spreading of vital education materials to all individuals, mainly students, across the globe. Online education is also being improved by the making of portable application that empowers students' access to educational material via their mobile phones.

In terms of agriculture, information technology plays a major part in advancing the agricultural sector. These days farmers can sell their products right from the homestead, utilizing the internet. All they need to do is create a site to advertise their products, orders will be placed directly by means of the site and the farmers will deliver fresh goods to the consumers once orders have been made. This gets rid of the middlemen who tend to increase cost of farming products with the aim of making profits. In this case, information technology benefits both the farmer and the consumer. The consumer gets the product at a low price when it is still fresh and the farmer gains additional income.

Somehow, most likely, the best point of interest of data innovation is the making of new and intriguing occupations. PC software engineers, system's analyzers, hardware and software designers, and Web fashioners are only a portion of the numerous new business opportunities made with the assistance of Information technology. In addition, job posting sites usually use information technology as a classification in their databases. The class incorporates an extensive variety of employments across architecture, engineering, and management functions. Individuals with occupations in these areas commonly have a progress on education mainly in software engineering and or ISs, which may also possess related industry certifications. Short courses in information technology fundamentals can also be discovered online and are particularly helpful for individuals who want to get some introduction to the field before focusing on it as a career. A career in information technology can include working in or leading information technology departments, product advancement teams, or research groups.

Headways in data innovation have had numerous huge points of interest on society, and might be the crown gem of our time and indicate the progression of humankind, be that as it may, this has not come without its disadvantages to information technology or ISs that leave individuals' thinking about whether the great exceeds the terrible. A number of weaknesses of data innovation incorporate while data innovation might have streamlined the business process it has additionally made occupation redundancies, cutting back, and outsourcing. This implies a ton of lower- and center-level occupations have been done away with bringing on more individuals to end up unemployed. Despite the fact that unemployment made and occupation made due to information technology is no place close in examination as information technology has unquestionably delivered inconceivable amount of employments. Change tragically is constant and in business terms, you need to move with movement or be abandoned.

Despite the fact that data innovation might have made correspondence speedier, less demanding and more advantageous, it has, in addition, purchased along with security issues. From wireless sign block attempts to email hacking, individuals are currently agonized over their once private data getting to be open information. Significant measure of individuals is uninformed of the endeavors substantial organizations go to gather information on individuals and the utilization and offering of this information. By and large thoughts, for example, online treats that promote the web client's hobbies can be seen as something to be thankful for yet one could think about whether faculty data processing is something worth being thankful for, in the hands of extensive organizations whose essential premium is to inspire you to spend your well-deserved cash.

On the other hand, industry specialists trust that the web has made professional stability a major issue as since alternation continues changing with every day. This suggests that one must be in a steady learning mode, on the off chance that he or she wishes for their business to be secured. This of the disservices of IS has been around since the presentation of all innovation and one must not overlook that life is a ceaseless learning cycle and that you should stick to it or be abandoned. Moreover, as computing systems and capabilities keep growing worldwide, information overload has turned into an undeniably critical issue for many information technology experts. Efficiently processing immense measures of information to produce beneficial business intelligence necessitates a lot of processing power, sophisticated software, and human analytic expertise.

As data innovation might have made the world a worldwide town, it has likewise added to one society overwhelming another weaker one. For instance, it is currently contended that US impacts how most youthful young people everywhere throughout the world now act, dress, and carry on. Dialects too have gotten to be dominated, with English turning into the essential method of correspondence for business and everything else. Loss of dialect and society is never something worth being thankful for, yet information technology or IS are, in addition, impressive at holding learning of society, dialect, and one of a kind practices, so it is hence down to individuals to hold their character or social personality.

Overdependence on innovation where computers and the internet have turned into a fundamental part of this current life, a few individuals, particularly youths who grow up with it, would not have the capacity to work without it. The internet is conceivably making individuals sluggish, especially with regards to task or venture research as opposed to perusing books in a library, individuals can simply do a Google search. Also, the usage of technology in an organization, company, or business decreases the number of hours that a human works at that company. This may even result in some people losing their work because technology is doing it for them. However, this is advantageous for the organization as their profit will increase because they do not need to pay their workers as much because they are not needed as much.

With the ever-growing variety of social networking sites, such as Facebook and Twitter, it is not impossible that the traditional communication skills will be lost. Especially children who always engrossed in these websites because exchange of information and responsive skills are not important with computers. Emails and instant messaging have replaced the old tradition of handwritten letters. Although this is advantageous considering time constraints a personal touch and sense of feeling are lost in comparison with consuming the time to sit down and handwrite a letter.

In terms of health, studies have proved that technology can create certain amount of problems with a person's health. Many scientists, doctors, and researchers are worried about potential links between technology and heart problems, eye strain, obesity, muscle problems, and deafness. Waste released from technology can contaminate the environment that not only makes people ill, but also harms the environment.

12.3.3 COMPUTER METHODS: CHEMINFORMATICS VIEW

The availability of open access and online chemical databases has made it easier for people to know more about what is going on in the chemistry world and to be updated with the recent findings. Integrating databases with other resources, including journal writing, it was important for advance scientific progress. Enhancement of data and information integration specifically in the scientific software system has become an issue of awareness among the chemists and the cheminformatics community for the past few years but the issue has been solved by the development of the Semantic Web techniques.

In the field of sciences, in order to develop new research, it is always based on the previous findings. Therefore, it is crucial to keep a record of the previous concepts of sciences, so that it can be used in the future whether for improvements on the finding or for the benefit of references.

The Semantic Web technique itself is the inclusion of machine-processable data in web documents and it is aimed to transform information that has not been structured or only semistructured into a fully organized web document that will be made accessible both to humans and machines. There are actually three major sections in the Semantic Web technique, they are the Dublin Core, Open Archives Initiative Object Reuse, and Exchange (OAI-ORE) and finally Simple Knowledge Organization (SKOS). In this modern world, IS plays a very important role especially in collecting, organizing, and storing data or information. It is almost impossible for an organization to not have IS department. IS also consists of a series of advanced technology, such as the latest software and hardware. In order to achieve information faster and in an organized manner, software, such as Microsoft Excel, might be of help in such situation. These sophisticated technologies can only be functioned to its full potential by people who have the accomplishment in handling those technologies. Basically, IS initially created for supporting operations, management, and decision making in an organization. It also helps an organization in terms of communication networking, processing, and interpreting data.

Science is the body of knowledge of the physical and natural world that often requires the support of technology. The advancement of science is largely maintained by the frequently updated technology. Especially in the field of research, technology is probably the most significant necessity in ensuring the success of that certain research. Even tools and equipment used in science cannot be developed without the help of technology. The results of science research, which is made possible by the existence of technology, are usually used for benefits of the society such as drug discovery in the medical world.

For instance, drug discovery nowadays requires proper management system and the enhancement of the accessibility of potentially useful data. This can only be achieved by the existence of an IS and technology. If IS and technology were to be combined in the advancement of science, surely, they will produce an efficient and a more effective finding in any research. This is due to the requirements needed for a research to be conducted, such as the complicated comprehensions of words in chemistry that can only be interpreted using a certain program, such as the Semantic Web techniques (Borkum and Frey, 2014).

Semantic as we know it is the study of meanings of words, phrases, signs, and symbols. In chemistry alone, there are plenty of words, phrases, signs, and symbols that are not familiar to the people who are not in the field of chemistry. Fortunately, with this newly founded technology, it is easier for the consumers to search for the meaning of a certain word, phrases, signs, or symbols. A Semantic Web technique is specially created for the purpose of smoothing of any chemistry research.

The Dublin Core specifically focuses on definitions of specifications, vocabularies, and best practice for the assertion of metadata on the web. Dublin core is an initiative to create a digital "library card catalog" for the Web the elements that offer extended categorize information and improved document indexing.

Similarly, OAI-ORE is when some resources have meaningful relationships with other resources or becoming a part of other resources, such as a figure or a table, that belong to another resource of a scientific publication. Another example is when a resource is being associated with another resource, such as when a review is made; it is related back to its original text of the scientific publication. The automated software system will then manipulate these sources as a whole instead of separating them. This technique can definitely make a certain research or reference a lot more efficient, effective, and less time-consuming.

Then finally, SKOS is a project that encouraged publication of controlled vocabularies on the Semantic Web, it is also highly dependent upon informal methods, including natural language. Another few important aspects of IS and technology in terms of supporting science are validity, accountability, and value proposition. Validity is deeply important, for instance in a laboratory environment an invalid risk assessment could have negative consequences, endangerment of human life inclusive. In the case of accountability, an organization or an individual is accountable or responsible for the validity of the information that they provide, whereas value proposition depends on individual perspective and also organizational perspective. From an individual perspective, it is less time-consuming in doing work and the data or information provided have been standardized so that it is easier to carry out a research. From an organizational perspective, it is actually risky to provide a source of information, as an unauthorized access can easily condemn or leak out valuable information on the website if the website has a weak safety system.

Science, innovation, and advancement each speak to a progressively bigger classification of exercises that are profoundly associated, however, particular. Science adds to innovation in no less than six ways.

First, new learning that serves as an immediate wellspring of thoughts for new mechanical potential outcomes. Second, wellspring of devices and methods for more proficient building outlines and an information base for assessment of attainability of plans. Third, research instrumentation, lab strategies, and expository techniques utilized as a part of examination that, in the end, discover their way into configuration or mechanical practices, frequently through middle of the road disciplines.

Next, routine of exploration as a hotspot for improvement and absorption of new human aptitudes and abilities, in the end, helpful for innovation. Moreover, formation of an information base that turns out to be progressively critical in the appraisal of innovation as far as its more extensive social and natural effects. Lastly, information base that empowers more productive methodologies of connected examination, advancement, and refinement of new innovations.

The opposite effect of innovation on science is of in any event measure up to significance that through giving a ripe wellspring of novel exploratory inquiries and in this manner additionally defending the assignment of assets expected to address these inquiries in an effective and auspicious way, amplifying the motivation of science and as a wellspring of generally inaccessible instrumentation and methods expected to address novel and more troublesome investigative inquiries more effectively.

Particular illustrations of each of these two-way connections are stated. Due to numerous backhanded and also coordinate associations in the middle of science and innovation, the exploration arrangement of potential social advantage is much more extensive and more looking so as to differ than would be recommended just at the immediate associations in the middle of science and innovation.

The institute has previously distributed contextual investigations, one of advancements that have been come regarding the interest-driven material science research. However, the studies introduced long-term scales that may lead handling the procedure between the first disclosures and the advancement of items for further stages of research observation. Four key advances highlighted in these distributions that empowered a number of the innovation-based developments found in the administration areas are portrayed beneath.

First example is fiber optics where the advancement of fiber optic advances has took into consideration broadband web associations and quick overall correspondence of data, empowering online developments in the administration areas, for example, virtual interfaces, online medicinal services checking, and remote system's administration. The innovation has its roots in material science research by John Tyndall in the 1800s and in later research into photonics.

Second, the utilization of lasers has taken into consideration both quick correspondence through broadband systems, furthermore for quick information stockpiling and recovery through compact disc in short CD and digital versatile disc or digital video disc, DVD advances. The guideline behind the laser was produced by Albert Einstein and it took over 40 years before the principal obvious wavelength laser was built. Analysts at the University of Surrey are, as of now, attempting to control quantum course lasers that could be utilized for restorative analysis, for instance, glucose observing for diabetics.

Third, liquid crystal display(LCD) innovation empowers cell phones through light-weight, low-control utilization, and ease screens. The first exploratory examination that supports LCD innovation was led over 100 years prior, with further advancement in the second 50% of the most recent century. Basic LCD showcases are found in watches and number crunchers with more intricate shows now in cell telephones, personal computer, or PC screens, and televisions.

Last is GPS, where the capacity to precisely decide the position of an article or individual has empowered advancements, for example, web-based robbery, following of autos, and satellite route. GPS is supported by an extensive variety of material science research, from nuclear timekeepers to the hypothesis of general relativity, consolidated with space science and innovation.

12.3.4   CHEMINFORMATICS: GENETIC ALGORITHM

Genetic algorithm, also called by partial-least-square methods with functions to find out and predicting comprehensive bioactive parts, most likely to be beneficial for the science field today, equally to the world, globally. The integration of chemistry and IS has proven to be advantages as it assists chemists to comprehended and gather all the feature of molecules with certain pharmacology has improved. This innovation and accessibility of database have helped to measure and analyze the discovery of potential new drug. Moreover, because of the unlimited accessibility in the field of cheminformatics, chemists have the ultimate pleasure in the use of enormous tools and methods that are advanced, safe, and bioavailable. Cheminformatics, and compound informatics, is the utilization of personal computer and instructive procedures connected to a scope of issues in the field of science. These in silico strategies are utilized as a part of, for instance, pharmaceutical organizations during the time spent medication disclosure. The utilization of cheminformatics instruments is picking up significance in the field of translational exploration from medicinal chemistry to neuropharmacology. Specifically, required for the investigation of substance data on huge datasets of bioactive mixes. These mixes frame vast multitarget complex systems; it is a drug-target interactome system, bringing about an extremely difficult information examination issue. Counterfeit Neural Network in short for CNN, calculations may offer some support with anticipating the collaboration of medications and focuses on CNS interactome. However, cheminformatics can be simple with the right devices, as it begins with database. It keeps all your basic data and delegate information readily available. In any case, all the more imperatively, a cheminformatics framework like CDD vault. CCD is a hosted database solution for secure analysis, management, and sharing chemical biological data, where it let intuitively organize chemical structure and biological structure data and allows collaboration within the organization internal or external partners through web interface and it incorporates every one of the instruments important for a simple to utilize, end-to-end arrangement. That registration, synthetic drawing, looking, and representation, and additionally SAR and different examinations. CDD vault even serves as the storehouse for more than 2 million open mixes and examine results. CDD vault gives a complete medication revelation informatics framework, effortlessly information for action, compound similitude, and selectivity.

CDD vault makes the testing objectives of cheminformatics less troublesome where commonplace web interface makes cheminformatics more available to learners. In addition, incorporated diagrams offer some assistance with visualizing the relationship in the middle of action and properties. Information-based instructional exercises also offer some assistance with navigating the cheminformatics scene such as submerge seek on exercises, structure, or naturally figured physical properties. What is more, as a protected, facilitated cloud application, CDD vault is a practical arrangement that is ideal for scholastic gatherings, charities, and little organizations. Current facilitated cloud design gives cost investment funds, in addition, makes it conceivable to fabricate more natural interfaces, better than legacy frameworks. At the point when cheminformatics work includes others, CDD vault makes coordinated effort simple with implicit correspondence and sharing capacities. The precise cut of the information with just the accomplices determined. For instance, cheminformatics analysis of organic substituents which identifies most of common substituents, calculation of substituent properties, and as well as mechanical recognition of drug-like bioisosteric groups.

Namely, enterprise capability as an industrial-strength database with all the benefits of the cloud that is easy to use. State the fact that there is zero footprint web interference run on all major browsers, served from the certified cloud for immediate turnkey deployment. As for the cloud technology provide private, multitenant architecture that is able to provide affordable security and has demonstrated 99.98% history availability. Moreover, it provides management tool that is easy to track in real-time changes, showing full details for new compounds and data. The fact that it is customizable, where it could define protocols, data fields, preferred graphs, and even chemical registration business rules, however, customization services are available for more complex requests.

The entire combinatorial science is likewise taking into account the idea of substituents, acting for this situation under the name building squares. The present study concentrates on natural substituents from the perspective of cheminformatics and tries to reply to questions about the aggregate number of substituents in known natural science space and the ramifications of this number for the span of virtual natural science space. The portrayal of substituents by ascertained properties is too examined, including a technique for computing substituent drug-resemblance in view of an examination of the dissemination of substituents in a vast database of medications versus an extensive database of nondrugs. At last an illustration of the application of a vast database of medication, such as substituents with computed properties, is presented a Web-based instrument for programmed distinguishing proof of bioisosteric gatherings.

12.4 CONCLUSIONS

The advancement of information technology achieved a turning point with the improvement of the internet. Through the course of its improvement, specialists started finding different utilizations for the network and utilization of the technology spread around the world. Access to the internet today by people, organizations, and institutions alike has created a worldwide business sector for internet service and has spurned an increase in productivity in the technological communication field. Information technology continuously developed in order to enhance today's organization system to be more manageable, productive and systematic. Constant improvements can possibly create new applications of information technology that can affect all the areas of the society that includes the economy, households, government, and private sectors. Therefore, it is important to always be aware of the latest update of the technology in order to use the IS and technology to their maximum potential; it will definitely result in improvement, not just in an organization but also among individuals of the society.

12.5   RECOMMENDATIONS

Although IS has already beneficial to almost all organizations, it is still open to embrace improvements. One of them is to create a technology usage agreement for the staff of that certain organization. For instance, an organization has the right to control its staff from browsing or visiting inappropriate web sites that are known to house viruses such as torrent and file sharing sites. In addition, the organization could also develop limitations for data or music downloading policy and include terms and conditions for data confidentiality. This is to assure that the organization's confidentiality is safely guarded by its staff.

The second improvement that can be made is having a back-up plan. It is impossible to know what will happen to an organization in the future, in order to be safe, hence, a backup plan is essential. One solution for this issue is to hire or consult one of the many firms that provide an off-site storage where an organization can keep record of the key documents and important databases. So that when an unexpected disaster happens, the organization can still recover by referring to the back-up plan.

The third betterment that can also be made is by setting up a schedule periodic maintenance downtime. Where the employees of the IS and technology department can manage updates, scan for viruses, backup all the data, fixing errors. This is due to the prevention of data loss and system malfunction.

Then the fourth method that can be considered is to contact the internet service provider of an organization. An organization can request to increase its bandwidth. Increasing bandwidth has a lot of advantages for the organization such as making multitasking much easier and it could also reduce application hang-ups caused by slow updating of the software of that certain application.

After that, the organization may also create a comprehensive technology plan. Comprehensive technology plan is where an organization can monitor the updates of technologies nowadays and, hence, use it for the benefit of the organization. The results of the monitoring on the latest technologies can be evaluated closely and then it can be adopted into the organization to replace aging workstations that are no longer efficient and effective in today's modern world.

Finally, an organization must at least make the effort to create a web site which is user-friendly. This can be done by considering software as a service. The organization then can provide many options for functions such as word processing, search engine, contact person or email and, most importantly, customer relationship management function.

KEYWORDS

- **cheminformatic**
- **dataset**
- **medical chemistry**
- **counterfeit neural network**
- **structure activity relationship**
- **genetic algorithms**

REFERENCES

1. Almunawar, M. N.; Anshari, M.; Susanto, H.; Chen, C. K. How People Choose and Use Their Smartphones. In *Management Strategies and Technology Fluidity in the Asian Business Sector*; IGI Global, 2018a; pp 235–252

2. Almunawar, M. N.; Anshari, M.; Susanto, H. Adopting Open Source Software in Smartphone Manufacturers' Open Innovation Strategy. In *Encyclopedia of Information Science and Technology*; *4th ed.* IGI Global, 2018b, pp 7369–7381.

3. Almunawar, M. N.; Anshari, M.; Susanto, H. Crafting Strategies for Sustainability: How Travel Agents Should React in Facing a Disintermediation. *Oper. Res.* **2013**, *13* (3), pp 317–342.

4. Almunawar, M. N.; Susanto, H.; Anshari, M. The Impact of Open Source Software on Smartphones Industry. In *Encyclopedia of Information Science and Technology,* 3rd ed; IGI Global, pp 5767–5776.

5. Almunawar, N. M.; Susanto, H.; Anshari, M. A cultural Transferability on IT Business Application: iReservation System. *J. Hosp. Tour. Technol.* **2013**, *4* (2), 155–176.
6. Bajdor, P.; Grabara, I. The Role of Information System Flows in Fulfilling Customers' Individual Orders. *J. Stud. Soc. Sci.* **2014**, *7* (2).
7. Brillouin, L. *Science and Information Theory*. Courier Corporation, 2013.
8. Chaves, C. V.; Moro, S. Investigating the Interaction and Mutual Dependence Between Science and Technology. *Res. Pol.* **2007**, *36* (8), 1204–1220.
9. Grabara, J.; Kolcun, M.; Kot, S. The Role of Information Systems in Transport Logistics. *Int. J. Edu. Res.* **2014**, *2* (2), 1–8.
10. Emmanuel, A.; Sanya, J. O.; Olubiyi, O. O. Preliminary Identification of Lactate Dehydrogenase Inhibitors Towards Anticancer Drug Development. *J. Develop. Drugs* **2015**, *4* (132), 2.
11. Jamal, S.; Scaria, V. Cheminformatic Models Based on Machine Learning for Pyruvate Kinase Inhibitors of Leishmania mexicana. *BMC Bioinform.* **2013**, *14* (1), 329.
12. Leu, F. Y.; Liu, C. Y.; Liu, J. C.; Jiang, F. C.; Susanto, H. S-PMIPv6: An intra-LMA Model for IPv6 Mobility. *J. Net. Comput. Appl.* **2015**, *58*, 180–191.
13. Leu, F. Y.; Ko, C. Y.; Lin, Y. C.; Susanto, H.; Yu, H. C. Fall Detection and Motion Classification by Using Decision Tree on Mobile Phone. In *Smart Sensors Networks*; 2017; pp. 205–237.
14. Liu, J. C.; Leu, F. Y.; Lin, G. L.; Susanto, H. An MFCC-based text-Independent Speaker Identification System for Access Control. *Concurr. Comput. Pract. Exp.* **2018**, *30* (2), e4255.
15. Loging, W.; Rodriguez-Esteban, R.; Hill, J.; Freeman, T.; Miglietta, J. Cheminformatic/ bioinformatic Analysis of Large Corporate Databases: Application to Drug Repurposing. *Drug Discov. Today: Ther Strat* **2012**, *8* (3), 109–116.
16. Power, D. J.; Sharda, R.; Burstein, F. *Decision Support Systems*. **2007**, *43* (3), 1044–1061.
17. Romero-Durán, F. J.; Alonso, N.; Yañez, M.; Caamaño, O.; García-Mera, X.; González-Díaz, H. Brain-inspired Cheminformatics of Drug-target Brain Interactome, Synthesis, and Assay of TVP1022 Derivatives. *Neuropharmacology* **2016**, *103*, 270–278.
18. Muchmore, S. W.; Edmunds, J. J.; Stewart, K. D.; Hajduk, P. J. Cheminformatic Tools for Medicinal Chemists. *J. Med. Chem.* **2010**, *53* (13), 4830–4841.
19. Stratton, C. F.; Newman, D. J.; Tan, D. S. Cheminformatic Comparison of Approved Drugs from natural Product Versus Synthetic Origins. *Bioorg. Med. Chem. Lett.* *25* (21), 4802–4807.
20. Susanto, H.; Almunawar, M. N. *Information Security Management Systems: A Novel Framework and Software as a Tool for Compliance with Information Security Standard*. CRC Press, 2018.
21. Susanto, H.; Chen, C. K. Macromolecules Visualization Through Bioinformatics: An Emerging Tool of Informatics. *Appl. Physi. Chem. Multidis Approach.* **2018**, 383.
22. Susanto, H.; Chen, C. K. Informatics Approach and Its Impact for Bioscience: Making Sense of Innovation. *Appl. Phys. Chem. Multidiscipl. Approach.* **2018**, 407.
23. Susanto, H. Smart Mobile Device Emerging Technologies: An Enabler to Health Monitoring System. *Kalman Filtering Techniques for Radar Tracking*; 2000; p 241.
24. Susanto, H.; Chen, C. K.; Almunawar, M. N. Revealing Big Data Emerging Technology as Enabler of LMS Technologies Transferability. In *Internet of Things and Big Data Analytics Toward Next-Generation Intelligence*; Springer: Cham, 2018, pp 235–252.

25. Susanto, H. Cheminformatics—The Promising Future: Managing Change of Approach Through ICT Emerging Technology. *Appl. Chem. Chem. Eng. Prin. Methodol Eval. Method.* **2017**, *2*, 313.

26. Susanto, H. Biochemistry Apps as Enabler of Compound and DNA Computational: Next-Generation Computing Technology. *Appl. Chem. Chem. Eng. Exp. Tech. Method. Develop.* **2017**, *2*, 181.

27. Susanto, H. Electronic Health System: Sensors Emerging and Intelligent Technology Approach. In *Smart Sensors Networks*; 2017; pp 189–203.

28. Susanto, H.; Chen, C. K. Information and Communication Emerging Technology: Making Sense of Healthcare Innovation. In *Internet of Things and Big Data Technologies for Next Generation Healthcare*; Springer: Cham, 2017; pp 229–250.

29. Susanto, H.; Almunawar, M. N.; Leu, F. Y.; Chen, C. K. Android vs iOS or Others? SMD-OS Security Issues: Generation Y Perception. *Int. J. Technol. Diffus. (IJTD)* **2016**, *7* (2), 1–18.

30. Susanto, H. Managing the Role of IT and IS for Suppoting Business Process Reengineering. *J. Systems Information Technol.* **2016**.

31. Susanto, H.; Kang, C.; Leu, F. Revealing the Role of ICT for Business Core Redesign.

32. Susanto, H.; Almunawar, M. N. Security and Privacy Issues in Cloud-Based E-Government. In *Cloud Computing Technologies for Connected Government*; IGI Global, pp 292–321.

33. Susanto, H.; Almunawar, M. N. Managing Compliance with an Information Security Management Standard. In *Encyclopedia of Information Science and Technology,* 3rd ed; IGI Global, pp. 1452–1463.

34. Tran, S. T.; Le Ngoc Thanh, N. Q. B.; Phuong, D. B. Introduction to Information Technology. In *Proc. of the 9th inter. CDIO Conf. (CDIO)*, 2013.

35. Wu, Y.; Cegielski, C. G.; Hazen, B. T.; Hall, D. J. Cloud Computing in Support of Supply Chain Information System Infrastructure: Understanding When to go to the Cloud. *J. Supp. Chain Manag.* **2013**, *49* (3), 25–41.

36. Zhang, X.; Zheng, N.; Rosania, G. R. Simulation-based Cheminformatic Analysis of Organelle-targeted Molecules: Lysosomotropic Monobasic Amines. *J. Comput.-Aid. Mol. Des.* **2008**, *22* (9), 629–645.

CHAPTER 13

IMPACT OF MOBILE CHEMISTRY APPS ON THE TEACHING-LEARNING PROCESS IN HIGHER EDUCATION

RAVINDRA S. SHINDE

Department of Chemistry, Dayanand Science College, Latur 413512, Maharashtra, India

E-mail: rshinde.33381@gmail.com

ABSTRACT

Electronic devices such as cellphones, smartphones, android, and tabs are valuable tools in teaching and learning process. Android mobiles are the daily needs in the life of education, teachers, and students. In the academic spectrum of chemistry, chemistry apps are powerful techniques used today in the teaching–learning process. According to the survey dating of 2017, the use of chemistry apps in the teaching–learning process helps both teachers and students to meet curriculum objectives and learning outcomes more meritoriously. The mobile apps in smartphones enhance student learning in and out of classrooms and laboratory. The accessibility of chemistry apps on smartphones and other portable electronic devices affords chemical experts and chemistry student's powerful and compact tools to solve problems.

13.1 INTRODUCTION

The incorporation of technology in our chemistry teaching is now ubiquitous but such is the amount of possibilities that it can be overwhelming for those new to the field to focus on areas of use to their teaching. Mobile chemistry apps technology has the potential to assist us in our role as teachers for teaching–learning process. One of such technology to study effective chemistry is through smartphones mobile apps learning. The chemistry "apps," on hand-held and portable touch-controlled computers such as smartphones and iPods are seeing dramatic growth with increasing adoption rates. In India, 60% of peoples use androids, iOS, smartphones, tabs, and other devices. The younger students are more likely to own androids. The potential of mobile learning is to provide education flexibly, most balk at the cost of providing students with mobile hardware. The habit of "bring your own device" is often mooted as a cost-effective alternative. The students are enthusiastically using mobile technologies such as androids, tablet computers, and smartphones to support their learning. The Indian government also promotes e-learning education like NPTEL, Swayam, Swayam Prabha, and e-PG Pathshala. The goal of mobile chemistry apps useful in teaching–learning process is to offer the project team with insights into how university students perceive mobile learning, the mobile technologies they own or access, the types of informal mobile learning they undertake and their mobile learning preferences. The teaching–learning process on mobile chemistry apps mostly depends on (1) student demographics; (2) practice of mobile devices by students to support learning or study; (3) technological resources such as internet access, Wi-Fi facility; (4) the accessibility and superiority of internet access; (5) ownership and access to mobile devices; (6) teacher attitudes and beliefs; (7) knowledge and skills; (8) classroom management; and (9) adapt technology and planning.

13.2 RESEARCH EXPERIMENTAL

All chemistry apps undergo support of a different operating system that you have on your smartphones.

1. Samsung
2. Redmi Mi
3. Apple iOS
4. Windows Phone
5. Google Android
6. Sony
7. Honor
8. Others

These are some of the smartphone and operating system used by student user, desktop, and tablet computers at different colleges and universities. The access of chemistry apps in androids, smartphones, and tablets is used in a large group of students. One of the simplest and most effective ways of incorporating technology into our teaching is to create podcasts and screencasts. Podcast (audio only) and screencast (audio with video or screen capture) allow students to recover material in their own time at their own pace. There are some useful resources for how these can be created in a chemistry context.

13.3 RESULT AND DISCUSSION

Chemistry is a complex topic involving a lot of terminologies, different vocabulary, and difficult to visualize molecule and dynamic of chemical bondings. The chemistry teacher recognizes these difficulties and uses various techniques to enhance the students. One such technique is the use of mobile chemistry apps into teaching classrooms, laboratory practical, and curriculum. Smartphones can aid as effective and suitable informative tools, which actually inspires learning. The android smartphones provide a crowd of applications that can be downloaded directly onto the cell phone. These mobile applications, or "apps," have an extensive array of functionalities and cover several disciplines. We have tested and use many of chemistry apps discussed here to give the students an objective outlook on the performance of each.

13.4 UTILIZATION OF MOBILE CHEMISTRY APPS IN TEACHING AND LEARNING CHEMISTRY

13.4.1 3D CHEMISTRY APPS

3D viewer apps can be a beneficial device in the chemistry classroom.

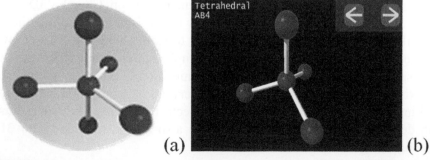

FIGURE 13.1 (a) 3D valence shell electron pair repulsion theory (VSEPR) and (b) screenshot of 3D VSEPR.

The 3D valence shell electron pair repulsion theory (VSEPR) app helps you to visualize the shapes of the VSEPR models in 3D such that you can understand more and you can sort out your confusion. This education app helps the students to learn chemistry in a smarter way. Students can see every parts of models by swiping their fingers on the screen. This app provides information of orbital, shape of orbitals, atoms, orbitals of all atoms, s orbital, p orbital, d orbital, f orbital, how orbitals looks, atomic orbitals, shapes of atomic orbitals, theory of orbitals, structure of atom, subshell, filling of orbitals in an atom, electronic density, arrangement of electrons in atoms, arrangements of orbitals, where are electrons, electronic configuration of atoms, see and observe orbitals, understanding orbitals, chemistry lovers, microworld, microscopic world, microscopic education, futuristic education, where electrons exist, molecular orbital theory, quantum physics, quantum numbers, molecular orbital diagrams, and orbital diagrams (Fig. 13.1).

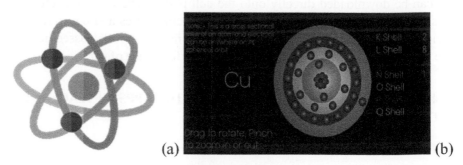

(a)   (b)

FIGURE 13.2   (a) ChemEx 3D; (b) screenshot of ChemEx 3D.

The other "ChemEx 3D" apps is a chemistry lab app. ChemEx 3D is all about adapting new ideas and creating new methods of learning chemistry. ChemEx 3D is a simplest and easiest way to learn chemistry. This allows for free check on periodic table and a lot information about 103 different atoms, such as atomic number, atomic name, atomic mass, atomic phase, electronic configuration, electron per cell, crystal structure, atomic density, melting and boiling point, critical point, electronegativity, covalent radius, van der Waals radius, and discovery that also allows to check their 3D structure (Fig. 13.2).

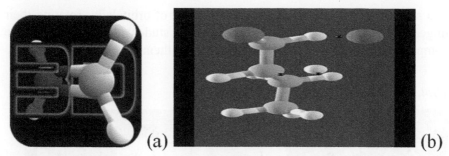

FIGURE 13.3   (a) Organic Chemistry Visualized and (b) screenshot of Organic Chemistry Visualized.

The "organic Chemistry Visualized" app provides a visual approach to organic chemistry. It is not meant to replace a textbook—it should be seen as a visual aid. The molecules and reactions are just briefly described. The main focus is the numerous animations of the molecules and reactions. Currently, alkanes, alkenes, and alkynes are described. You can test your knowledge through a quiz of 50 questions. The hardware requirements—screen size >3.7 in. and dual-core CPU. The animations are strongly compressed to keep the app size small. This is why the animations only run on newer smartphones. The Organic Chemistry Visualized is a free software application from the teaching and training tools subcategory, part of the Education category. The app is currently available in English. The program can be installed on android (Fig. 13.3).

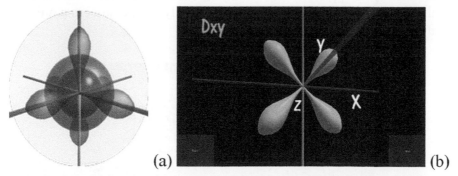

FIGURE 13.4   Virtual Orbitals 3D Chemistry and (b) screenshot of virtual d_{xy} orbitals.

The Virtual Orbitals 3D Chemistry app helps you to visualize the shapes of the orbitals in 3D such that you can understand more and you can sort out your confusions. This education app helps the students to learn chemistry in a smarter way. Students can see every parts of orbitals by rotating their fingers to screen. It is a useful app for chemistry students to understand some complex parts of orbitals and help them to feel them (Fig. 13.4).

13.4.2 CLASSIC PERIODIC TABLE APPS

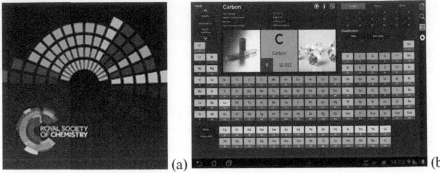

(a)   (b)

FIGURE 13.5 (a) Periodic Table Royal Society of Chemistry (RSC) and (b) screenshot of Periodic Table.

The "Periodic Table" app is useful for learning general chemistry and supporting several aspects of elemental periodicity. In the Periodic Table application, you will find a huge amount of data about chemical elements for free. The chemistry falls into the number of the most important sciences and is one of the main school objects. Its studying begins with the Periodic Table. Interactive approach to a training material is more effective than classical, as in it technologies that became the family for the modern pupils are used (Fig. 13.5).

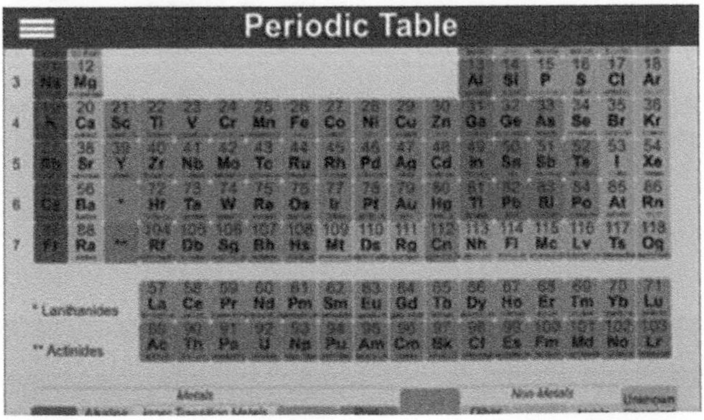

FIGURE 13.6 Screenshot of Periodic Table.

Periodic Table—is a free application for android that displays the entire periodic table at startup interface. The table has a long form approved by the International Union of Pure and Applied Chemistry as the core. Besides the Periodic Table of chemical elements, you can use the table of solubility. Did you know that neodymium is used in microphones? Or europium in Euro banknotes to help stop counterfeiting? These are just two of the absorbing facts in our customizable app, based on our popular and well-respected Royal Society of Chemistry Periodic Table website (Fig. 13.6).

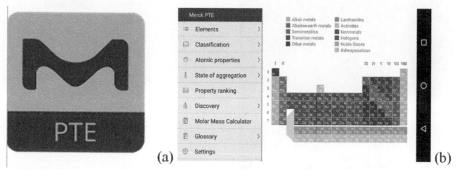

(a)   (b)

FIGURE 13.7   (a) "Merck PTE" app and (b) screenshot images of "Merck PTE" app.

The "Merck PTE" app is the ultimate tool for every friend of chemistry—whether pupil or teacher, student or professor, amateur or expert, hobbyist or technician. The app is a must-have of digital periodic tables. Get informed with mobile reference work, any time, with ease, offline and in detail (with 1 million downloads). Even more features; even more improvements; even more user-friendliness. It can be downloaded free of charge and start experimenting.

Features are as follows:

- All important information about the elements, such as atomic number, valence electrons, oxidation state, electronegativity according to Allred–Rochow and Pauling, atomic mass, boiling point, melting point, atomic radius, density, history, discoverer, classification, crystalline structure type, electron configuration, basic state, ionization energy, isotopic composition, state of matter, hardness according to Mohs, oxidation numbers, percentage of mass in Earth's crust, and year of discovery.
- Visualized element properties: atomic radius, atomic radius graphic, electronegativity (according to Allred–Rochow and Pauling),

ionization energy, relative atomic mass, state of matter, ranking list of properties, discovery, and classifications.

- Molar mass calculator, simple entry field for chemical formulas; calculate molar mass simply and quickly (Fig. 13.7).

13.4.3 CHEMISTRY NOTES APPS

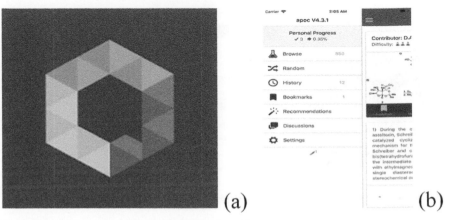

FIGURE 13.8   (a) apoc apk and (b) screenshot of apoc.

apoc—advanced problems in organic chemistry—is your fully interactive virtual set of enhanced flashcards. The app allows students of university-level organic chemistry classes to test their proficiency with chemical mechanisms and transformations taught in advanced chemistry courses. For students of organic chemistry around the world, the Evans database of organic chemistry problems has become an extremely useful learning resource. Hundreds of unique questions focusing on organic synthesis, conformational analysis, and many other categories can be browsed. Additionally, more than a hundred subcategories help refine the browsing process and tailor the results. The work done in this mobile project transforms this rich database into an easy-to-use and free mobile app, putting hundreds of organic chemistry problems at your fingertips. Additional features like shake for random questions, direct access to suggested solutions and the primary literature, and being able to keep track of your learning process and favorite exercises by bookmarking make apoc (Fig. 13.8).

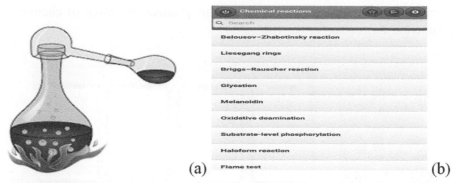

(a)　　　　　　　　　　　　　　　　　　　　　(b)

FIGURE 13.9 (a) Chemical Reaction app and (b) screenshot image of Chemical Reaction app.

The "Chemical Reaction" app describes the chemical reactions. The application includes a handy search and the ability to reproduce the text aloud (Fig. 13.9).

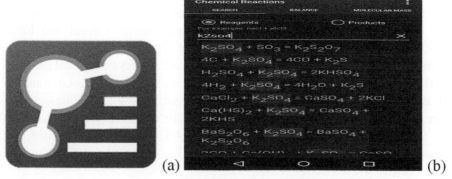

(a)　　　　　　　　　　　　　　　　　　　　　(b)

FIGURE 13.10 (a) Chemical Reaction apps and (b) screenshot image of Chemical Reaction.

This application provides an information about important inorganic chemistry reactions help to balance chemical reactions and to calculate molecular masses of chemical compounds. The application is free and works in offline mode.

However, the database of chemical reactions will be updated periodically. The key features of this app are searching for chemical reactions by reagents and by-products, search functionality comes together with the autocomplete functionality, balancing chemical reactions, calculation of molecular masses of chemical compounds and convenient mechanism for input of chemical compound formulas (Fig. 13.10).

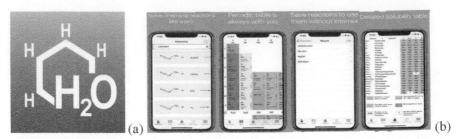

(a)   (b)

FIGURE 13.11   (a) Chem./Homework apps and (b) screenshot of Chemistry & Homework apple.

This app allows us to discover chemical reactions and to solve the chemical equations with one and or unknown variables. You will always have Mendeleev's Periodic Table and Solubility table handy, and even the calculator of molar masses. The app discovers the equations of chemical reactions even if the right or left part is unknown, helps you with organic and inorganic chemistry. The discovered reactions in a usual and ionic aspect will be mapped and formulas of organic chemistry are drawn for you. Mendeleev's periodic convenient interactive table: press a chemical element in the table to look at the information, the calculator of molar masses. Enter correctly a chemical compound and it will show molar masses and percentage of elements. The table of solubility of substances is added in the app. "Now your textbooks become waste! The best solver of chemical equations for iPhone and iPad" (Fig. 13.11).

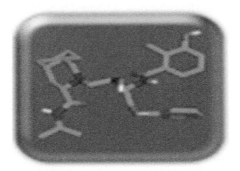

FIGURE 13.12   Organic Chemistry Nomenclature apps.

You are downloading the "Organic Chemistry Nomenclature" apk file for Android: Organic Chemistry Nomenclature is the ultimate way for chemistry students to study and memorize the names and structures of all the important chemical function (Fig. 13.12).

FIGURE 13.13   Chemistry Cheat Sheets apk.

The "Chemistry Cheat Sheets Free" apk covers the important topics in general chemistry and organic chemistry in a brief mode with several summary tables and figures. It covers maximum entry-level knowledge in all branches of chemistry, for example, organic compounds, salts, inorganic acids, gases, and biomolecules under "chemicals," study of various chemical, physical, and other properties (Fig. 13.13).

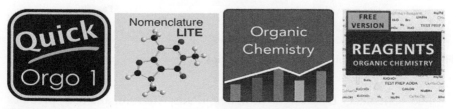

FIGURE 13.14   Apks for more study in teaching–learning process for students and teachers.

13.4.4 CHEMISTRY DESIGN APK

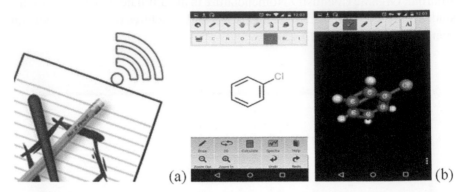

FIGURE 13.15   (a) ChemDoodle Mobile and (b) app screenshot.

ChemDoodle Mobile content rating is everyone. This app is listed in Education category of app store and has been developed by www.chem-doodle.com. You could visit iChemLabs, LLC.'s website to know more about the company/developer who developed this. ChemDoodle Mobile can be downloaded and installed on android devices supporting 10 api and above. Download the app using your favorite browser and click on install to install the app. Please note that we provide original and pure apk file and provide faster download speed than ChemDoodle Mobile apk mirrors. You could also download apk of ChemDoodle Mobile and run it using popular android emulators (Fig. 13.15).

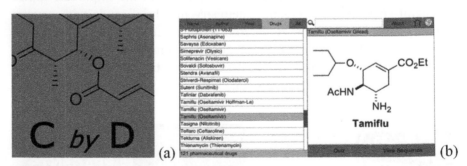

FIGURE 13.16   Chemistry By Design and (b) apk screenshot.

The "Chemistry By Design" (iPhones, iPads, Android, and iOS) précises the entire synthesis paths of the pharmaceutical compounds. The paths are branded by name, author, year, and drugs can be searched within the app.

This apk allows students to test their skills using known synthetic sequences. The goal is to display the reagents, starting materials and products for every single step used in constructing a natural product or pharmaceutical. A continuously growing database, currently featuring hundreds of syntheses of natural products and pharmaceuticals, is now available for browsing and testing your organic chemistry skills. You can choose to start your browsing by selecting from various lists or using the search box. Chemistry By Design is a free software application from the teaching and training tools subcategory, part of the Education category (Fig. 13.16).

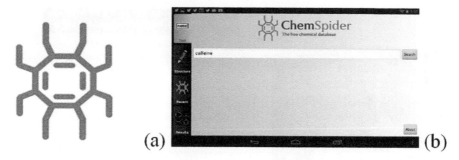

(a)   (b)

FIGURE 13.17   (a) ChemSpider and (b) screenshot image of ChemSpider.

ChemSpider Mobile allows you to search the ChemSpider chemical database provided by the Royal Society of Chemistry. Compounds can be searched by structure or by name and browsed within the app. Results can be examined by jumping to the webpage (Fig. 13.17).

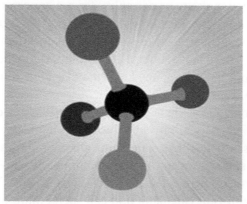

FIGURE 13.18   Chirys Draw apk.

Chirys Draw complex molecular structures and reaction schemes using your fingertips. An easy circle gesture (US Patent No. 9754085) makes even complicated polycyclic rings easy and fun to draw. Multiple double bonds or multiple functional groups are simple to add all at one time. Annotation and reaction condition text is chemically aware and properly subscripts numbers. With over 175k downloads, Chirys Draw is speeding the way scientists draw and communicate. Imagine what you will create (Fig. 13.18).

13.4.5 CHEMISTRY LABORATORY APK

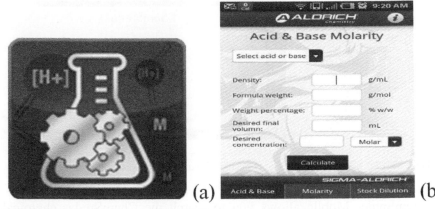

FIGURE 13.19   (a) Molarity App and (b) screenshot of Molarity App.

The Molarity App from Sigma-Aldrich(R) is a chemistry calculator tool that generates lab-ready directions describing how to prepare an acid or base solution of specified molarity or normality from a concentrated acid or base solution. A second tab includes a general molarity function that calculates the mass of any reagent needed to prepare a given volume of solution of desired molarity. A third tab features a stock dilution function that calculates how to dilute a stock solution of any known molarity to your desired volume and molarity (Fig. 13.19).

"Solution Calculator" is a handy tool for students taking chemistry classes or researchers/scientists working in a biology, chemistry, or biochemistry laboratory. It has a convenient calculator for making chemical solutions and for diluting solutions using a stock solution. It helps you to quickly determine how much chemical/stock solution you need. You do not need to mess around with your calculator and can spend more time in your study or research. It contains a handy tool to calculate molecular weight (MW) of commonly used chemicals in the lab. You do not need to enter the name or molecular formula of the chemical, you just need to press a few buttons to get the MW of the chemical instantaneously (Fig. 13.20).

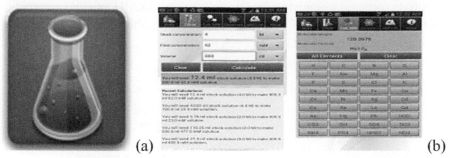

(a)   (b)

FIGURE 13.20   (a) Solution Calculator and (b) screenshot image of Solution Calculator.

13.4.6   RESEARCH APPS

FIGURE 13.21   Chemistry apk.

A simple app designed as a quick reference for chemistry students; includes a periodic table with links to Wikipedia, a tool to calculate molecular masses of compounds (with a button at the top to perform simple grams/ moles calculations, calculate mass percent, and do stoichiometry with that compound), a table of polyatomic ions, constants, solubility rules, tools for calculating molarity and volume of solutions, tools for converting between units commonly used in chemistry, compound lookup that links to major chemical search engines (and Wikipedia, of course), a table of reduction potentials, organic chemistry functional groups, IR, ^1H nuclear magnetic resonance (NMR) and ^{13}C NMR spectroscopy tables and a table with common ligands (Fig. 13.21).

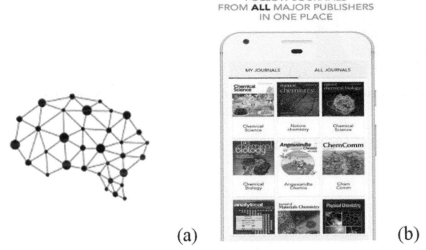

(a)   (b)

FIGURE 13.22   (a) Researcher apk and (b) screenshot image of Researcher apk.

In the Researcher app, you can find all the top impact factor journals and papers from Nature Springer, ACS, Elsevier, Wiley, Taylor and Francis, RSC, BMJ, IEEE, arXiv, PubMed, PLOS, PNAS, PubMed, F1000, Cell, Science AAAS, and more. The university students and researchers find the app invaluable when writing literature reviews, composing their thesis, and learning what's new in the world of science. It's free and easy to use, allowing you stay on top of your academic research by enabling you to follow, filter and save papers from all journals relevant to your research, and then sync them with your reference manager (Fig. 13.22).

13.5   CONCLUSIONS

The mobile apps are the beautiful technology for increasing student's engagement in chemistry. The mobile apps along with the multimedia features introduced in this chapter will potentially attract students to obtain effective and interactive learning experience in the fields of chemical science, chemical technology, chemical engineering, and other subjects. The apps create opportunities for collaborative activities among students. The mobile chemistry apps on androids and other portable electronic devices give chemistry students and chemical professionals effective and chemical tools to solve problems appropriately with no tension and good interest from electronic media, high-price heavy books, and large personal computer system (PCs)

ACKNOWLEDGMENTS

The author is thankful to Mr. Mahesh Ramrao Kularni, Ph.D. Student, K. T. H. M. College, for helpful discussion during the preparation of this manuscript. The authors would like to thank all app developers for their permission to let us publish the app icons and screenshots.

COMPLIANCE WITH ETHICAL STANDARDS

This study was not funded by any agency.

ETHICAL APPROVAL

This article does not contain any studies with human participants or animals performed by any of the authors.

CONFLICT OF INTEREST

The author declares no conflict of interest.

KEYWORDS

- **m-learning**
- **m-classrooms technology**
- **m-chemistry apps**
- **m-learning**
- **m-effective teaching**

REFERENCES

1. Bennett, J.; Pence, H. E. Managing Laboratory Data Using Cloud Computing as an Organizational Tool. *J. Chem. Educ.* **2011,** *88*, 761–763.
2. Alrasheedi, M.; Capretz, L. F. Determination of Critical Success Factors Affecting Mobile Learning: A Meta-Analysis Approach. *Turk. Online J. Educ. Technol.* **2015,** *14*, 41–51.
3. Clark, A. M.; Ekins, S.; Williams, A. J. Open Drug Discovery Teams: A Chemistry Mobile App for Collaboration. *Mol. Inform.* **2012,** *31*, 585–597.
4. Williams, A. J.; Pence, H. E. Smart Phones, a Powerful Tool in the Chemistry Classroom. *J. Chem. Educ.* **2011,** *88*, 683–688.
5. Williams, A. J.; Ekins, S.; Clark, A. M.; Jack, J. J.; Apodaca, R. L. Mobile Apps for Chemistry in the World of Drug Discovery. *Drug Discov. Today* **2011,** *16*, 928–939.
6. Evans-Cowley, J. Planning in the Real-Time City: The Future of Mobile Technology. *J. Plann. Lit.* **2010,** *25*, 136–149.
7. Ismail, I.; Azizan, S. N.; Azman, N. Mobile Phone as Pedagogical Tools: Are Teachers Ready. *Int. Educ. Stud.* **2013,** *6*, 36–47.
8. Murphy, A.; Farley, H.; Lane, M.; Hafeez, B. A.; Carter, B. Mobile Learning Anytime, Anywhere: What Are Our Students Doing. *Australas. J. Inf. Syst.* **2014,** *18*, 331–345.
9. Lampe, C.; Wohn, D. Y.; Vitak, J.; Ellison, N. B.; Wash, R. Student Use of Facebook for Organizing Collaborative Classroom Activities. *Int. J. Comput.-Support. Collabor. Learn.* **2011,** *6*, 329–347.
10. Clark, A. M.; Ekins, S.; Williams, A. J. Redefining Cheminformatics with Intuitive Collaborative Mobile Apps. Mol. Inform. **2012,** *31*, 569–584.
11. Kobus, M.; Rietveld, P.; van Ommeren, J. N. Ownership versus On-Campus Use of Mobile IT Devices by University Students. Comput. Educ. **2013,** *68*, 29–41.
12. Mobile   Reagents.   http://itunes.apple.com/us/app/mobilereagents/id395953310 (accessed November 3, 2018).
13. ChemSpider   Mobile.   http://itunes.apple.com/us/app/chemspider/id458878661 (accessed November 3, 2018).
14. Android. www.android.com/ (accessed November 3, 2018).
15. Ekins, S.; Clark, A. M.; Williams, A. J. Incorporating Green Chemistry Concepts into Mobile Chemistry Applications and Their Potential Uses. *ACS Sustain. Chem. Eng.* **2013,** *1*, 8–13.

PART 3
Perspectives on Technological Innovations

CHAPTER 14

NUCLEAR SCIENCE AND TECHNOLOGY

FRANCISCO TORRENS[1,*] and GLORIA CASTELLANO[2]

[1]Institut Universitari de Ciència Molecular, Universitat de València, Edifici d'Instituts de Paterna, PO Box 22085, E-46071 València, Spain

[2]Departamento de Ciencias Experimentales y Matemáticas, Facultad de Veterinaria y Ciencias Experimentales, Universidad Catylica de Valencia San Vicente Mártir, Guillem de Castro-94, E-46001 València, Spain

*Corresponding author. E-mail: torrens@uv.es

ABSTRACT

The atom consists of a nucleus presenting more than 99% mass surrounded by electrons. In Einstein's *Theory of Special Relativity*, energy and mass are equivalent in the equation: $E = mc^2$. In fission, a heavy nucleus splits into two smaller others. Thermal reactors need to moderate fast neutrons from fission to sustain a chain reaction. Residual heat occurs in nuclear fuel once when the fissions are finished because of fission-product radio-activity. Most nuclear power plants use U as nuclear fuel to obtain heat. Nuclear fuel cycle follows mining, extraction, enrichment, making, reactor, and management as nuclear waste. Nuclear fuel elements are rod bundles of piled U-ceramic tablets that gather with different designs. The main structural components of an element are bolsters, grids, guide tubes, and fuel rods. Reserves of U will continue rising as a result of the new exploitation efforts. Nuclear plants need a culture shared by all workers: workers, society, and environmental safety. The prior objective of nuclear energy is the safe exploitation of its installations. To compensate mechanical faults and human errors, the principle of *defense in depth* is used. Dangerous nuclear waste is not expelled but confined via successive hermetic barriers. Factors, organizational and personnel management, play a relevant role in nuclear plants' safety. Council of Nuclear Safety watches the fulfillment of safety objectives and radiological protection. Radiations cause *stochastic* and *deterministic* damages interacting with deoxyribonucleic acid. Received dose is measured in Sieverts, being obtained multiplying Grays by arbitrary coefficients. Detectors present characteristics that validate them for different measurement types.

14.1 INTRODUCTION

Setting the scene: nuclear science and technology.

In earlier publications, it was informed nuclear fusion, American nuclear cover-up in Spain after Palomares (Almería) disaster (1966),[1] *Manhattan Project, Atoms for Peace*, nuclear weapons and accidents.[2] The aim of this work is to initiate a debate by suggesting a number of questions (Q), which can arise when addressing subjects of basic nuclear science and technology, in different fields, and providing, when possible, answers (A), hypotheses (H), and experiments (E). Specific fields of the debate are as follows: the principles of nuclear physics and radiation, nuclear power plants (NPPs), nuclear fuel (NF), nuclear safety (NS), radiological protection (RP), radioactive waste (RW), dismantling and close of radioactive and nuclear installations, NPPs of the future, another applications of nuclear technology, nuclear technology, and sustainable development (SD).

14.2 PRINCIPLES OF NUCLEAR PHYSICS AND RADIATION

Pérez Martín organized *Basic Course on Nuclear Science/Technology*.[3] Mesado Meliá proposed H/E/Q/A on nuclear physics/radiation principles.

H1. (Democritus, ca. −400). Matter is an entity formed by *atoms*, which are indivisible entities.

H2. (Dalton, 1808). Matter is formed by atoms indivisible, equal in every element, uncharged.

E1. (Becquerel, 1896). Some materials that can be found in nature emit particles (*radioactivity*).

E2. (Thompson, 1897). He discovers *electron* (e^-), particle smaller than atom/electrically charged.

H3. (Thompson, 1898). Atomic model: a positively charged mass that has lodged inside e^-.

E3. (Curie, Curie, 1898). *Radioactive* Ra/Po emit α-*particles* [2 *protons* (p^+) + 2 *neutrons* (n^0)].

E4. (Rutherford). Bombarding Au sheets with α-particles most passed, some deviated, few bounced.

H4. (Rutherford, 1911). Nuclear model: atom formed by positively charged nucleus with e^- around (cf. Fig. 14.1).

H5. (Bohr, 1913). Nuclear model that takes into account quantum physics.

H6. (Schrödinger, 1926). Quantum mechanics model.

E5. (Chadwick, 1932). Experiments verifying the n^0.

E6. Nucleus is formed by nucleons: uncharged n^0 and positively charged p^+.

E7. Nucleons are composed of smaller particles called *quarks*.

E8. (Fermi). He bombarded the heaviest material U with n^0.

E9. (Hahn, Meitner, Strassmann, 1939). Nuclear fission: bombarding U with n^0, Ba nuclei appeared.

E10. (Fermi, 1942). Chicago Pile-1: First self-maintained nuclear *chain reaction* (CR).

E11. (Manhattan Project, 1944). Hanford (WA): First nuclear reactor (NR) operating continuously.

E12. (Los Alamos National Laboratory, 1951). EBR-1: The first NPP.

H7. (Eisenhower, 1953). Discourse *Atoms for Peace*: The United States opens nuclear technology for its civil use.

E13. [Calder Hall (UK), 1956]. The first NPP in Europe.

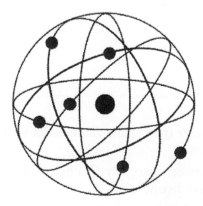

FIGURE 14.1 Scheme of the atom according to Rutherford's model.
Source: Ref. [3].

Table 14.1 lists the typical dimensions of different objects.

H8. Matter is almost empty: H nucleus (p^+) is 100,000 times smaller than H atom.

E14. *Isotopes*: Elements with different number of n^0 [H (1 p^+): ^{1}H (0 n^0); ^{2}H (D, 1 n^0); ^{3}H (T, 2 n^0)].

E15. Another example, U (92 p^+): ^{238}U (146 n^0); ^{235}U (143 n^0); ^{233}U (141 n^0).

TABLE 14.1 Typical Dimensions of Different Objects.

Object	Dimension (m)
Galaxy	10^{22}
Light year	10^{16}
Solar system	10^{14}
Earth orbit	10^{11}
Moon orbit	10^9
Earth	10^7
Madrid–Segovia distance	10^5
1 kilometer	10^3
A tree	10^1
A table	10^0
A pencil	10^{-1}
A fly	10^{-2}
Pencil tip	10^{-3}
Human cell	10^{-4}
Cell nucleus	10^{-6}
Chromosome	10^{-8}
DNA diameter	10^{-9}
H atom	10^{-10}
Pb nucleus	10^{-11}
Proton and neutron	10^{-15}

Source: Ref. [3].

E16. *Stable* and *unstable* (turn stable via radioactive processes) *isotopes* of the same element.

H9. Equilibrium between two forces: charges Coulomb repulsion and attraction at short distances.

H10. For small nuclei, 1 n^0 by p^+; for big nuclei, more n^0 than p^+ are necessary to maintain stability.

E17. Unstable nuclei develop strategies to be stable (*radioactive processes*).

E18. Radioactive processes: α/β^- ($n^0 \rightarrow p^+ + e^-$)/$\beta^+$ [$p^+ \rightarrow n^0$ + positron (e^+)]/γ-disintegration (electromagnetic radiation); fission ($\rightarrow 2$ nuclei).

Table 14.2 shows disintegration times of some radioactive isotopes.

TABLE 14.2 Disintegration Times of Some Radioactive Isotopes.

Isotope	Semidisintegration time ($T_{1/2}$)
^{238}U	4470,000,000 years (natural)
^{235}U	704,000,000 years (natural)
^{239}U	23.5 min (artificial)
^{233}Th	22.2 min (artificial)

Source: Ref. [3].

In nuclear fission heavy nucleus splits emitting particles (n^0, $\gamma/\alpha/\beta^-$ radiation, cf. Fig. 14.2).

E19. In Einstein's *Theory of Special Relativity*, mass excess transforms into energy: $E = mc^2$.

E20. Only some heavy nuclei are capable of being fissioned by any energy n^0; for example, $^{233/235}U/^{239}Pu$.

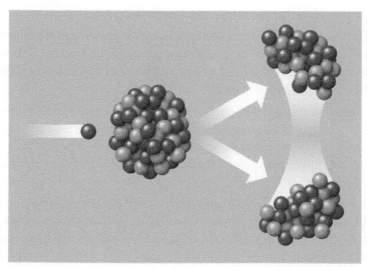

FIGURE 14.2 Scheme of a fission reaction.
Source: Ref. [3].

Q1. How do we know which reaction will occur when a particle interacts with an atom?

A1. The concept of *effective section*.

Q2. What is a CR?

A2. If in fissioned-nucleus surroundings other nuclei exist able to be fissioned absorbing emitted n^0.

Q3. How is a CR controlled?

A3. In an NR, CR is self-maintained during the operation cycle between NF recharges.

E21. Generated n^0 are fast ($E \sim 1$ MeV) but energy for the greatest probability to fission ^{235}U is 0.025 eV.

E22. *Thermalization/moderation*: Moderator (H_2O, $C_{graphite}$) reduces n^0 energy via atomic collisions.

E23. Fast n^0 improve NF yield; nonfissile ^{238}U is converted into fissile ^{239}Pu:

$$^{238}U + n \rightarrow {}^{239}U \rightarrow {}^{239}Np \rightarrow {}^{239}Pu \qquad (14.1)$$

E24. *Residual heat*: When CR is stopped NF continues heating during years decaying exponentially.

Q4. Why cannot an NPP explode like a nuclear bomb?

He provided the following conclusions (Cs).

C1. The atom consists of a central nucleus that presents more than 99% of its mass, surrounded by e^- that move around the nucleus. The nucleus of an atom is formed by n^0 and p^+, which are called *nucleons* (because they conform to the nucleus).

C2. Energy and mass are equivalent, as Einstein postulated in his equation: $E = mc^2$. Fusion and fission are two types of nuclear reactions in which energy is released because the nuclei resultant of the reaction present lower mass than the initial nuclei. The mass difference is transformed into energy.

C3. In fusion, two light nuclei join into other of greater mass but lower than the sum of the original nuclei. In fission, a heavy nucleus splits into two others of lower mass. In it, it is released, among other particles, n^0, which can, in turn, fission new nuclei and create a CR, which is the base of NPPs working. The first NR was built by Fermi (1942).

C4. Thermal reactors need to moderate the fast n^0 coming from fission to sustain a CR. Fast reactors use directly the fast n^0 for their work, without the necessity of slowing them down.

C5. Residual heat occurs in NF once finished the fissions, because of the radioactivity of the fission products and other transuranic products, elements heavier than U, generated when this absorbs n^0.

14.3 NUCLEAR POWER PLANTS

Gómez Zarzuela proposed the following question, answer, experiment and hypothesis on NPPs.

Q1. How is the calorific energy obtained from the U atoms?

A1. To free and dispose of nuclear energy (NE), it is essential to split (fission) the nucleus.

E1. Energetic equivalences: 1 U tablet = 565 L of petroleum = 810 kg of coal = 480 m^3 of natural gas.

H1. Reactors: light [pressurized (PWR) and boiling water reactor (BWR)]; Canada deuterium U (CANDU); VVER .

She provided the following conclusion.

C1. The great majority of NPPs use U as NF to obtain heat. To produce the heat, n^0 cause that U-nuclei break (fission), releasing a great amount of energy. The generated heat is used to heat water and produce steam, which is converted into mechanical energy in turbines, which is finally transformed into electricity in the alternator. In Spain, two types of NPPs exist: PWR and BWR.

14.4 NUCLEAR FUEL

Olmo Juan proposed the following hypotheses, questions, answers, and experiments on NF.

H1. *Fissionable*: That can fission faced with an energetic enough neutron (n^0).

H2. *Fissile*: That fissions independently of the energy of the incidental n^0; for example, $^{233/235}U/^{239}Pu$.

Q1. Why is U enriched?

A1. ^{235}U is fissile but ^{238}U is not.

E1.   Natural U (0.7% fissile ^{235}U and 99.3% nonfissile ^{238}U) is enriched till 1–4.5% fissile ^{235}U.

E2. $UF_{6(g)}$: ^{235}U enrichment methods: diffusion through a porous membrane and centrifugation.

Q2.   What aspect has NF?

E3. $UO_{2(s)}$ is compacted in tablets and sintered (1700°C, tablets of 1 cm diameter × 1 cm height).

E4.   Heavy water D_2O reactors (HWRs) absorb lesser n^0 than H_2O and do not need U enrichment.

E5.   Types of HWRs: CANDU; pressurized HWR.

He provided the following conclusions.

C1.   Cycle of NF comprises the following phases: mining, extraction, U-enrichment, NF manufacture, its use in the reactor, and its management as nuclear waste.

C2.   Elements of NF are bundles of rods of piled U-ceramic tablets that gather with different designs, because of PWR–BWR differences.

C3.   The main structural components of an element are (higher and lower) bolsters, grids, guide tubes, and NF rods. Component functions are to provide structural integrity and allow and favor the pass of cooler to transfer the produced energy.

C4.   Reserves of U will continue rising as a result of the new exploitation efforts; in addition, new technologies are in development that will allow a greater NF use. Therefore, U supplies will be suitable to provide NPPs source.

14.5   NUCLEAR SAFETY

González Navarro proposed the following questions, answers, and hypotheses on NS.

Q1.   What is risk?

H1.   Equation:

$$Risk = Probability \times Consequences \qquad (14.2)$$

Q2. What have we learned from all the accidents that we have had in nuclear industry?

Q3. Objective of NS?

A3. Lower probability, fewer consequences, lower *ionizing radiations* (IRs) risks.

Q4. How do we measure risk?

H1. (Kemeny, 1979). *Report of President's Commission on the Accident at Three Mile Island.*[4]

Q5. How can it be achieved that workers act safely in every moment?

A5. The NPPs should have a strong *culture of safety*.

Q6. How do workers never take a short cut to be more productive when doing their job?

Q7. What to do to guarantee NPP safety in the case of an emergency?

He provided the following conclusion on human factor and *culture of safety*.

C1. To become an NPP free of accidents requires something more than the design and implementation of safe technologies. Every NPP needs a culture shared by all the workers, which cornerstone and main objective be workers, society, and environmental safety. Culture of safety is *to do well the things even when nobody is looking.*

He provided the following conclusions.

C2. The prior NE objective is the safe exploitation of its installations. The NS allows guaranteeing the defense versus the pernicious effects of IRs, without renouncing to the benefits that NE use reports to humanity.

C3. To compensate mechanical faults and human errors, the principle of *defense in depth* is used, which is focused on several protection levels: multibarrier protection, technological safeguards, and emergency plans. The NPPs are designed to face the different phenomena that could give rise to an accident with IR consequences, for example, earthquakes, floods, explosions, fires, hurricane winds, toxic clouds, accidents inside installation (breaking of the reactor cooling circuit), and so on.

C4. The dangerous NPPs waste is not expelled to the environment but applied its confinement via successive hermetic barriers, which treatment is different from that received by other industrial (and domestic) waste of contrasted negative impact on the environment, to which the *principle of dispersion and dilution* in the ecosystem is applied, for example, CO_2.

C5. Factors, organizational and personnel management, exist that play a relevant role in NPPs safety. Every NPP should count on not only reliable and robust technologies but also a *culture of safety*, which guarantees workers, society, and environmental protection.

C6. An independent regulating organism exists, Council of Nuclear Safety, which watches the fulfillment of NS and RP objectives.

14.6 RADIOLOGICAL PROTECTION

Sáez Muñoz proposed the following questions, answers, and experiment on RP.

Q1. The IRs, how do they affect us?

Table 14.3 presents all the possibilities that can be given rise versus a caused damage.

TABLE 14.3 Summary of Cell Damages.

Cell damage	Repaired			Without consequences
	Badly repaired	Fundamental zone	High reproductive potential	Probabilistic risk of damages
			Low reproductive potential	Without consequences
		Non-fundamental zone		Without consequences
	Not repaired	Fundamental zone		Deterministic risk of damages
		Nonfundamental zone		Without consequences

Source: Ref. [3].

Q2. How can they affect us?

A2. Effects: *deterministic* (dose threshold, instantaneous); *stochastic* (any dose, time deferred).

Q3. How can their presence be detected?

Q4. Which class of IR does one want to measure?

Q5. How do we measure them?

E1. Risks: IR (radioactive field) and pollution (fixed and disperse).

Q6. How to protect oneself from IRs?

She provided the following conclusions.

C1. The IRs cause damages in biosystems on interacting with deoxyribonucleic acid (DNA), which damage can be of *stochastic* (vs. any dose, deferred in time) or *deterministic* (with a dose threshold, instantaneous) character.

C2. The received dose is measured in Sieverts (Sv), which are obtained multiplying Grays (Gy = J/kg) by arbitrary coefficients, dependent each of them on the type of incident particle and organ where it affects.

C3. The IR cannot be perceived by the human senses. Detectors present different characteristics that do them valid for different types of measurements: detection efficiency and energy resolution.

C4. It is essential to distinguish between IR (which ceases on leaving the zone) and pollution (which continues, can affect third persons).

C5. The dose limit for the public in general is 1 mSv/year, while that for exposed workers is 100 mSv in five years without ever exceeding 50 mSv/year. For zones control, barriers are established between them.

14.7 RADIOACTIVE WASTE (DISMANTLING RADIOACTIVE/ NUCLEAR INSTALLATIONS)

Verdú Company proposed Q/A/H on RW (dismantling/close of radioactive/ nuclear installations).

Q1. What is RW?

Q2. How is RW classified?

A2. Very low-level waste (VLLW) < 100; low/intermediate-level waste (LILW) < 4000; high-level waste (HLW) > 4000 Bq/g.

Q3. Which activities do they generate RW?

A3. Normal NPPs operation, NPPs dismantling, NF factory, CIEMAT dismantling, and others.

H1. RW-management stages: segregation/reception, previous storage, treatment, solidification/immobilization, packing, temporal storage, and definitive storage.

H2. Definitive storage via barriers: chemical, physical, engineered, and geological barriers.

H3. Transport via containers guaranteeing integrity versus free fall, free fall on steel punch, fire resistance, locomotive impact, and immersion tests.

H4. Inequality: To dismantle ≠ demolish.

H5. Equation: To dismantle = deconstruct.

Q4. What is it needed?

H6. Three levels of dismantling.

Q5. Which one to select?

He provided the following conclusions.

C1. The RW is a dangerous waste because of the isotopes' contents that generate IRs. They are classified into VLLW/LILW/HLW. Waste management consists of the stages of segregation and reception, previous storage, treatment, solidification or immobilization, packing, previous storage of the conditioned RW, and definitive storage.

C2. In Spain, RW management is carried out by ENRESA. The VLLW/ LILW is stored in El Cabril (Córdoba, Spain) Store Centre. The HLW is temporally stored in pools, individualized temporal stores and centralized temporal stores (CTSs). After that, they are managed according to NF cycle: open [deep geological store (DGS)], closed (reelaboration, DGS), or advanced closed (transmutation, DGS).

C3. For HLW, in Spain, a CTS exists and a DGS is foreseen (2050). The NPPs are designed for a certain working lifetime, generally, 40 years. Real NPP working and good operation and maintenance practices allow testing that the extension of its working lifetime is possible more than the initially estimated 40 years.

C4. Numerous NPPs in all the world are implanting lifetime extension programs to secure their long-term working.

14.8 NUCLEAR POWER PLANTS OF THE FUTURE

Labarile proposed the following hypotheses on NPPs of the future.[5]

H1. Generation III: EPR (AREVA), AP1000 (Westinghouse), advanced/ economic simplified BWR, and small modular reactor.

H2. Generation IV: sodium/lead fast/melted salts/very high-temperature reactors.

She provided the following conclusions.

C1. Since the start of commercial NRs, their safety and working have not stopped improving *via* technological advances in NFs, materials, systems, and components.

C2. Generations III/III+ NPPs reunite evolutionary improvements that affect above all safety systems, reliability, plant-working capability, costs, and design standardization.

C3. Generations IV NPPs will offer great advantages with regard to present NPPs in the fields of sustainability, economy, safety and reliability, nonproliferation, and physical protection.

C4. Nuclear fusion NPPs suppose a great technological challenge, which will achieve to generate electricity via fuels abundant in nature.

14.9 ANOTHER APPLICATIONS OF NUCLEAR TECHNOLOGY

Morató Rafet proposed hypotheses, Qs and As on another application of nuclear technology.

Q1. What is NE?

Q2. Which are their applications?

A2. Electricity production, nuclear medicine (NM), industry, agriculture/ food, art, civil safety, space exploration, and radioisotopes (RITs).

A3. NM applications: image diagnose, diseases treatment, equipment sterilization, and RIT production.

A4. Image diagnose: X-ray (including computerized axial tomography, CAT), p^+ emission tomography (PET), nuclear magnetic resonance, gammagraphy (γ-ray), radioimmunoanalysis.

A5. Treatment of diseases: radiotherapy, monoclonal antibodies (mAbs), n^0 therapy.

A6. Industrial applications: based on IR energy decay through matter, based on radiotracers.

Q3. H production, what do I need?

Q4. Why, to irradiate food?

A4. It prevents germination, production of buds, and so on.

Q5. Did you know that astronauts in the space eat irradiated meat?

Q6. Which is NE importance?

Table 14.4 shows a list of the main RITs of civil interest and their applications.

TABLE 14.4　Main Radioisotopes of Civil Interest and Their Applications.

Isotope	Radiation	Semidisintegration time	Use(s)
^{60}Co	γ	5.3 years	Gamma therapy, metal-thickness measure
^{125}I	γ	60 days	Cancer: prostate, eyes
^{131}I	β	25 min	Thyroid cancer treatment
^{241}Am	α	460 years	Smoke detectors, neutron calibration
^{14}C	β	5570 years	Fossils dating
^{89}Sr	β	50 days	Bone cancer metastasis
^{90}Sr	β	28 years	Radiotherapy, thermoelectric generators
^{99}Tc	γ	6 h	Bone cancer diagnosis, cardiac monitoring, thrombosis diagnosis, liver diseases study, cerebrovascular accidents, etc.
^{45}Ca	β	165 days	Ca absorption study
^{133}Xe	γ	2.3 days	Air flux in lungs
^{24}Na	β and γ	15 h	Study of the mineral content of a body
^{42}K	β	12.5 h	Study of the mineral content of a body
^{15}O	β⁺	124 s	Brain study via PET
^{18}F	β⁺	110 min	PET
^{99}Mo	β	67 h	^{99}Tc production

TABLE 14.4 *(Continued)*

^{192}Ir	β and γ	73.83 days	Cervical, neck, mouth, tongue, lung cancer; to prevent stenosis after an angioplasty; gammagraphies
^{153}Sm	β	1.95 days	Bone cancer metastasis
^{186}Re	β	3.78 days	Bone cancer metastasis
^{90}Y	β	2.67 days	Arthritis
^{169}Er	β	9.4 days	Arthritis
^{177}Lu	β	6.71 days	Several cancer types
^{36}Cl	β	301000 years	Groundwater dating
^{210}Pb	β	22.6 years	Sand and soil layers dating
^{137}Cs	β	30 years	Thickness measure; erosion and deposits detection in soil
^{198}Au	β	2.7 days	Pollution detection; sand tracking in runnings
^{169}Yb	γ	32 days	Gammagraphies

Source: Ref. [3].

He provided the following conclusions.

C1. Many and different applications of nuclear technology are important in people's day-to-day: (a) in industry, the development, automation, improvement, quality control of industrial processes, and resources mining exploration; (b) in agriculture/food, cultivations improvement, food protection versus plagues, and extension of their preservation time; (c) diseases diagnostic via techniques of image obtaining/treatment versus diseases, for example, cancer; (d) maintenance, dating, and authentication of artworks; (e) in safety/civil protection, smoke detectors; (f) space exploration.

14.10 NUCLEAR TECHNOLOGY AND SUSTAINABLE DEVELOPMENT

García García proposed questions and A on nuclear technology and SD.

Q1. What is SD?

A1. Reasonable and nonpollutant energy; action for the climate.

Q2. Can we do without NE in Spain?

Q3. Which will the consequences of doing without NE in Spain be?

He provided the following conclusions on both present and future energy perspectives.

C1. The necessity of powering the development of all the sources that could supply energy in conditions safe, reliable, economical, and of respect to environment, for example, NE.

C2. People will have to continue using fossil fuels (FFs) in the long term so that new techniques should be developed orientated to their clean combustion.

He provided the following conclusions.

C3. In previous years, the world and EU energy scenes changed substantially. An important increment occurred in energy demand, particularly electric, increased spectacularly because of the development of emergent countries.

C4. A climate-change threat emerged caused by the rise in the emission of *greenhouse effect* (GHE) gases, especially CO_2, coming from FFs combustion. Consider that despite the calls to energy saving, energy demand will continue its escalation, driven by the population rise and accelerated incorporation of emerging societies to the consumption levels of developed countries. People will have to use rising energy amounts, to substitute that used in the transport sector by other not emitting GHE gases, and for drinking-water production by seawater desalination. One should wait for a strong electricity penetration as a substitute for FFs direct use.

C5. Faced with the situation above, in the future, it will be necessary to count with all available sources, for example, NE, in an energy mix the most balanced as possible, so that the necessary sustainability criteria be reached simultaneously.

C6. The NE offers, via the analysis of the parameters that condition the covering of demand, solutions, which convert it into one of the basic energies in the present and future world energy scene, according to the international organisms experts in the matter pick up (World Energy Council, International Energy Agency, Organisation for the Economic Cooperation and Development). Spain should not be outside the considerations of the organisms if it does not want to lose the train of future competitiveness and development.

C7.   In Spain, the characteristics of people's energy system, the high dependence on foreign, the distance of fulfillment of people's environmental obligations, the scarce efficiency, and competitiveness make necessary a long-time stable framework to which it is necessary that all economic, political, and social agents reach a Pact of State in energy matter.

C8.   In the sense above, it is needed to take into account the challenges that should be faced in the medium and long time, so that an energy model sustainable in time could be established, in which all available technologies be incorporated, maximizing their advantages and minimizing their disadvantages, in which NE should be a fundamental option in search for the solution to the challenges created in the present and future.

14.11   FINAL REMARKS

From the present discussion, the following final remarks can be drawn:

(1)   The atom consists of a nucleus presenting more than 99% mass surrounded by electrons.

(2)   Energy and mass are equivalent, as Einstein postulated in his equation: $E = mc^2$.

(3)   In fission, a heavy nucleus splits into two others of lower mass.

(4)   Thermal reactors need to moderate fast neutrons from fission to sustain a CR.

(5)   Residual heat occurs in fuel once finished fissions due to fission-product radioactivity.

(6)   The great majority of NPPs use uranium as fuel to obtain heat.

(7)   Fuel cycle: mining, extraction, enrichment, making, reactor, and management as nuclear waste.

(8)   Fuel elements are rod bundles of piled U-ceramic tablets that gather with different designs.

(9)   The main structural components of an element are bolsters, grids, guide tubes, and fuel rods.

(10) Reserves of uranium will continue rising as a result of the new exploitation efforts.

(11) Plants need a culture shared by all workers: workers, society, and environmental safety.

(12) The prior objective of NE is the safe exploitation of its installations.

(13) To compensate mechanical faults and human errors, the principle of *defense in depth* is used.

(14) Dangerous nuclear waste is not expelled but confined via successive hermetic barriers.

(15) Factors, organizational and personnel management, play a relevant role in plants' safety.

(16) Council of Nuclear Safety watches safety objectives/radiological-protection fulfillment.

(17) Radiations cause *stochastic/deterministic* damages interacting with deoxyribonucleic acid.

(18) Received dose is measured in Sieverts, being obtained multiplying Grays by coefficients.

(19) Detectors present characteristics that validate them for different measurement types.

(20) Distinguish between irradiation (which ceases on leaving zone) and pollution (continues).

(21) Dose limit for public is 1 mSv/year, while for exposed workers, 100 mSv in 5 years and 50 mSv/year.

(22) Radioactive waste is dangerous due to isotopes' contents that generate IRs.

(23) Very low- and low/intermediate-level waste are stored in El Cabril (Córdoba, Spain).

(24) For high-level waste, centralized temporal store exists/deep geological store is foreseen.

(25) Numerous NPPs are implanting lifetime extension programs.

(26) Since the start of commercial NRs their safety/working did not stop improving.

(27) Generation III/III+ plants reunite improvements that affect safety systems, and so on.

(28) Generations IV plants will offer great advantages in sustainability, economy, safety, and so on.

(29) Nuclear fusion plants suppose a great technological challenge to generate electricity.

(30) Applications: industry, agriculture, diseases diagnostic, artworks, smoke detectors, and so on.

(31) Need of powering development of all sources supplying energy in conditions safe, and so on.

(32) New techniques should be developed orientated to the clean combustion of FFs.

(33) Increment occurred in energy demand that increased due to emergent countries' development.

(34) A climate-change threat emerged by the rise in the emission of *greenhouse effect* gases.

(35) Faced with the situation above, it will be necessary to count on all available sources.

(36) NE offers positive solutions that convert it into one of the basic energies.

(37) In Spain, it is necessary a Pact of State in energy matter.

(38) It is necessary to face challenges so that a sustainable energy model could be established.

ACKNOWLEDGMENTS

The authors thank the support from Generalitat Valenciana (Project No. PROMETEO/2016/094) and Universidad Católica de Valencia *San Vicente Mártir* (Project No. UCV.PRO.17-18.AIV.03).

KEYWORDS

- nuclear physics
- radiation
- nuclear power plant
- nuclear fuel
- nuclear safety
- radiological protection
- radioactive waste
- radioactive installation dismantling
- future nuclear power plant
- nuclear technology application
- sustainable development

REFERENCES

1. Torrens, F.; Castellano, G. Nuclear Fusion and the American Nuclear Cover-Up in Spain: Palomares Disaster (1966). In *Engineering Technology and Industrial Chemistry with Applications*; Haghi, A. K., Torrens, F., Eds.; Apple Academic–CRC: Waretown, NJ, in press .
2. Torrens, F.; Castellano, G. Manhattan Project, Atoms for Peace, Nuclear Weapons, and Accidents. In *Molecular Chemistry and Biomolecular Engineering: Integrating Theory and Research with Practice*; Pogliani, L., Torrens, F., Haghi, A. K., Eds.; Apple Academic–CRC: Waretown, NJ, in press.
3. Pérez Martín, S., Ed. *Curso Básico de Ciencia y Tecnología Nuclear*; Jóvenes Nucleares–Sociedad Nuclear Española: Madrid, Spain, 2013.
4. Kemeny, J. G., Ed. *Report of the President's Commission on the Accident at Three Mile Island. The Need for Change: The Legacy of TMI*; US Government: Washington, DC, 1979.
5. Fernández-Cosials, K.; Barbas Espa, A., Eds. *Curso Básico de Fusión Nuclear*; Jóvenes Nucleares–Sociedad Nuclear Española: Madrid, Spain, 2017.

CHAPTER 15

CAPITAL VS NATURE CONTRADICTION AND INCLUSIVE SPREADING OF SCIENCE

FRANCISCO TORRENS[1*] and GLORIA CASTELLANO[2]

[1]Institut Universitari de Ciència Molecular, Universitat de València, Edifici d'Instituts de Paterna, P. O. Box 22085, E-46071 València, Spain

[2]Departamento de Ciencias Experimentales y Matemáticas, Facultad de Veterinaria y Ciencias Experimentales, Universidad Catylica de Valencia San Vicente Mártir, Guillem de Castro-94, E-46001 València, Spain

*Corresponding author. E-mail: torrens@uv.es

ABSTRACT

The trap of diversity is how neoliberalism fragmented the identity of the *working class*. In 21st century, together with *the fundamental contradiction* described by Marx, at least two others exist without which the *political subjects* question cannot be tackled: one faces capital *vs* nature; the other political and geographically opposes actions *vs* representations. To take the capital/nature contradiction forces to accept, *vs* classical Marxism, that *the productive forces are at the same time destructive forces* and their development, far from securing without brake emancipation, should be controlled and perhaps stopped/reverted. The universal *consumer middle class*, cracked today, is inseparable from the two other contradictions. Benjamin compared capitalism to a runaway locomotive without engine driver, on the point of being derailed by speed, saying that revolution consists in finding as soon as possible the emergency break. The *consume proletarianization* of the imaginary *consumer middle class* accelerates planetary resources destruction, but it comes accompanied of no conservative *conscience* in species terms. At the beginning of 21st century, it is urgent to build that universalizing subject that, however, seems incompatible with the subjectivity of this consumer, which is individually connected with only merchandises. The concepts of inclusive spreading are *integration, segregation, exclusion,* and *inclusion*. The pillars of science inclusive spreading are *cognitive accessibility, easy lecture,* and *associative fabric*. A questionnaire is proposed to measure the results of a presentation. Some points in the questionnaire need some explanations.

15.1   INTRODUCTION

Setting the scene: The outer contradiction of capital *vs* nature and the paradigm of inclusive spreading of science. The trap of diversity was discussed as how neoliberalism fragmented the identity of the *working class*.[1] Scale, the universal laws of growth, innovation, sustainability, and the pace of life in organisms, cities, economies, and companies were examined.[2] *Supercities* as the intelligence of territory were analyzed.[3] The following hypotheses (H) were proposed on biofuel economic and ethical risks.

H1.   (Fidel Castro, 2003). The use of food to produce biofuel will increase the price of food.

H2.   (Fidel Castro, 2007). The use of food to produce biofuel will affect region food security.

At the beginning of 21st century, together with *the fundamental contradiction* described by Marx (the one that faces capital *vs* work) at least two others exist, crossed, and nonassimilable, without which the *political subjects* question cannot be tackled: the one that faces capital *vs* nature and the other that opposes, in the political and geographic plane, actions *vs* representations. To take the capital/nature contradiction seriously forces to accept, *vs* classical Marxism and according to Sacristán, that *the productive forces are at the same time destructive forces* and their development, far from securing without brake emancipation, should be controlled, and perhaps stopped and till reverted. The universal *consumer middle class*, cracked today, is inseparable from the two other contradictions. With a surprising anticipatory view, in a moment in which the left wing let itself charm by an emancipating concept of progress, Benjamin compared capitalism to a runaway locomotive without engine driver and control, on the point of being derailed by speed, saying that revolution had to consist, not in speeding up the march, but in finding as soon as possible the emergency break. The *consume proletarianization* of the imaginary *consumer middle class* accelerates the planet resources destruction, but it came accompanied, by its proper nature, of no conservative *conscience* in terms now of species. At the beginning of 21st century, it is urgent to build that universalizing subject that, however, seems incompatible with the subjectivity of this consumer (unsuccessful or not) that is individually connected with only merchandises (and not with other humans). The key concepts of inclusive spreading follow: *integration, segregation, exclusion,* and *inclusion.* The pillars of inclusive spreading of science follow: *cognitive accessibility, easy lecture,* and *associative fabric.* A questionnaire was proposed to measure the results of a presentation. Some points in the questionnaire needed some explanations.

In earlier publications, it was reported the periodic table of the elements (PTE)[4-6], quantum simulators[7-15], science, ethics of developing sustainability *via* nanosystems, devices[16], *green nanotechnology* as an approach toward environment safety[17], molecular devices, machines as hybrid organic–inorganic structures[18], PTE, quantum biting its tail, sustainable chemistry[19], quantum molecular *spintronics*, nanoscience, and graphenes[20]. It was informed cancer, its hypotheses[21], precision personalized medicine from theory to practice, cancer[22], how human immunodeficiency virus/acquired immunodeficiency syndrome (HIV/AIDS) destroy immune defences, hypothesis[23], 2014 emergence, spread, uncontrolled Ebola outbreak[24,25], Ebola virus disease, questions, ideas, hypotheses, models[26], metaphors that made history, reflections on philosophy, science and deoxyribonucleic acid (DNA)[27], scientific integrity, ethics, science communication, and psychology.[28] In the present report, it is reviewed some reflections on biofuel increasing the price of food and decreasing regional food security, political subjects, civilizational relief, the outer capital *vs* nature contradiction, a new ecological movement for 21st century, the devaluation of nature, the *negavalue,* and the paradigm of the inclusive spreading of science. The aim of this work is to initiate a debate by suggesting a number of questions (Q), which can arise when addressing subjects of biofuel economic and ethical risks, the outer capital *vs* nature contradiction, the ecological movement, *devalue, negavalue* and the paradigm of the inclusive spreading of science, and providing, when possible, hypotheses (H) on the key concepts of inclusive spreading (*integration, segregation, exclusion,* and *inclusion*) and the pillars of the inclusive spreading of science (*cognitive accessibility, easy lecture,* and *associative fabric*), and explanations (E) on a questionnaire to measure the results of a presentation.

15.2 POLITICAL SUBJECTS AND CIVILIZATIONAL RELIEF

At the beginning of 21st century, together with *the fundamental contradiction* described by Marx (the one that faces capital *vs* work) at least two others exist, crossed, and nonassimilable, without which the *political subjects* question cannot be tackled: the one that faces capital *vs* nature and the other that opposes, in the political and geographic plane, actions *vs* representations.[29] To take the capital/nature contradiction seriously forces to accept, *vs* classical Marxism and according to Sacristán, that *the productive forces are at the same time destructive forces*[30] and their development, far from securing without brake emancipation, should be controlled, and perhaps stopped and till reverted.

15.3 THE OUTER CONTRADICTION: CAPITAL VS NATURE

The universal *consumer middle class*, cracked today, is inseparable from the two other contradictions above. The first, each time more pressing, is the one that opposes capital *vs* nature. In order to notice till which point people continue living in a material production universe, which not only exploits human beings but also their conditions themselves of *outer* existence, it is enough to refer not to the data of the gross domestic product of India or China (which are also eloquent) but to the ecological consequences of the cultural capitalism transformations in the last years. In 2010, for example, the generation rhythm of domestic waste was multiplied by ten. The planet produces 8,000,000,000 t rubbish per day. With regard to *electronic waste*, only United States scraps 60,000,000 t computers per year while the global number is 400,000,000–500,000,000 t of the residues every year. The 70% of this waste, otherwise, ends up in poor countries, Pakistan and Ghana above all, where it produces devastating ecological and healthy effects. The *consumer imagery*, functional to capitalism, is incompatible enough with human existence. One should remember Walter Benjamin and his advertences on progress and ruins. With a surprising anticipatory view, in a moment in which the proper left wing let itself charm by an emancipating concept of progress (according to which one will be always *going with the flow*, so that it would be enough to let oneself to get carried away and let do to change the model) Walter Benjamin compared capitalism to a runaway locomotive, with a train with neither engine driver nor control, on the point of being derailed by speed, and said that revolution had to consist, not in speeding up the march, not, but in finding as soon as possible the emergency break. The *consume proletarianization* of the imaginary *consumer middle class* accelerates the planet resources destruction, but it came accompanied, by its proper nature, of no conservative *conscience* in terms now of species. No collective subject exists called humanity. Marx joked it in a moment in which neither nuclear weapons nor climate change (CC) threatened the human species as a whole. Capitalism threatened *many men*, above all the *working-class* members but not humanity itself. It was in 19th century. At the beginning of 21st century, it is urgent to build that universalizing subject that, however, seems incompatible with the subjectivity of this consumer (unsuccessful or not) that is individually connected with only merchandises (and not with other humans).

15.4   A NEW ECOLOGICAL MOVEMENT FOR 21ST CENTURY

The 1% richest persons in the world, and a few at their service, get rich further every day exploiting the human being and its environment and, at the same time, are the least affected by the environmental degradation.[31] They use power and violence monopoly (*via* State) and *vs* global ecological crisis (GEC) they have no exit but a blind flight into front in the madness of getting more benefits than their competitors. In the context of polyhedral crisis at planetary level (ecological, economic, humanitarian, *etc.*), the modern ecological movement that were born in 1960s, drinking in the hippy movement, seems exhausting its possibilities as an agent of significant social change, if sometime it was it. Different ecological nongovernmental organizations (NGOs) of all sizes and colours never stop doing lobbyism and trying to raise awareness the population while GEC does not do but worsening. If in the beginning of capitalism, who first were mobilized, *vs* environmental degradation, were the workers most affected by pollution in their working centres and residential areas near to factories[32], nowadays a new ecological movement seems to be emerging, which links narrowly social and environmental questions in its fights. The claims union is similar to the one that used the personnel that were mobilized in the beginning of First Industrial Revolution, mixing working-improvements demands (to work lesser hours, earn more, have lesser accidents, *etc.*) with environmental claims (*e.g.*, no to breathe polluted air). Now, in a moment of economic crisis, which is used by neoliberal governments to impose social austerity, at the time that the different socio-environmental problems worsen, a movement emerges that drinks in fight waters of the last lustrums, from the antiglobalization movement to 15M, Occupy, Arab Springs, Greek strikes waves, *via* the international movement *vs* war. Economic and ecological crises union seems having shaped *via* the fights of the last years a noosphere common to millions persons, better connected than ever by Internet and its social nets. Persons that claim new forms of organizing society in its entirety, including the relationship with their environment and, so, entering fully, consciously or not, in the working sphere; for example, environmental fights in Arab Springs framework (paralyzation of building two new fertilizing plants in Egyptian coast, fight that emerged in Taksim Gezi Park of Istanbul defending a green zone, and extended by Turkey as a movement for democratic rights and *vs* social inequality and politic corruption), protests in Brazil *vs* the price rise of public transport and building society speculation driven by the Olympics and Soccer World, direct blockings net *vs* great extractive projects in the natural environment, Blocklandia[33], movement *vs* fracking, *and so on*. Environmental fights are and will be, now more than ever, social and, especially, working struggles. A phantom goes over the world: that of fight *vs* CC and social injustices. Observing the composition of the new ecological movement with strong social component, one proves that its militants are fed, greatly, with natives, small landowners, and working people that, however, do not habitually mobilize from their paid working post. In order that the ecological movement go much further, it is necessary that its activism rest on collaborative production and correct with working fights so that, in the course of the struggles, workers movement develop deep ecological roots; *for example*, green bans to inspire the class, social, and environmental interests union.

15.5 DEVALUATION OF NATURE

The more productive forces productivity rises, the more is needed that it rise to try to save benefit.[34] More commodities should be manufactured and permanently consumed, so that their lifetime should be shortened anyway. In order to try to connect the loss of *value* and give incessant production a free hand, capitalist class tries by all means to make cheaper constant capital *via* circulating capital (raw materials, inputs) depreciation; *that is*, it needs a *cheaper nature*. So that its urge to carry out a nature perception and social construction like an infinite place of free resources and inexhaustible rubbish dump, which bears natural environmental costs scorn in economic processes and *externalization* the life set of their negative consequences, which is, in short, *nature devaluation* as it is *built* as totally outside *value* (and so without intrinsic own value). Its mercantile *reification* is performed, however, as mere *circulating capital*. Nature passed so from being for so many humanity cultures life source, and the milieu that provides and makes possible it, which so should be cared and with which one should interact sustainably (use with replacement) to be another merchandise, or a milieu facilitating merchandises and waste receiver, which is a new indication that *value* is carried out increasingly *vs* real wealth, in first place natural wealth.

15.6   THE *NEGAVALUE*: THE EXTREME FACE OF THE *DEVALUE*

The *negavalue* shows as a result of capitalist sociosphere interaction with the rest of ecosphere. The use of chemicals that arrive at infant food and are found in most homes, pesticides, public water pollution, exposure to an indeterminate but rising number of environmental toxins, and so on, which as a whole and mixed are associated with malformations from birth, premature cancers (World Health Organization, WHO, foresees a rise in 75% cancers in 10 years, which is *more or less* the trend that people have since 1990s), chronic and allergic diseases incapacitating in different grades, neurological disorders in infant population, hyperactivity and attention deficits, reproductive problems proliferation, *and so on*, affect directly *human capital*. If *devalue* denies persons, *negavalue* destroys them as individuals and finally kill them. Gain rate priority above health is a eugenics form.

15.7   THE PARADIGM OF INCLUSIVE SPREADING OR *TELL IT TO ME EAS-I-LY*

Perales Guillen proposed H/Q on key concepts of the inclusive spreading of science.[35]

H1.   The key concepts of inclusive spreading follow: *integration*; *segregation*; *exclusion*; *inclusion*.

Q1.   What is *integration*?

Q2.   What is *segregation*?

Q3.   What is *exclusion*?

Q4.   What is *inclusion*?

   She raised the following hypothesis/questions on the pillars of inclusive spreading.

H2.   Pillars of inclusive spreading follow: *cognitive accessibility*; *easy lecture*; *associative fabric*.

Q5.   What is *cognitive accessibility*?

Q6.   What is *easy lecture*?

Q7.   What is *associative fabric*?

    She raised a questionnaire to measure the results of a presentation (*cf.* Fig. 15.1).

Q8.   Do you think that a correct way of spreading exists?

Q9.   In which part of the talk did you get most bored?

Q10.   Explain what *cognitive accessibility* is.

Q11.   Would you recommend this talk to a friend?*

    Three marks (*) in the questionnaire above need some explanations.

E1.   Would you recommend this talk to a friend?*: Asking something that has not happened is error.

E2.   PERSONAL DATA*: They should be included at the questionnaire bottom rather than at top.

E3.   Gender*: Female, Male, and Others.

IMPORTANTE: Puedes usar el conocimiento colectivo y
preguntar a otras personas asistentes

1. ¿Crees que existe una forma de divulgar correcta?

2. ¿En qué parte de la charla te has aburrido más?

3. Explica qué es la accesibilidad cognitiva

4. ¿Recomendarías esta charla a un amigo/a? *

ESPACIO DE LIBRE EXPRESIÓN

DATOS PERSONALES*

Género*

☐ Mujer   ☐ Hombre   ☐ Otros

Edad _____

Nivel formativo _____

Sector profesional _____

FIGURE 15.1   Questionnaire to measure the results of a presentation.
Source: Ref. [35].

15.8   FINAL REMARKS

From the present results, the following final remarks can be drawn.

1. The trap of diversity is how neoliberalism fragmented the identity of the *working class*. Biofuel has economic and ethical risks increasing food price and decreasing regional food security.
2. At the beginning of 21st century, together with *the fundamental contradiction* described by Marx (the one that faces capital *vs* work) at least two others exist, crossed and nonassimilable, without which the *political subjects* question cannot be tackled: the one that faces capital *vs* nature and the other that opposes, in the political and geographic plane, actions *vs* representations.
3. To take the capital/nature contradiction seriously forces to accept, *vs* classical Marxism, that *the productive forces are at the same time destructive forces* and their development, far from securing without brake emancipation, should be controlled and stopped/reverted. The universal *consumer middle class*, cracked today, is inseparable from the two other contradictions above.
4. With an anticipatory view, in a moment in which the left wing let itself charm by an emancipating concept of progress, Benjamin compared capitalism to a runaway locomotive without engine driver and control, on the point of being derailed by speed, saying that revolution had to consist, not in speeding up the march, but in finding as soon as possible the emergency break.
5. The *consume proletarianization* of the imaginary *consumer middle class* accelerates the planet resources destruction, but it came accompanied, by its proper nature, of no conservative *conscience* in terms now of species. At the beginning of 21st century, it is urgent to build that universalizing subject that, however, seems incompatible with the subjectivity of this consumer (unsuccessful or not) that is individually connected with only merchandises.
6. The key concepts of inclusive spreading follow: *integration, segregation, exclusion,* and *inclusion*. The pillars of inclusive spreading follow: *cognitive accessibility, easy lecture,* and *associative fabric*. A questionnaire was proposed to measure the results of a presentation. Some points in the questionnaire needed some explanationss.

ACKNOWLEDGMENTS

The authors acknowledge support from Generalitat Valenciana (Project No. PROMETEO/2016/094) and Universidad Católica de Valencia *San Vicente Mártir* (Project No. UCV.PRO.17-18.AIV.03.)

KEYWORDS

- **working class**
- **consumer middle class**
- **consume proletarianization**
- **political subject**
- **civilizational relief**
- **outer contradiction**
- **ecological movement**
- **biofuel**
- **food price**
- **food security**
- **cognitive accessibility**
- **easy lecture**
- **associative fabric**
- **integration**
- **segregation**
- **exclusion**

REFERENCES

1. Bernabé, D. *La Trampa de la Diversidad: Cómo el Neoliberalismo Fragmentó la Identidad de la Clase Trabajadora*; Akal: Tres Cantos, Madrid, Spain, 2018.
2. West, G. *Scale: The Universal Laws of Growth, Innovation, Sustainability, and the Pace of Life in Organisms, Cities, Economies, and Companies*; Penguin: New York, NY, 2017.
3. Vergara Gómez, A. *Supercities: The Intelligence of Territory*; Fundación Metrópoli: Pamplona (Spain), 2016.
4. Torrens, F.; Castellano, G. Reflections on the Nature of the Periodic Table of the Elements: Implications in Chemical Education. In *Synthetic Organic Chemistry*; Seijas,

J. A., Vázquez Tato, M. P., Lin, S. K., Eds.; MDPI: Basel, Switherland, 2015; Vol. 18; pp 1–15.

5. Torrens, F.; Castellano, G. Nanoscience: From a Two-Dimensional to a Three-Dimensional Periodic Table of the Elements. In *Methodologies and Applications for Analytical and Physical Chemistry*; Haghi, A. K., Thomas, S., Palit, S., Main, P., Eds.; Apple Academic–CRC: Waretown, NJ, 2018; pp 3–26.

6. Torrens, F.; Castellano, G. Periodic Table. In *New Frontiers in Nanochemistry: Concepts, Theories, and Trends*; Putz, M. V., Ed.; Apple Academic–CRC: Waretown, NJ, in press.

7. Torrens, F.; Castellano, G. Ideas in the History of Nano/Miniaturization and (Quantum) Simulators: Feynman, Education and Research Reorientation in Translational Science. In *Synthetic Organic Chemistry*; Seijas, J. A., Vázquez Tato, M. P., Lin, S. K., Eds.; MDPI: Basel, Switzerland, 2015; Vol. 19; pp 1–16.

8. Torrens, F.; Castellano, G. Reflections on the Cultural History of Nanominiaturization and Quantum Simulators (Computers). In *Sensors and Molecular Recognition*; Laguarda Miró, N., Masot Peris, R., Brun Sánchez, E., Eds.; Universidad Politécnica de Valencia: València, Spain, 2015; Vol. 9; pp 1–7.

9. Torrens, F.; Castellano, G. Nanominiaturization and Quantum Computing. In *Sensors and Molecular Recognition*; Costero Nieto, A. M., Parra Álvarez, M., Gaviña Costero, P., Gil Grau, S., Eds.; Universitat de València: València, Spain, 2016; Vol. 10; pp 31–1–5.

10. Torrens, F.; Castellano, G. Nanominiaturization, Classical/Quantum Computers/Simulators, Superconductivity, and Universe. In *Methodologies and Applications for Analytical and Physical Chemistry*; Haghi, A. K., Thomas, S., Palit, S., Main, P., Eds.; Apple Academic–CRC: Waretown, NJ, 2018; pp 27–44.

11. Torrens, F.; Castellano, G. Superconductors, Superconductivity, BCS Theory and Entangled Photons for Quantum Computing. In *Physical Chemistry for Engineering and Applied Sciences: Theoretical and Methodological Implication*; Haghi, A. K., Aguilar, C. N., Thomas, S., Praveen, K. M., Eds.; Apple Academic–CRC: Waretown, NJ, 2018; pp 379–387.

12. Torrens, F.; Castellano, G. EPR Paradox, Quantum Decoherence, Qubits, Goals and Opportunities in Quantum Simulation. In *Theoretical Models and Experimental Approaches in Physical Chemistry: Research Methodology and Practical Methods*; Haghi, A. K., Ed.; Apple Academic–CRC: Waretown, NJ, 2018; Vol. 5, pp 317–334.

13. Torrens, F.; Castellano, G. Nanomaterials, Molecular Ion Magnets, Ultrastrong and Spin–Orbit Couplings in Quantum Materials. In: *Physical Chemistry for Chemists and Chemical Engineers: Multidisciplinary Research Perspectives*; Vakhrushev, A. V., Haghi, R., de Julián-Ortiz, J. V., Allahyari, E., Eds.; Apple Academic–CRC: Waretown, NJ, in press.

14. Torrens, F.; Castellano, G. Nanodevices and Organization of Single Ion Magnets and Spin Qubits. In *Chemical Science and Engineering Technology: Perspectives on Interdisciplinary Research*; Balköse, D., Ribeiro, A. C. F., Haghi, A. K., Ameta, S. C., Chakraborty, T., Eds.; Apple Academic–CRC: Waretown, NJ, in press.

15. Torrens, F.; Castellano, G. Superconductivity and Quantum Computing via Magnetic Molecules. In *New Insights in Chemical Engineering and Computational Chemistry*; Haghi, A. K., Ed.; Apple Academic–CRC: Waretown, NJ, in press.

16. Torrens, F.; Castellano, G. Developing Sustainability via Nanosystems and Devices: Science–Ethics. In *Chemical Science and Engineering Technology: Perspectives on*

Interdisciplinary Research; Balköse, D., Ribeiro, A. C. F., Haghi, A. K., Ameta, S. C., Chakraborty, T., Eds.; Apple Academic–CRC: Waretown, NJ, in press.

17. Torrens, F.; Castellano, G. Green Nanotechnology: An Approach towards Environment Safety. In *Advances in Nanotechnology and the Environmental Sciences: Applications, Innovations, and Visions for the Future*; Vakhrushev, A. V.; Ameta, S. C.; Susanto, H., Haghi, A. K., Eds.; Apple Academic–CRC: Waretown, NJ, in press.

18. Torrens, F.; Castellano, G. Molecular Devices/Machines: Hybrid Organic–Inorganic Structures. In *Research Methods and Applications in Chemical and Biological Engineering*; Pourhashemi, A., Deka, S. C., Haghi, A. K., Eds.; Apple Academic–CRC: Waretown, NJ, in press.

19. Torrens, F.; Castellano, G. The Periodic Table, Quantum Biting its Tail, and Sustainable Chemistry. In *Chemical Nanoscience and Nanotechnology: New Materials and Modern Techniques*; Torrens, F., Haghi, A. K., Chakraborty, T., Eds.; Apple Academic–CRC: Waretown, NJ, in press.

20. Torrens, F.; Castellano, G. Quantum Molecular Spintronics, Nanoscience and Graphenes. In *Molecular Physical Chemistry*; Haghi, A. K., Ed.; Apple Academic–CRC: Waretown, NJ, in press.

21. Torrens, F.; Castellano, G. Cancer and Hypotheses on Cancer. In *Molecular Chemistry and Biomolecular Engineering: Integrating Theory and Research with Practice*; Pogliani, L., Torrens, F., Haghi, A. K., Eds.; Apple Academic–CRC: Waretown, NJ, in press.

22. Torrens, F.; Castellano, G. Precision Personalized Medicine from Theory to Practice: Cancer. In *Molecular Physical Chemistry*; Haghi, A. K., Ed.; Apple Academic–CRC: Waretown, NJ, in press.

23. Torrens, F.; Castellano, G. AIDS Destroys Immune Defences: Hypothesis. *New Front. Chem.* **2014**, *23*, 11–20.

24. Torrens-Zaragozá, F.; Castellano-Estornell, G. Emergence, Spread and Uncontrolled Ebola Outbreak. *Basic Clin. Pharmacol. Toxicol.* **2015**, *117* (Suppl. 2) 38–38.

25. Torrens, F.; Castellano, G. 2014 Spread/Uncontrolled Ebola Outbreak. *New Front. Chem.* **2015**, *24*, 81–91.

26. Torrens, F.; Castellano, G. Ebola Virus Disease: Questions, Ideas, Hypotheses and Models. *Pharmaceuticals* **2016**, *9*, 14–6–6.

27. Torrens, F.; Castellano, G. Metaphors That Made History: Reflections on Philosophy/Science/DNA. In *Molecular Physical Chemistry*; Haghi, A. K., Ed.; Apple Academic–CRC: Waretown, NJ, in press.

28. Torrens, F.; Castellano, G. Scientific Integrity/Ethics: Science Communication and Psychology. In *Molecular Physical Chemistry*, Haghi, A. K. Ed.; Apple Academic–CRC, Waretown (NJ), 2019, pp 1–13.

29. Alba Rico, S. Sujetos Politicos y Relevo Civilizacional. In *La Clase Trabajadora: ¿Sujeto de Cambio en el Siglo XXI?*; Tarín Sanz, A., Rivas Otero, J. M., Eds.; Siglo XXI: Tres Cantos, Madrid, Spain, 2018; pp 53–69.

30. Sacristán, M. *Pacifismo, Ecologismo, y Política Alternativa*; Público, Madrid, Spain, 2009.

31. Castillo, J. M. Clase Trabajadora y Ecología del Trabajo. In *La Clase Trabajadora: ¿Sujeto de Cambio en el Siglo XXI?*; Tarín Sanz, A., Rivas Otero, J. M., Eds.; Siglo XXI: Tres Cantos, Madrid, Spain, 2018; pp 221–241.

32. Castillo, J.M. *Trabajadores y Medio Ambiente: La Lucha contra la Degradación Ambiental desde los Centros de Trabajo*; Atrapasueños: Sevilla, Spain, 2013.
33. Klein, N. *Esto lo Cambia Todo: El Capitalismo contra el Clima*; Paidós: Barcelona, Spain, 2015.
34. Piqueras, A. *Las Sociedades de las Personas sin Valor: Cuarta Revolución Industrial, Des-substanciación del Capital, Desvalorización Generalizada*; El Viejo Topo: Vilassar de Dalt, Barcelona, Spain, 2018.
35. Perales Guillen, I. Book of Abstracts. IV Jornadas de Divulgación Inclusiva de la Ciencia, València, Spain, November 26–27, 2018, Universitat de València: València, Spain, 2018; O–7.

CHAPTER 16

ROLE OF HALIDE ION IN GOLD NANOPARTICLES SYNTHESIS USING SEED GROWTH METHOD

LAVANYA TANDON and POONAM KHULLAR*

Department of Chemistry, B.B.K.D.A.V College for Women, Amritsar 143001, Punjab, India

Corresponding author. E-mail: virgo16sep2005@gmail.com

ABSTRACT

Nanoparticles are being used widely in the field of catalysis, adsorption, and photonoics due to their unique chemical and physical properties. Template method effectively controls particle size, structure, and morphology. Surfactants lower the surface tension between two liquids or between liquid and a solid. On replacing citrate-stabilized seeds with CTAB-stabilized seeds, nanorods are formed. Halide ions also affect the morphology of nanoparticles. Br^- ion act as shape-directing agent in the formation of nanoparticles and when Br^- ion is replaced by the Cl^-, then uniform or spherical structures are obtained. Addition of high concentration of iodide ion results in homogeneous formation of nanorods, while low concentration of I^- results in nanorods with shorter aspect ratios. The degree of binding of the different halides to the gold surfaces has a strong impact on the structure of the surfactant layer. On increasing the amount of the chloride ion, there occurs faster and more isotopic growth of nanoparticles.

16.1 INTRODUCTION

Nanotechnology is the branch of science that deals with the manipulation of matter on atomic, molecular, and superamolecular scale. Advancement in the field of nanotechnology has resulted due to the development of the engineered nanoparticles. Thousands of synthetic protocols are being reported for the synthesis of metal nanoparticles, but the mechanisms and the physiochemical processes that are used in the formation of metal nanoparticles are known scarcely. At present, two processes are involved in the formation of metal nanoparticles, which are: nucleation and growth. Self-nucleation takes place very fast and the growth is diffused controlled. In case of metallic systems, nucleation is continous[1-8] and growth is autocatalytic[1-8] and aggressive[9] Other processes like digestive ripening,[10] ostwald ripening,[11,12] and oxidative etching[13] have been identified. These factors contribute to the shape, size, and stability of the nanoparticles. For the synthesis planning, one of the important keys to control the dispersion of size is to separate the stages of nucleation and growth on time[14] and other important requirements are long-term stability, functionalization, and suspension media. In synthesizing very small nanoparticles, strong reducing agents—like borohydride ion—and stabilizing agents that bound tightly to the surface of nanoparticles—like thiols on surface of gold—favors the nucleation over the growth.[15] Another method is the hot injection method in which the separation of growth and nucleation takes place through application of temperature ramps. The most extensively employed method is the seed-mediated approach that have initiated this separation through use of different chemical environments.[16-18] The Murphy's group[19-21] has firstly reported the method of quaternary ammonium-based method of seed growth. This group have synthesized successfully anisotropic nanoparticles of gold possessing different shapes like nanocubes,[22] nanorods,[23-25] jackshaped,[26] dumbbell-like,[27] nanoprisms,[28] and bipyramids.[29] This method involves the seeding of small gold nanoparticles into the reactive solutions; gold nanoparticles are shown to grow in the absence of self-nucleation. These reactive solutions are also known as growth baths, which contains the precursor Au (III), quaternary ammonium salt like hexadecyltrimethylammonium bromide (CTAB), and reducing agent like ascorbic acid (AA). Gold nanoparticles are being used for labeling, delivering, heating, sensing, therapy,[30] drug and gene delivery,[31] probes,[32] sensors,[33] diagnostics,[34] and photocatalyst.[35] Much interest has been attracted by gold nanoparticles as these possess certain specific features. By changing the characterization of the particles like size, shape, and environment, their intrinsic features like electronic, optical, physiochemical, and surface Plasmon resonance (SPR) can be altered.

By using different physical, chemical, and biological methods, AuNP's of different sizes and shapes have been synthesized. At near-IR, absorbing nanoparticles of gold can be excited by light and due to this these becomes contrasting agents of the next generation for phototherapeutic applications and diagnosis.

16.2   STRATEGIES TO SYNTHESIZE NANOPARTICLES

There are two ways to synthesize nanomaterials.

1.   Bottom-up approach
2.   Top-down approach

In bottom-up approach, there is miniaturization of materials components, which, further through self-assembly process, leads to formation of nano-structure. The materials and devices are built from molecular components, which assemble themselves chemically through molecular recognition. The term *molecular recognitions* means specific interaction between two or more molecules through non-covalent bonding like hydrogen bonding, metal coordination, hydrophobic forces, vanderwaal forces, pi-pi interactions, and electrostatic and electromagnetic effects. During the self-assembly, physical forces operating at the nanoscale combine the basic units into stable nanostructures.

The top-down approach uses larger macroscopic initial structures, which are externally controlled in the process of nanostructures like etching through mask, ball milling, and application of severe deformation. Creating nanostructure materials based on bottom-up approach is a fast-growing field of research. Two-dimensional arrays of nanocrystals are of great interest, which shows unique optoelectronic, magnetic, or catalytic properties, and they can be tuned by varying the size and interparticle separation distance. Advanced functional materials are used in many areas, including miniaturized nanoelectronics,   quantum dots (QDs) in quantum computers, and high-density memory.

16.2.1   TEMPLATE METHOD

Template method controls the morphology, structure, and particle size during the formation of nanoparticles. Morphology plays an important role in the characterization of the materials. Template method is easy to operate and implement; and it controls the crystal growth and nucleation by changing the morphology. According to the structure, templates may be regarded as hard templates or soft templates. Hard templates are the rigid material whose structure determines the morphology and size of the particles. Few examples are ion exchange resins, carbon fiber, and porous anodic aluminum oxide (AAO).[36,37]

Template method involves three steps:

1. Preparation of template
2. Use of precipitation, sol-gel, and hydrothermal method for synthesizing the target under the template function
3. Removal of template

Through dissolution, sintering, and etching, template can be removed. Hard templates are used for the synthesis of nanowires, nanotubes, and nanobelts. These templates are very stable and are thus used as microreactors during the synthesis in which the precursor is filled in the pores of the templates through impregnation.

Rigid structure is not possessed by the soft templates. An aggregate with specific structural features are formed due to the intermolecular forces of interactions, hydrogen bonding, chemical bonding, and so on, and act as the templates; and on their surfaces, inorganic species are deposited through precipitation or electrochemical method. Soft templates can be polymer, biopolymer, or surfactants.

16.2.2 SURFACTANTS

Surfactants are used in the formation of nanoparticles and are regarded as amphiphilic molecules. These have hydrophilic head and hydrophobic tail. Surfactants are categorized according to the charge on the hydrophilic portion of the molecule, that is,

- Anionic surfactants
- Nonionic surfactants
- Cationic surfactants
- Amphoteric surfactants

They are the compounds that lower the surface tension between two liquids or between liquid and a solid. Surfactants may be detergents, wetting agents, emulsifiers, and foaming agents. Surfactants diffuse in water and are adsorbed at interface between air and water or at interface between oil and water. The water-soluble hydrophobic group extends out of the bulk water phase, into the air or into the oil phase, whereas the water-soluble head group remains in the water phase. Some of the surfactants are cetyl trimethyl ammonium bromide (CTAB), cetyl trimethyl ammonium chloride (CTAC), didodecyldimethylammonium bromide (DDAB), and N-dodecyl-beta-maltoside. The morphology of the mesoporus material is decided on the basis of the interaction between the organic–inorganic interface and liquid crystal phase of the surfactant.[38,39] In acidic environment, weak interactions takes place, and in alkaline environment, strong interactions takes place.

16.2.3   HALIDE ION EFFECT

Nanorods of gold are prepared through seed-mediated growth technique. To the aqueous solution of cetyltrimethylammonium bromide, reducing agent like ascorbic acid is added for the reduction of Au^{3+} to Au^+ and gold seeds are added and these catalyzes the Au^+ reduction on the surface. The yield of nanoparticles is increased to 100% by adding silver nitrate to growing solution. In the synthesis of gold nanoparticles, polymers, stabilizing surfactants, and halide ions like iodide, bromide, or chloride are present in water.[40-48] Halide ions affects the morphology of the gold nanoparticles.[49,50] Halides have strong tendency to adsorb on surface of the metal. The nanoparticles grow in the very complex mixture of surfactants and salts in water.[51]

On the addition of bromide ions to the solution of gold chloride, an intensity increase is observed. It is due to the exchange of chloride and hydroxyl ions by the bromide ions in the coordination sphere of the Au(III). Increase in concentration of the bromide ions results in increase in absorbtion due to the ligand-exchange process. Faster growth reaction is observed due to the presence of larger gold nanoparticles.

The addition of iodide ions to the Au(III), the tri-iodide ions are formed along the ligand exchange reactions because of the oxidation of iodide ions, which has been caused by the Au(III) particles.

16.3   DISCUSSION

16.3.1   ROLE OF BROMIDE IONS IN SEED GROWTH

It is observed (Jin 2010) that even when CTAB is used below its *cmc* and Br⁻ ion are added in the form of NaBr, a good yield of nanorods is obtained.

An appropriate control over the externally added Br⁻ ions, results in good control of dimensions nanorods. Thus, Br⁻ ions serve as an important shape-directing agent for the good nanorod formation in the seed-growth process.

Jin et al. synthesized nanorods using two-step seed growth method in the presence of $AgNO_3$. They found that the concentration of CTAB is a very important parameter for the formation of typical Au nanorods with aspect ratio approximately 3:5 that shows LSPR at approximately 800 nm, a concentration lower than 0.1 M. Comparison of different reactions using CTAC in place of CTAB depicts that Cl⁻ ion is less efficient in inhibiting radial growth of nanorods as compared to Br⁻ ion. Also, sufficient concentration of Br⁻ ion is necessary to bring higher anisotropy.

(Some studies showed that I⁻ ion strongly bind to Au and seem to destroy particles.)

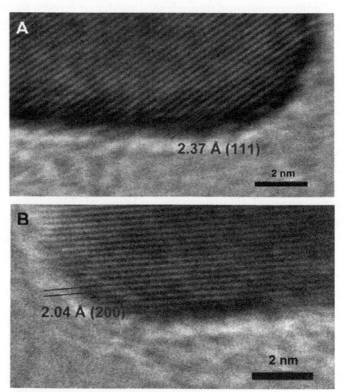

FIGURE 16.1 HR-TEM of Au nanorods effected by NaBr in the presence of minimal CTABr.

Source: Reproduced with permission from Ref. 53. © 2010 American Chemical Society.

HR-TEM study analysis results in observed lattice spacing of 2.37 for [111] plane, which is quiet closed to the bulk, Au(III) spacing (2.35 A⁰).The rotation of rods leads to (200) lattice showing spacing value of 2.05A⁰ quiet close to the bulk spacing of 2.04 A⁰. Further results indicated that extrinsic Br⁻ ions as NaBr can also result in desired anisotropy even at low concentration of CTAB. It clearly indicates very important role of Br⁻ ion as shape-directing agent in addition to CTA⁺. And, as long as required amount of Br⁻ ion are available in the growth solution, nanorods are observed even at low concentration of CTA⁺ (i.e., below its cmc (1mM)).[52]

The counter anions play a significant role in the synthesis of gold nanorods and nanowires.[53] Considering CTAB, CTA⁺ ion provides steric stabilization and Br⁻ ion act as shape-directing agent.

Relative comparison of various studies of synthesis revealed that the mode of interaction of different halide ions follows the order F⁻< Cl⁻< Br⁻< I⁻.[54] Among these, I⁻ bind too strongly so that no nanorod formation takes place. F⁻ ion adsorb nonspecifically on the gold surface, and a weak interaction is observed between Cl⁻ and Au surface, hence, leaving only Br⁻ ion, which interact neither too strongly nor too weakly with gold. Some studies have shown that the nanorod structure is dependent on the types of seeds used, for example, on replacing citrate stabilized seeds with CTAB-stabilized seeds, nanorods are formed.

In the previous study by Murphy et al., it is observed that Au nanorods adopt a penta –twinned structure, while the rod sides are bound by the {110} and {100}facets and the other two ends are enclosed by {111} facet. HR-TEM studies of externally added NaBr indicated that the nanorods are perfectly face centered cubic single crystalline predominantly [55]{111} planes are observed.

This is further supported by the theoretical calculations of Barnard and Curtiss that growth along {110} is energetically favorable. The structural model (scheme1) indicates that nanorod growth[56] is along {110} plane rather than along {100} or {111} plane.

{111} facet possess lowest surface energy (most stable facet) among the three types of low index facet followed by {100} facet and {110} facet possess the highest surface energy.

Since {110} is the least stable facet, seeds grow along {110} direction to maximize {111} facet.

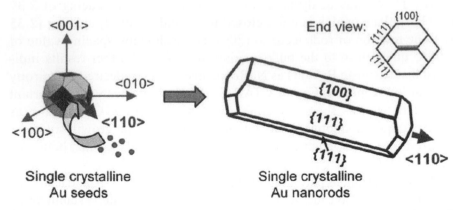

FIGURE 16.2   Structural model of single crystalline Au nanorods and proposed growth mechanism.
Source: Reproduced with permission from Ref. 53. © 2010 American Chemical Society.

Scheme 1 (model) indicates a structural model that shows that the side faces consist of four {111} facets and two {100} facets. However, the two {100} facets are due to longitudinal truncation; this has been shown by various studies.

16.3.2   ROLE OF IODIDE ION IN SEED GROWTH PROCESS

Iodide in CTAB prevents gold nanoparticle formation. In silver-assisted approach, synthesis of seed-mediated methods using CTAB suffers from the major drawback, that is, in some studies, it is observed that iodide >40 ppm present as an impurity in various samples of CTAB prevents nanorod formation. Iodide gets adsorbed on the Au {111} surface and, therefore, prevents interference in the reduction of Au^{3+} to Au^0.

The presence of nanorod is indicated by the presence of longitudinal surface plasmon peak at 700–800 nm and the absence of this wavelength indicates the absence of nanorods. To detect or to determine the quality of the sample, the quantity "absorbance max ratio" A_{LPB} / A_{TPB} can be used. Large value of the ratio indicates larger population of nanorods. It is observed that as more and more KI is added to the reaction, the nanorod aspect ratio decreases and LSPR was absent when I^- concentration reaches 0.57 μm.

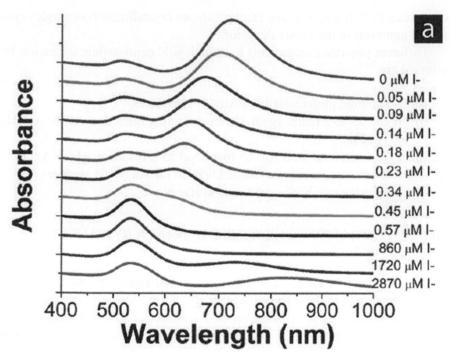

FIGURE 16.3 The iodide concentration in the growth solution is indicated next to each spectrum.

Source: Reproduced with permission from Ref. 61. © 2009 American Chemical Society.)

Iodide addition to CTAB-stabilized AuNPs is found to disrupt dispersion of Au nanorods, which results in the transformation of nanorods into spherical particles. It occurs because of the competition of adsorption between CTAB and I⁻ for the Au surface. Therefore, I⁻ displaces the CTAB capping ligand, which causes destabilization of nanorod shape.

The role of I⁻ vary depending upon the method adopted. As in stepwise additive growth method, I⁻ is required for the formation of AuNPs.[57] However, in Ag(I)-assisted method, the presence of I⁻ ion in the form of the contaminant inhibits the formation of Au nanorods. The nanorod-aspect ratio obtained is approximately 5 in Ag(I)-assisted method where as it is > 1 g in stepwise-additive growth method.

In Ag(I)-assisted method, nanorods extended in the {100} direction, whereas in stepwise additive growth method, nanorods grow in the {110} direction. Also, the seeds in the former method are pentafold twinned, while in later are single crystalline.[58] Au facets promotes crystallization of single crystalline nanorods in the absence of facet selective Ag(I) under potential deposition (VPD) on {110} Au facet promotes crystallization of single crystalline nanorods in the {100} direction.

Different possible mechanisms by which gold nanoparticle formation be affected are:

(a) I^- can bind preferentially to Au {111}surfaces.
b) It can act as a redox agent, hence, affecting reduction rate of Au(III) to Au(0).
c) I^- can combine with Ag^+ to form AgI as a result of which Ag^+ ions will become less available for UPD on the Au {110} surface.
d) Lastly, it can etch the gold seed surfaces.

16.3.3 FORMATION OF SPHERICAL NANOCRYSTALS

Even the presence of I^- impurities clearly indicates the effective nature of CTAB as capping agent and that it do not disrupted by the I^- etching. I^- etching significantly affects the nanorod formation in nanorod photosynthesis protocol as chemisorption of I^- displaces CTAB-capping ligand, which causes destabilization of the nanorod shape and hence, promotes aggregation and coalescence.

Low concentration of KI on the performed nanorod results in dumbbell-shaped nanorods, which clearly indicates that gold salt reduction took place preferentially at the nanorods tips. Higher concentration of I^- ion results in homogeneous rod, which clearly indicates that AgI gets deposited on all surfaces. In some studies, the formation of AgI leads to reduce availability of Ag for UPD on Au {110}, which results in isotropic nanocrystal shorter nanorods formation as well as lower redox potential[59] of Au(III) and increase in reaction kinetics.

In some studies (Ha et al.), it is observed that I^- ions bound strongly on Au {111}, which cause inhibition of Au deposition. In both the methods, that is, stepwise additive and Ag-assisted approach {111}, surfaces are exposed at the tips of the gold nanorods as shown in the Figure 16.4.

And Au should deposit onto these surfaces for nanorods to grow. The deposition of the I^- on {111} facet significantly slow down the Au deposition and this factor significantly slows down the Au deposition; this factor is increased when high concentration of CTAB is used.[60] However, presence of low concentration of I^- ion can lead to only partial surface coverage of Au {111} surface that results in nanorods with shorter aspect ratios. Hence, these results clearly indicate that how concentration impurities play a critical role in deciding the final shape of the nanocrystal.

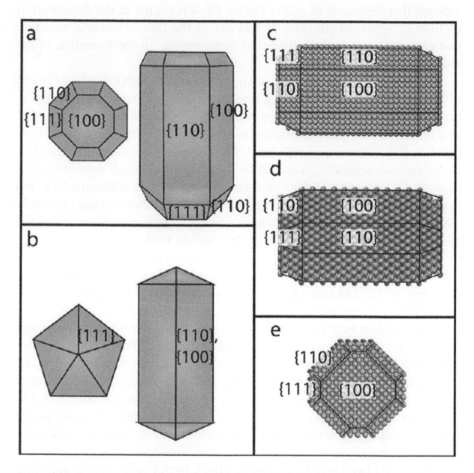

FIGURE 16.4 Top (left) and side (right) view of an Au nanorod synthesized with the (a) Ag(I)-assisted method or (b) stepwise additive method with the exposed facets labeled. Crystallographic models of an Au nanorod synthesized with the Ag(I)-assisted growth method on its (c) {100} side facet, (d) {110} side facet, or (e) {100} bottom facet.
Source: Reproduced with permission from Ref. 61. © 2009 American Chemical Society.)

16.3.4 ROLE Br⁻ ION IN CTAB

Sulphezi et al. studied the role of halide ions in controlling the shape of AuNPs in seed growth method using computational and experimental study. It was observed that not only Br⁻ ion played an active role in surface passivation, but also it acted as driving force for CTAB-micelle adsorption and stabilization of the gold surface in a facet-dependent way. Results also indicate that replacement of Br⁻ ion by Cl⁻ ion results in the formation of uniform or spherical structure. It was due to the lack of micellar structure that can protect the gold surface and hence results in the formation of the uniform structure.

The rational control of the size and shape of the AuNP synthesis is one of the main targets of the ongoing research.[61] The gold nanorods are usually prepared using a seed-mediated growth technique. In this technique, ascorbic acid (a mild reducing agent) is added to an aqueous cetyltrimethylammonium bromide (CTAB) solution of $HAuCl_4$ for the selective reduction of the Au^{3+} to Au^0 (growth solution), followed by the addition of small gold seeds that catalyze the reduction of Au⁺ on their surface.[62-64] The addition of Ag^+ ion to the growing solution can increase the yield of the short nanorod to nearly 100%.[62] The presence of halide ions significantly influence the morphology of the gold nanoparticles[65, 66] due to their strong tendency to adsorb on the metallic surfaces.[67,68] MD simulations showed that on all the surface, CTAB forms distorted cylindrical micelles spaced by water channels containing Br⁻ ions , which can provide a path for the diffusion of the gold reactants toward the gold surface.[69]

These studies help to understand the structure of gold/surfactants/electrolytes interface charges when Cl⁻ replaces Br⁻ at different concentration. It was observed that the degree of binding of the different halides to the gold surfaces has a strong impact on the structure of the surfactant layer at the interface and replacement of Br⁻ by Cl⁻; only a few disordered CTA^+ molecules can bind to the gold surface, confirming that CTAC is unable to exert a protecting action on the gold surface and is unable to inhibit the growth of any gold surface.

The CTAB head groups favorably attach to the gold surface and form a distorted cylindrical micelle in which the head groups of the CTAB molecule arrange in the outer layer of the micelle and CTAB tail arrange in the core of the micelle. It is observed that with the increasing amount of the chloride ion, there occurs faster and more isotopic nanoparticle growth and also CTAC results in considerably thinner layer on the gold than CTAB.

CTAB is able to form a compact micellar layer on the gold surface, which is actually denser on Au {100} and Au {110} with respect to Au {111}. The key element in the micelle adhesion to the surface is bromide propensity for the gold surface. So, when Br⁻ ion is replaced by the Cl⁻, the micelle prefers to diffuse into the electrolyte solution leaving the gold surface unprotected.

16.4 CONCLUSION

Two processes are involved in the formation of metal nanoparticles, which are: nucleation and growth. One of the important key to control the dispersion of size is to separate the stages of nucleation and growth on time The most extensively employed method is the seed-mediated approach that has initiated this separation through use of different chemical environments. Template method controls the morphology, structure, and particle size during the formation of nanoparticles. Nanorods of gold are prepared through seed-mediated growth technique. Faster growth reaction is observed due to the presence of larger gold nanoparticles. The concentration of CTAB is very important parameter for the formation of typical Au nanorods. It has been found that Cl^- ion is less efficient in inhibiting radial growth of nanorods as compared to Br^- ion; and I^- ion prevents the gold nanorod formation.

KEYWORDS

- **template method**
- **surfactant**
- **halide ions**
- **morphology**
- **nanotechnology**

REFERENCES

1. Watzky, M. A.; Finke, R. G. J. *Am. Chem. Soc.* **1997,** *119*, 10382.
2. Papp, S. Z.; Patakfalvi, R.; Dékány, I. *Croat. Chem. Acta* **2007,** *80*, 493.
3. Patakfalvi, R.; Papp, S. Z.; Dékány, I. *J. Nanopart. Res.* **2007,** *9*, 353.
4. Pérez, M. A.; Moiraghi, R.; Coronado, E. A.; Macagno, V. A. *Cryst. Growth Des.* **2008,** *8*, 1377.
5. Tatarchuk, V. V.; Sergievskaya, A. P.; Druzhinina, I. A.; Zaikovsky, V. I. *J. Nanopart. Res.* **2011,** *13*, 4997.
6. Streszewskia, B.; Jaworskia, W.; Pacławskia, K.; Csapó, E.; Dékány, I.; Fitzner, K. *Colloid. Surface. A* **2012,** *397*, 63.
7. Zhou, Y.; Wang, H.; Lin, W.; Lin, L.; Gao, Y.; Yang, F.; Du, M.; Fang, W.; Huang, J.; Sun, D.; Li, Q. *J. Colloid. Int. Sci.* **2013,** *407*, 8.
8. Sidhaye, D. S.; Prasad, B. L. *New J. Chem.* **2011,** *35*, 755.
9. Wang, F.; Richards, V. N.; Shields, S. P.; Buhro, W. E. *Chem. Mater.* **2014,** *26*, 5.

10. Gentry, S. T.; Kendra, S. F.; Bezpalko, M. W. Oxidative Etching and Its Role in Manipulating the Nucleation and Growth of Noble-Metal Nanocrystals, *J. Phys. Chem. C* **2011**, *115*, 12736.

11. Sidhaye, D. S.; Prasad, B. L. *New J. Chem.* **2011**, *35*, 755.

12. Redmond, P. L.; Hallock, A. J.; Brus, L. E. *Nano Lett.* **2005**, *5*, 131.

13. Zheng, Y.; Zeng, J.; Ruditskiy, A.; Liu, M.; Xia, Y. *Chem. Mater.* **2014**, *26*, 22.

14. Pérez, M. A. [ch. 6]. In *Recent Advances in Nanoscience*; Mariscal, M. M., Dassie, S. A., Eds.; Research Signpost: Trivandrum, Kerala, India, 2007; pp. 143.

15. Brust , M.; Fink, J.; Bethell, D.; Schiffrin, D. J.; Kiely, C. J. *J. Chem. Soc. Chem. Commun.* **1995**, 1655.

16. Gao, C.; Goebl, J.; Yin, Y. *J. Mater. Chem. C* **2013**, *1*, 3898.

17. Niu, W.; Zhang, L.; Xu, G. *Nanoscale* **2013**, *5*, 3172.

18. Ziegler, C.; Eychmüller, A. *J. Phys. Chem. C* **2011**, *115*, 4502.

19. Jana, N. R.; Gearheart, L.; Murphy, C. J. *Langmuir* **2001**, *17*, 6782.

20. Jana, N. R.; Gearheart, L.; Murphy, C. J. *J. Phys. Chem. B* **2001**, *105*, 4065.

21. Jana, N. R.; Gearheart, L.; Murphy, C. J. *Adv. Mater.* **2001**, *13*, 1389.

22. Sau, T. F.; Murphy, C. J. *J. Am. Chem. Soc.* **2004**, *126*, 8648.

23. Jana, N. R.; Gearheart, L.; Murphy, C. J. *J. Phys. Chem. B* **2001**, *105*, 4065.

24. Jana, N. R.; Gearheart, L.; Murphy, C. J. *Adv. Mater.* **2001**, *13*, 1389.

25. Nikoobakht, B.; El-Sayed, M. A. *Chem. Mater.* **2003**, *15*, 1957–1962.

26. Sánchez, A.; Díez, P.; Villalonga, R.; Martínez-Ruiz, P.; Eguiláz, M.; Fernández, I.; Pingarrón, J. M. *Dalton Trans.* **2013**, *42*, 14309.

27. Grzelczak, M.; Sánchez-Iglesias, A.; Rodríguez-González, B.; Alvarez-Puebla, R.; Pérez-Juste, J.; Liz-Marzán, L. M. *Adv. Funct. Mater.* **2008**, *18*, 3780.

28. Millstone, J. E.; Park, S.; Shuford, K. L.; Qin, L.; Schatz, G. C.; Mirkin, C. D. *J. Am. Chem. Soc.* **2005**, *127*, 5312.

29. Liu, M.; Guyot-Sionnest, P. *J. Phys Chem. B* **2005**, *109*, 22192.

30. Linic, S.; Asiam, U.; Boerigter, C.; Morabito, M. *Nat. Mat.* **2015**, *14*, 567–576.

31. Tauran, Y.; Brioude, A.; Coleman, A. W.; Rhimi, M.; Kim, B. *World J Biol Chem.* **2013**, *4*, 35–63.

32. Yong, I.; Li, J.; Shim, Y. *Clin Endosc.* **2013**, *46*, 7–23.

33. Ali, M. E.; Hashim, U.; Mustafa, S.; Che Man, Y. B.; Islam, Kh. N. *J. Nanomat.* **2012**, *2012*, Article ID 103607.

34. Perrault, S. D.; Chan, W. C. W. *Proc. Nat. Acad. Sci. USA* **2010**, *107*, 11194–11199.

35. Zhuang, M.; Ding, C.; Zhu, A.; Tian, Y. *Anal. Chem.* **2014**, *86*, 1829–1836.

36. Ren, N.; Tang, Y. *Petrochem. Technol.* **2005**, *34* (5), 405–411.

37. Masuda, H.; Fukuda, K. *Science*, **1995**, *268* (5216), 1466–1468.

38. Zhao, D.; Peidong, Y.; Qisheng, H.; Bradley, C. F.; Galen, D. *Curr. Opin. Solid State Mat. Sci.* **1998**, *3* (1), 111–121.

39. Lin, -P.; Mou, C.-Y. *Acc. Chem. Res.* **2002**, *35* (11), 927–935.

40. Smith, D. K.; Miller, N. R.; Korgel, B. A. *Langmuir* **2009**, *25* (16), 9518–9524.

41. Soejima, T.; Kimizuka, N. *J. Am. Chem. Soc.* **2009**, *131*, 14407–14412.

42. Fan, X.; Guo, Z. R.; Hong, J. M.; Zhang, Y.; Zhang, J. N.; Gu, N. *Nanotechnology* **2010**, *21*, 105602–105609.

43. Zhang, J.; Langille, M. R.; Personick, M. L.; Zhang, K.; Li, S.; Mirkin, C. A. *J. Am. Chem. Soc.* **2010**, *132* (40), 14012–14014.

44. Jiao, Z.; Xia, H.; Tao, X. *J. Phys. Chem. C* **2011**, *115* (16), 7887–7895.

45. DuChene, J. S.; Niu, W.; Abendroth, J. M.; Sun, Q.; Zhao, W.; Huo, F.; Wei, W. D. *Chem. Mater.* **2013**, *25*, 1392–1399.

46. Langille, M. R.; Personick, M. L.; Zhangand, J.; Mirkin, C. A. *J. Am. Chem. Soc.* **2012**, *134*, 14542–14554.

47. da Silva, M. G. A.; Meneghetti, M. R.; Denicourt-Nowicki, A.; Roucoux, A. *RSC Adv.* **2014**, *4*, 25875–25879.

48. Kedia, A.; Kumar, P. S. *RSC Adv.* **2014**, *4*, 4782–4790.

49. Sau, T. K.; Murphy, C. J. *Philos. Mag.* **2007**, *87*, 14–15.

50. Lohse, S. E.; Burrows, N. D.; Scarabelli, L.; Liz-Marz'an, L. M.; Murphy, C. J. *Chem. Mater.* **2014**, *26* (1), 34–43.

51. Murphy, C. J.; Thompson, L. B.; Alkilany, A. M.; Sisco, P. N.; Boulos, S. P.; Sivapalan, S. T.; Yang, J. A.; Chernak, D. J.; Huang, J. *J. Phys. Chem. Lett.* **2010**, *1* (19), 2867–2875.

52. Garg, N., Scholl, C. Mohanty, A., and Jin, R. The Role of Bromide Ions in Seeding Growth of Au Nanorods. *Langmuir* 2010, 26, 10271–10276.

53. Sau, T. K.; Murphy, C. J. *Philos. Mag.* **2007**, *87*, 2143–2158.

54. Filankembo, A.; Giorgio, S.; Lisiecki, I.; Pileni, M.-P. *J. Phys. Chem. B* **2003**, *107*, 7492–7500.

55. Barnard, A. S.; Curtiss, L. A. *J. Mater. Chem.* **2007**, *17*, 3315–3323.

56. Murphy, C. J. *ACS Nano* **2009**, *3*, 770–774.

57. Millstone, J. E.; Wei, W.; Jones, M. R.; Yoo, H. J.; Mirkin, C. A. *Nano Lett.* **2008**, *8*, 2526–2529.

58. Liu, M. Z.; Guyot-Sionnest, P. *J. Phys. Chem. B* **2005**, *109*, 22192–22200.

59. Sibbald, M. S.; Chumanov, G.; Cotton, T. M. *J. Phys. Chem.* **1996**, *100*, 4672–4678.

60. Smith, D. K.; Miller, N. R.; Korgel, B. A. Iodide in CTAB Prevents Gold Nanorod Formation. *Langmuir* 2009, 25 (16), 9518–9524.

61. Grzelczak, M.; Perez-Juste, J.; Mulvaney, P.; Liz-Marzan, L. M. *Chem. Soc. Rev.* **2008**, *37*, 1783–1791.

62. Jana, N. R.; Gearheart, L.; Murphy, C. J. *J. Phys. Chem. B* **2001**, *105*, 4065–4067.

63. Murphy, C. J.; Sau, T. K.; Gole, A. M.; Orendorff, C. J.; Gao, J.; Gou, L.; Hunyadi, S. E.; Li, T. *J. Phys. Chem. B* **2005**, *109* (29), 13857–13870.

64. Nikoobakht; El-Sayed, M. A. *Chem. Mater.* **2003**, *15* (10), 1957–1962.

65. Sau, T. K.; Murphy, C. J. *Philos. Mag.* **2007**, *87*, 14–15.

66. Lohse, S. E.; Burrows, N. D.; Scarabelli, L.; Liz-Marz'an, L. M.; Murphy, C. J. *Chem. Mater.* **2014**, *26* (1), 34–43.

67. Meena, S. K.; Sulpizi, M. *Langmuir* **2013**, *29* (48), 14954–14961.

68. Vivek, J. P.; Burgess, I. J. *Langmuir* **2012**, *28* (11), 5040–5047.

69. Yu, C.; Varghese, L.; Irudayaraj, J. *Langmuir* **2007**, *23*, 9114–9119.

INDEX

P

Printed and bound by CPI Group (UK) Ltd, Croydon, CR0 4YY

23/10/2024

01777703-0008